Progress in Mathematical Physics
Volume 47

Einstein, 1905–2005

Poincaré Seminar 2005

Thibault Damour
Olivier Darrigol
Bertrand Duplantier
Vincent Rivasseau
Editors

Birkhäuser Verlag
Basel · Boston · Berlin

Editors:

Thibault Damour
IHÉS
35, Route de Chartres,
91440 Bures-sur-Yvette
France
e-mail: damour@ihes.fr

Bertrand Duplantier
Service de Physique Théorique
Orme des Merisiers
CEA - Saclay
91191 Gif-sur-Yvette Cedex
France
e-mail: bertrand.duplantier@cea.fr

Olivier Darrigol
Université Denis Diderot (Paris 7)
2, place Jussieu
75251 Paris Cedex 05
France
e-mail: darrigol@paris7.jussieu.fr

Vincent Rivasseau
Laboratoire de Physique Théorique
Université Paris XI
91405 Orsay Cedex
France
e-mail: Vincent.Rivasseau@th.u-psud.fr

2000 Mathematics Subject Classification 83-03, 83A05, 83B05, 83Cxx, 60Hxx

A CIP catalogue record for this book is available from the Library of Congress, Washington D.C., USA

Bibliographic information published by Die Deutsche Bibliothek
Die Deutsche Bibliothek lists this publication in the Deutsche Nationalbibliografie; detailed bibliographic data is available in the Internet at <http://dnb.ddb.de>.

ISBN 3-7643-7435-7 Birkhäuser Verlag, Basel – Boston – Berlin

Part of Springer Science+Business Media
Printed on acid-free paper produced of chlorine-free pulp. TCF ∞
Printed in Germany
ISBN-10: 3-7643-7435-7
ISBN-13: 978-3-7643-7435-8

9 8 7 6 5 4 3 2 1

www.birkhauser.ch

Contents

Foreword

This book is the fourth in a series of lectures of the *Séminaire Poincaré*, which is directed towards a large audience of physicists and of mathematicians.

The goal of this seminar is to provide up-to-date information about general topics of great interest in physics. Both the theoretical and experimental aspects are covered, with some historical background. Inspired by the Bourbaki seminar in mathematics in its organization, hence nicknamed "Bourbaphi", the Poincaré Seminar is held twice a year at the Institut Henri Poincaré in Paris, with contributions prepared in advance. Particular care is devoted to the pedagogical nature of the presentations so as to fulfill the goal of being readable by a large audience of scientists.

This volume contains the seventh such Seminar, held in 2005. It is devoted to Einstein's 1905 papers and their legacy. After a presentation of Einstein's epistemological approach to physics, and the genesis of special relativity, a centenary perspective is offered. The geometry of relativistic spacetime is explained in detail. Single photon experiments are presented, as a spectacular realization of Einstein's light quanta hypothesis. A previously unpublished lecture by Einstein, which presents an illuminating point of view on statistical physics in 1910, at the dawn of quantum mechanics, is reproduced. The volume ends with an essay on the historical, physical and mathematical aspects of Brownian motion.

We hope that the publication of this series will serve the community of physicists and mathematicians at the graduate student or professional level.

We thank the Commissariat à l'Énergie Atomique (Division des Sciences de la Matière), the Centre National de la Recherche Scientifique (Sciences Physique et Mathématiques), and the Daniel Iagolnitzer Foundation for sponsoring the Seminar. Special thanks are due to Chantal Delongeas for the preparation of the manuscript.

Thibault Damour
Olivier Darrigol
Bertrand Duplantier
Vincent Rivasseau

Institut Henri Poincaré · Amphi Hermite
11, rue Pierre et Marie Curie · 75005 Paris
O. Darrigol : La genèse de la Relativité · 10h00
C. M. Will : Tests of Special Relativity · 11h00
B. Duplantier : Le mouvement brownien · 14h00
Ph. Grangier : Expériences à un seul photon · 15h00
T. Damour : Einstein épistémologue · 16h00
Einstein,
1905-2005
Séminaire Poincaré
Samedi 9 avril 2005
Crédit photos :
Julie Mehretu, excerpt (suprematist evasion), 2003
ink and acrylic on canvas, 32"x54"
Archives Lotte Jacobi, Université du New Hampshire
www.lpthe.jussieu.fr/poincare
CENTRE NATIONAL
DE LA RECHERCHE
SCIENTIFIQUE
Fondation
Iagolnitzer

Einstein, 1 – 31

Poincaré Seminar 2005

The Genesis of the Theory of Relativity

Olivier Darrigol

The most famous of Albert Einstein's papers of 1905 is undoubtedly the one concerning the theory of relativity. Any modern physicist knows that this theory imposes a strict and general constraint on the laws of nature. Any curious layman wonders at the daring reform of our ancestral concepts of space and time. As often happens for great conceptual breakthroughs, the theory of relativity gave rise to founding myths whose charm the historian must resist.

The first of this myth is that Einstein discovered the theory of relativity in a single stroke of genius that defies any rational analysis. Some of Einstein's reminiscences favor this thesis, for instance his allusion to a conversation with Michele Besso in which he would have suddenly realized that a reform of the concept of time solved long standing paradoxes of electrodynamics. One could also argue that the historical explanation of a deep innovation is by definition impossible, since a radically new idea cannot be derived from received ideas. In the case of Einstein's relativity the rarity of pre-1905 sources further discourages historical reconstruction, and invites us to leave this momentous discovery in its shroud of mystery.

This romantic attitude does not appeal to teachers of physics. In order to convey some sort of logical necessity to relativity theory, they have constructed another myth following which a few experiments drove the conceptual revolution. In this empiricist view, the failure of ether-drift experiments led to the relativity principle; and the Michelson-Morley experiment led to the constancy of the velocity of light; Einstein only had to combine these two principles to derive relativity theory.

As a counterpoise to this myth, there is a third, idealist account in which Einstein is supposed to have reached his theory by a philosophical criticism of fundamental concepts in the spirit of David Hume and Ernst Mach, without even knowing about the Michelson-Morley experiment, and without worrying much about the technicalities of contemporary physics in general.

A conscientious historian cannot trust such myths, even though they may contain a grain of truth. He must reach his conclusions by reestablishing the contexts in which Einstein conducted his reflections, by taking into account his education and formation, by introducing the several actors who shared his interests, by identifying the difficulties they encountered and the steps they took to solve them. In this process, he must avoid the speculative filling of gaps in documentary sources. Instead of rigidifying any ill-founded interpretation, he should offer an open spectrum of interpretive possibilities. As I hope to show in this paper, this sober method allows a fair intelligence of the origins of relativity.

A first indication of the primary context of the early theory of relativity is found in the very title of Einstein's founding paper: "On the electrodynamics of moving bodies." This title choice may seem bizarre to the modern reader, who defines relativity theory as a theory of space and time. In conformity with the latter view, the first section of Einstein's paper deals with a new kinematics meant to apply to any kind of physical phenomenon. Much of the paper nonetheless deals with the application of this kinematics to the electrodynamics and optics of moving bodies. Clearly, Einstein wanted to solve difficulties he had encountered in this domain of physics. A survey of physics literature in the years 1895-1905 shows that the electrodynamics of moving bodies then was a widely discussed topic. Little before the publication of Einstein's paper, several studies with similar titles appeared in German journals. Much experimental and theoretical work was being done in this context. The greatest physicists of the time were involved. They found contradictions between theory and experience or within theory, offered mutually incompatible solutions, and sometimes diagnosed a serious crisis in this domain of physics.

Since Heinrich Hertz's experiments of 1887-8 on the electric production of electromagnetic waves, Maxwell's field theory was the natural frame for discussing both the electrodynamics and the optics of moving bodies. In order to understand the evolution of this subject, one must first realize that the theory that Maxwell offered in his treatise of 1873 widely differed from what is now meant by "Maxwell's theory."

1 Maxwell's theory as it was

Like most of his contemporaries, Maxwell regarded the existence of the ether as a fundamental and undeniable fact of physics. He held this medium responsible for the propagation of electromagnetic actions, which included optical phenomena in his view. His theory was a phenomenological theory concerned with the macroscopic states of a continuous medium, the ether, which could combine with matter and share its velocity $\mathbf{v}$. These states were defined by four vectors $\mathbf{E}$, $\mathbf{D}$, $\mathbf{H}$, $\mathbf{B}$ that obeyed a few general partial differential equations as well as some relations depending on the intrinsic properties of the medium. In the most complete and concise form later given by Oliver Heaviside and Heinrich Hertz, the fundamental equations read

$$\nabla \times \mathbf{E} = -D\mathbf{B}/Dt\ , \quad \nabla \times \mathbf{H} = \mathbf{j} + D\mathbf{D}/Dt$$
$$\nabla \cdot \mathbf{D} = \rho\ , \nabla \cdot \mathbf{B} = 0\ , \tag{1}$$

where $\mathbf{j}$ is the conduction current and D/Dt is the convective derivative defined by

$$D/Dt = \partial/\partial t - \nabla \times (\mathbf{v}\times\ ) + \mathbf{v}(\nabla\cdot\ ). \tag{2}$$

In a linear medium, the "forces" $\mathbf{E}$ and $\mathbf{H}$ were related to the "polarizations" $\mathbf{D}$ and $\mathbf{B}$ by the relations $\mathbf{D} = \epsilon\mathbf{E}$ and $\mathbf{B} = \mu\mathbf{H}$, and the energy density $(1/2)(\epsilon E^2 + \mu H^2)$ of the medium had the form of an elastic energy. For Maxwell and his followers, the charge density and the conduction current $\mathbf{j}$ were not primitive concepts: the former corresponded to the longitudinal gradient of the polarization or "displacement" $\mathbf{D}$, and the latter to the dissipative relaxation of this polarization in a conducting medium. The variation $D\mathbf{D}/Dt$ of the displacement constituted another form of current. Following Michael Faraday, Maxwell and his disciples regarded the electric fluids of earlier theories as a naïvely substantialist notion.[1]

The appearance of the convective derivative D/Dt in Maxwell's theory derives from his understanding of the polarizations $\mathbf{D}$ and $\mathbf{B}$ as states of a single medium made of ether and matter and moving with a well-defined velocity $\mathbf{v}$ (that may vary from place to place): the time derivatives in the fundamental equations must be taken along the trajectory of a given particle of the moving medium. The resulting law of electromagnetic induction,

$$\nabla \times \mathbf{E} = -D\mathbf{B}/Dt = -\partial\mathbf{B}/\partial t + \nabla \times (\mathbf{v} \times \mathbf{B}) \,, \tag{3}$$

contains the $(\mathbf{v} \times \mathbf{B})$ contribution to the electric field in moving matter. By integration around a circuit and through the Kelvin-Stokes theorem, it leads to the expression

$$\oint \mathbf{E} \cdot d\mathbf{l} = -\frac{d}{dt} \iint \mathbf{B} \cdot d\mathbf{S} \tag{4}$$

of Faraday's law of induction, wherein the integration surface moves together with the bordering circuit. When the magnetic field is caused by a magnet, the magnetic flux only depends on the relative position of the magnet and the circuit so that the induced current only depends on their relative motion.

In sum, the conceptual basis of Maxwell's original theory widely differed from what today's physicists would expect. Electricity and magnetism were field-derived concept, whereas modern electromagnetism treats them as separate entities. A quasi-material ether was assumed. The fundamental equations (1) only correspond to our "Maxwell equations" in the case of bodies at rest, for which the velocity $\mathbf{v}$ is zero and the convective derivative D/Dt reduces to the partial derivative $\partial/\partial t$. One thing has not changed, however: the theory's ability to unify electromagnetism and optics. In a homogenous insulator at rest, Maxwell's equations imply the existence of transverse waves propagating at the velocity $c = 1/\sqrt{\epsilon\mu}$. Having found this electromagnetic constant to be very close to the velocity of light, Maxwell identified these waves with light waves. The resulting theory automatically excludes the longitudinal vibrations that haunted the earlier, elastic-solid theories of optics.

Within a few years after Maxwell's death (in 1879), a growing number of British physicists saluted this achievement and came to regard Maxwell's theory as

[1] J.C. Maxwell, *A treatise on electricity and magnetism*, 2 vols. (Oxford, 1973); H. Hertz, "Über die Grundgleichungen der Elektrodynamik für bewegte Körper," *Annalen der Physik*, 41 (1890), 369-399.

philosophically and practically superior to earlier theories. The Germans had their own theories of electricity and magnetism, based on electric and magnetic fluids (or Amperean currents) directly acting at a distance. They mostly ignored Maxwell's theory until in 1888 Heinrich Hertz demonstrated the emission of electromagnetic waves by a high-frequency electric oscillator. After this spectacular discovery was confirmed, a growing number of physicists adopted Maxwell's theory in a more or less modified form.

Yet this theory was not without difficulties. Maxwell had himself noted that his phenomenological approach led to wrong predictions when applied to optical dispersion, to magneto-optics, and to the optics of moving bodies. In these cases he suspected that the molecular structure of matter had to be taken into account.

2 Flashback: The optics of moving bodies

Maxwell's idea of a single medium made of ether and matter implied that the ether was fully dragged by moving matter, even for dilute matter. Whereas this conception worked very well when applied to moving circuits and magnets, it was problematic in the realm of optics. The first difficulty concerned the aberration of stars, discovered by the British astronomer James Bradley in 1728: the direction of observation of a fixed star appears to vary periodically in the course of a year, by an amount of the same order as the ratio (10^{-4}) of the orbital velocity of the earth to the velocity of light.[2]

The old corpuscular theory of light simply explained this effect by the fact that the apparent velocity of a light particle is the vector sum of its true velocity and the velocity of the earth (see Fig. 1). In the early nineteenth century, the founders of the wave theory of light Thomas Young and Augustin Fresnel saved this explanation by assuming that the ether was completely undisturbed by the motion of the earth through it. Indeed, rectilinear propagation at constant velocity is all that is needed for the proof.[3]

Fresnel's assumption implied an ether wind of the order of 30km/s on the earth's surface, from which a minute modification of the laws of optical refraction ought to follow. As Fresnel knew, an earlier experiment of his friend François Arago had shown that refraction by a prism was in fact unaffected by the earth's annual motion. Whether or not Arago had reached the necessary precision of 10^{-4}, Fresnel took this result seriously and accounted for it by means of a partial dragging of the ether within matter. His theory can be explained as follows.

According to an extension of Fermat's principle, the trajectory that light takes to travel between two fixed points (with respect to the earth) is that for

[2] J. Bradley, "A new apparent motion discovered in the fixed stars; its cause assigned; the velocity and equable motion of light deduced," Royal Society of London, *Proceedings*, 35 (1728), 308-321.

[3] A. Fresnel, "Lettre d'Augustin Fresnel à François Arago sur l'influence du mouvement terrestre dans quelques phénomènes d'optique," *Annales de chimie et de physique*, 9(1818), also in *Oeuvres complètes*, Paris (1868), vol. 2, 627-636.

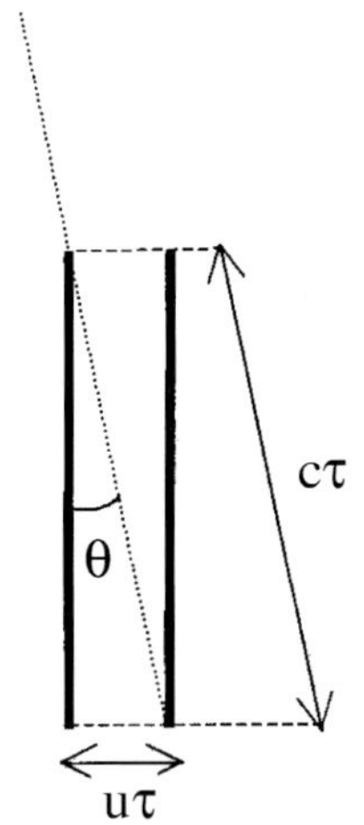

Figure 1: *Stellar aberration.* Suppose that the position of a fixed star in the sky is judged by the orientation of a narrow straight tube through which it can be seen. If the earth is moving with respect to the fixed stars at the velocity **u**, the latter sweeps the distance $u\tau$ during the time τ that the light from the star takes to travel from the beginning to the end of the tube. Therefore, the true light path makes a small angle with the direction of the tube. When the velocity of the earth is perpendicular to the tube, this angle is $\theta \approx \tan\theta = u/c$. Owing to the annual motion of the earth, the apparent position of the star varies with a period of one year.

which the traveling time is a minimum, whether the medium of propagation is at rest or not. The velocity of light with respect to the ether in a substance of optical index n is c/n, if c denotes the velocity of light. The absolute velocity of the ether across this substance is $\alpha\mathbf{u}$, where α is the dragging coefficient and $\mathbf{u}$ is the absolute velocity of the substance (the absolute velocity being that with respect to the remote, undisturbed parts of the ether). Therefore, the velocity of light along the element $d\mathbf{l}$ of an arbitrary trajectory is $c/n + (\alpha - 1)\mathbf{u} \cdot d\mathbf{l}/ds$ with respect to the substance (with $ds = \|d\mathbf{l}\|$). To first order in u/c, the time taken by light during this elementary travel is

$$dt = (n/c)ds + (n^2/c^2)(1 - \alpha)\mathbf{u} \cdot d\mathbf{l} \ . \tag{5}$$

Note that the index n and the dragging coefficient in general vary along the path, whereas the velocity $\mathbf{u}$ has the same value (the velocity of the earth) for the whole optical setting. The choice $\alpha = 1$ (complete drag) leaves the time dt and the trajectory of minimum time invariant, as should obviously be the case. Fresnel's choice,

$$\alpha = 1 - 1/n^2 \tag{6}$$

yields

$$dt = (n/c)ds + (1/c^2)\mathbf{u} \cdot d\mathbf{l} \ , \tag{7}$$

so that the time taken by light to travel between two fixed points of the optical setting differs only by a constant from the time it would take if the earth were not moving. Therefore, the laws of refraction are unaffected (to first order) under Fresnel's assumption.[4]

In 1846, the Cambridge professor George Gabriel Stokes criticized Fresnel's theory for making the fantastic assumption that the huge mass of the earth was completely transparent to the ether wind. In Stokes' view, the ether was a jelly-like substance that behaved as an incompressible fluid under the slow motion of immersed bodies but had rigidity under the very fast vibrations implied in the propagation of light. In particular, he identified the motion of the ether around the earth with that of a perfect liquid. From Lagrange, he knew that the flow induced by a moving solid (starting from rest) in a perfect liquid is such that a potential exists for the velocity field. From his recent derivation of the Navier-Stokes equation, he also knew that this property was equivalent to the absence of instantaneous rotation of the fluid elements. Consequently, the propagation of light remains rectilinear in the flowing ether, and the apparent position of stars in the sky is that given by the usual theory of aberration.[5]

In order to account for the absence of effects of the earth's motion on terrestrial optics, Stokes further assumed that the ether adhered to the earth and had a negligible relative velocity at reasonable distances from the ground.

To sum up, before the middle of the century, there were two competing theories of the optics of moving bodies that both accounted for stellar aberration and for the absence of effects of the earth's motion on terrestrial optics. Fresnel's theory assumed the stationary character of the ether everywhere except in moving refractive media, in which a partial drag occurred. Stokes' theory assumed complete ether drag around the earth and irrotational flow at higher distances from the earth.

[4]Cf. E. Mascart, *Traité d'optique*, 3 vols. (Paris, 1893), vol. 3, chap. 15. Fresnel justified the value $1 - 1/n^2$ of the dragging coefficient by making the density of the ether inversely proportional to the square of the propagation velocity c/n (as should be in an elastic solid of constant elasticity) and requiring the flux of the ether to be conserved. As Mascart noted in the 1870s, this justification fails when double refraction and dispersion are taken into account.

[5]G.G. Stokes, "On the aberration of light," *Philosophical magazine*, 27(1845), 9-55; "On Fresnel's theory of the aberration of light," ibid., 28 (1846), 76-81; "On the constitution of the luminiferous ether, viewed with reference to the phenomenon of the aberration of light," ibid., 29 (1846), 6-10. This result can be obtained from Fermat's principle, by noting that to first order the time taken by light to travel along the element of length $d\mathbf{l}$ has the form $dt = (1/c)ds - (1/c^2)\mathbf{v} \cdot d\mathbf{l}$ ($\mathbf{v}$ denoting the velocity of the ether), so that its integral differs only by a constant (the difference of the velocity potentials at the end points) from the value it would have in a stationary ether.

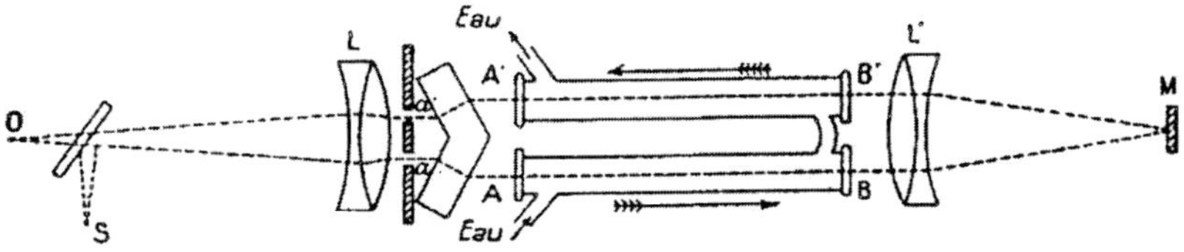

Figure 2: *Fizeau's experiment.* After reflection on a semi-reflecting blade, the light from the source S is divided into two beams. The upper beam travels against the water stream in A′B′, crosses the lens L′, is reflected on the mirror M, crosses L′ again, travels *against* the water stream in AB, and returns to the semi-reflecting blade. The lower beam does the symmetrical trip, which runs twice along the water stream. The phase difference between the two beams is judged from the interference pattern in O.

In 1850 Hippolyte Fizeau performed an experiment in which he split a light beam into two beams, had them travel through water moving in opposite directions, and measured their phase difference by interference (see fig. 2). The result confirmed the partial drag of light waves predicted by Fresnel. Maxwell knew about Fizeau's result, and, for a while, wrongly believed that it implied an alteration of the laws of refraction by the earth's motion through the ether. In 1864, he performed an experiment to test this modification. The negative result confirmed Arago's earlier finding with improved precision. As Stokes explained to Maxwell, this result pleaded for, rather than contradicted the Fresnel drag. Yet Maxwell remained skeptical about the validity of Fizeau's experiment. In 1867 he wrote:

> This experiment seems rather to verify Fresnel's theory of the ether; but the whole question of the state of the luminiferous medium near the earth, and of its connexion with gross matter, is very far as yet from being settled by experiment.

In this situation, it was too early to worry about an incompatibility between the electromagnetic theory of light and the optics of moving bodies. In 1878, one year before his death, Maxwell still judged Stokes' theory "very probable."[6]

---

[6] H. Fizeau, "Sur les hypothèses relatives à l'éther lumineux, et sur une expérience qui paraît démontrer que le mouvement des corps change la vitesse avec laquelle la lumière se propage dans leur intérieur," Académie des Sciences, *Comptes-rendus*, 33 (1851), 349-355; J.C. Maxwell, "On an experiment to determine whether the motion of the earth influences the refraction of light," unpub. MS, in Maxwell, *The scientific letters and papers*, ed. Peter Harman, vol. 2 (Cambridge, 1995), 148-153; Maxwell to Huggins, 10 Jun 1867, ibid., 306-311; "Ether," article for the *Encyclopedia Britannica* (1878), reproduced ibid., 763-775.

In the 1870s a multitude of experiments confirmed the absence of effect of the earth's motion on terrestrial optics. In 1874, the author of the best of those, Eleuthère Mascart, concluded:

> The translational motion of the earth has no appreciable influence on optical phenomena produced by a terrestrial source, or light from the sun, so these phenomena do not provide us with a means of determining the absolute motion of a body, and relative motions are the only ones that we are able to determine.

Mascart and other continental experts interpreted this finding by means of Fresnel's theory. British physicists mostly disagreed, as can be judged from a British Association report of 1885 in which a disciple of Maxwell criticized "Fresnel's somewhat violent assumptions on the relation between the ether within and without a transparent body."[7]

In 1881 the great American experimenter Albert Michelson conceived a way to decide between Fresnel's and Stokes' competing theories. Through an interferometer of his own, he compared the time that light took to travel the same length in orthogonal directions (see fig. 3). If the ether was stationary, he reasoned, the duration of a round trip of the light in the arm parallel to the earth's motion was increased by a factor $[l/(c-u)+l/(c+u)]/(2l/c)$, which is equal to $1/(1-u^2/c^2)$. The corresponding fringe shift was about twice what his interferometer could detect. From the null result, Michelson concluded that Fresnel's theory had to be abandoned.[8]

A French professor at the Ecole Polytechnique, Alfred Potier, told Michelson that he had overlooked the increase of the light trip by $1/\sqrt{1-u^2/c^2}$ in the perpendicular arm of his interferometer. With this correction, the experiment became inconclusive. Following William Thomson's and Lord Rayleigh's advice and with Edward Morley's help, Michelson first decided to repeat Fizeau's experiment with his powerful interferometric technique. In 1886 he thus confirmed the Fresnel dragging coefficient with greatly improved precision.[9]

At this critical stage, the Dutch theorist Hendrik Lorentz entered the discussion. He first blasted Stokes' theory by noting that the irrotational motion of an incompressible fluid around a sphere necessarily involves a finite slip on its surface.[10] The theory could still be saved by integrating Fresnel's partial drag, but only at the price of making it globally more complicated than Fresnel's. Lorentz therefore favored Fresnel's theory, and called for a repetition of Michelson's experiment

[7] E. Mascart, "Sur les modifications qu'éprouve la lumière par suite du mouvement de la source et du mouvement de l'observateur," *Annales de l'Ecole Normale*, 3 (1874), 363-420, on 420; R.T. Glazebrook, Report on "optical theories," British Association for the Advancement of Science, *Report* (1885), 157-261.

[8] A. Michelson, "The relative motion of the earth and the luminiferous ether," *American journal of science*, 22 (1881), 120-129.

[9] A. Michelson and E. Morley, "Influence of the motion of the medium on the velocity of light", *American journal of science*, 31 (1886), 377-386.

[10] It seems dubious that Stokes, as an expert on potential theory in fluid mechanics, could have overlooked this point. More likely, his jelly-like ether permitted temporary departures from irrotationality.

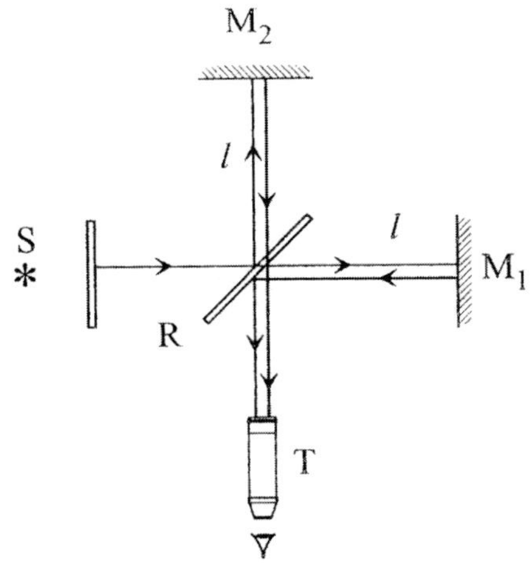

Figure 3: *The Michelson-Morley experiment.* The light from the source **S** is divided into two beams by the semi-reflecting blade **R**. After reflection on the mirrors $\mathbf{M}_1$ and $\mathbf{M}_2$, the two beams return to **R**. Their interference pattern is observed through the telescope **T**.

of 1881 after noting the error already spotted by Potier. Michelson and Morley fulfilled this wish in 1887 with an improved interferometer. The result was again negative, to every expert's puzzlement: while Fizeau's experiment confirmed Fresnel's theory, the new experiment contradicted it.[11]

3 Lorentz's theory

When in the early 1890s Hertz and Heaviside perfected Maxwell's electrodynamics of moving bodies, they noted that it was incompatible with Fresnel's theory of aberration, but decided to postpone further study of the relation between ether and matter. Unknown to them, Lorentz had long ago reflected on this relation and reached conclusions that sharply departed from Maxwell's original ideas. Unlike Maxwell's British disciples, Lorentz learned Maxwell's theory in a reinterpretation by Hermann Helmholtz that accommodated the continental interpretation of charge, current, and polarization in terms of the accumulation, flow, and

[11] H.A. Lorentz, "De l'influence du mouvement de la terre sur les phénomènes lumineux," *Archives néerlandaises* (1887), also in *Collected papers*, 9 vols. (The hague, 1934-1936), vol. 4, 153-214; Michelson and Morley, "On the relative motion of the earth and the luminiferous ether," *American journal of science*, 34 (1887), 333-345.

displacement of electric particles. In 1878 he gave a molecular theory of optical dispersion based on the idea of elastically bound charged particles or "ions" that vibrated under the action of an incoming electromagnetic wave and thus generated a secondary wave. For the sake of simplicity, he assumed that the ether around the molecules and ions had exactly the same properties as the ether in a vacuum. He could thus treat the interactions between ions and electromagnetic radiation through Maxwell's equations in a vacuum supplemented with the so-called Lorentz force.[12]

Using lower-case letters for the microscopic fields and Hertzian units, these equations read

$$\begin{aligned} &\nabla\times\mathbf{e} = -c^{-1}\partial\mathbf{b}/\partial t\ , \quad \nabla\times\mathbf{b} = c^{-1}[\rho_m\mathbf{v}+\partial\mathbf{e}/\partial t]\ , \\ &\nabla\cdot\mathbf{e} = \rho_m\ , \quad \nabla\cdot\mathbf{b} = 0\ , \\ &\mathbf{f} = \rho_m[\mathbf{e}+c^{-1}\mathbf{v}\times\mathbf{b}]\ , \end{aligned} \tag{8}$$

where ρ_m denotes the microscopic charge density (confined to the ions) and $\mathbf{f}$ denotes the density of the force acting on the ions. Note that there are only two independent fields $\mathbf{e}$ and $\mathbf{b}$ since the constants ϵ and μ are set to their vacuum value. Although from a formal point of view these equations can be seen as a particular case of the Maxwell-Hertz equations (1), they were unthinkable to true Maxwellians who regarded the concepts of electric charge and polarization as emergent macroscopic concepts and believed the molecular level to be directly ruled by the laws of mechanics.

Using his equations and averaging over a macroscopic volume element, Lorentz obtained the first electromagnetic theory of dispersion. In 1892, he realized that he could perform similar calculations when the dielectric globally moved through the ether at the velocity $\mathbf{u}$ of the earth. He only had to assume that the ions and molecules moved through the ether without disturbing it. Superposing the incoming wave and the secondary waves emitted by the moving ions, he found that the resulting wave traveled at the velocity predicted by Fresnel's theory. The partial ether drag imagined by Fresnel was thus reduced to molecular interference in a perfectly stationary ether.[13]

Notwithstanding with their global intricacy, Lorentz's original calculations contained an interesting subterfuge. In order to solve equations that involved the wave operator $\partial^2/\partial x^2 - c^{-2}(\partial/\partial t - u\partial/\partial x)^2$ in a reference frame bound to the transparent body, Lorentz introduced the auxiliary variables

$$x' = \gamma x\ , \quad t' = \gamma^{-1}t - \gamma ux/c^2 \tag{9}$$

[12]Lorentz, "Over het verband tusschen de voortplantings sneldheit en samestelling der midden stofen," Koninklijke Akademie van Wetenschappen, *Verslagen* (1878), transl. as "Concerning the relation between the velocity of propagation of light and the density and composition of media" in *Collected papers* (ref. 11), vol. 2, 3-119.

[13]Lorentz, "La théorie électromagnétique de Maxwell et son application aux corps mouvants," *Archives néerlandaises* (1892), also in *Collected papers* (ref. 11), vol. 2, 164-321.

that restored the form of the operator in the ether-bound frame for

$$\gamma = 1/\sqrt{1 - u^2/c^2} \ . \tag{10}$$

He thus discovered the Lorentz transformation for coordinates (up to the Galilean transformation $x = \bar{x} - ut$, where $\bar{x}$ is the abscissa in the ether frame).[14]

A few months later, Lorentz similarly realized that to first order in u/c the field equations in a reference frame bound to the earth could be brought back to the form they have in the ether frame through the transformations

$$t' = t - ux/c^2 \ , \quad \mathbf{e}' = \mathbf{e} + c^{-1}\mathbf{u} \times \mathbf{b} \ , \quad \mathbf{b}' = \mathbf{b} - c^{-1}\mathbf{u} \times \mathbf{e} \ . \tag{11}$$

In other words, the combination of these transformations with the Galilean transformation $x = \bar{x} - ut$ leaves the Maxwell-Lorentz equations invariant to first order. Lorentz used this remarkable property to ease his derivation of the Fizeau coefficient and to give a general proof that to first order optical phenomena were unaffected by the earth's motion through the ether.[15]

It is important to understand that for Lorentz the transformed coordinates and fields were mathematical aids with no direct physical significance. They were only introduced to facilitate the solution of complicated differential equations. The "local time" t' was only called so because it depended on the abscissa. The true physical quantities were the absolute time t and the fields $\mathbf{e}$ and $\mathbf{b}$ representing the states of the ether. In order to prove the first-order invariance of optical phenomena, Lorentz considered two systems of bodies of identical constitution, one at rest in the ether, the other drifting at the velocity $\mathbf{u}$. He first noted that to a field pattern $\mathbf{e}_0 = F(x, y, z, t)$, $\mathbf{b}_0 = G(x, y, z, t)$ for the system at rest corresponded a field pattern $\mathbf{e}, \mathbf{b}$ for the drifting system such that $\mathbf{e}' = F(x, y, z, t')$, $\mathbf{b}' = G(x, y, z, t')$ (the abscissa x being measured in a frame bound to the system). He then noted that $\mathbf{e}'$ and $\mathbf{b}'$ vanished simultaneously if and only if $\mathbf{e}$ and $\mathbf{b}$ did so. Consequently, the borders of a ray of light or the dark fringes of an interference pattern have the same locations in the system at rest and in the drifting system. The change of the time variable is irrelevant, since the patterns observed in optical experiments are stationary. We may conclude that Lorentz's use of the Lorentz invariance was quite indirect and subtle.

There remained a last challenge for Lorentz: to account for the negative result of the Michelson-Morley experiment of 1887. As George Francis FitzGerald had already done, Lorentz noted that the fringe shift expected in a stationary ether theory disappeared if the longitudinal arm of the interferometer underwent a contraction by the amount $\gamma^{-1} = \sqrt{1 - u^2/c^2}$ when moving through the ether. In order to justify this hypothesis, Lorentz first noted that in the case of electrostatics the field equations in a frame bound to the drifting body could be brought back to those for a body at rest through the transformation $x' = \gamma x$. He further assumed that the equilibrium length or a rigid rod was determined by the value

[14] Ibid. : 297

[15] Lorentz, "On the reflexion of light by moving bodies," Koninklijke Akademie van Wetenschappen, *Verslagen* (1892), also in *Collected papers* (ref.11), vol. 4, 215-218.

of intermolecular forces and that these forces all behaved like electrostatic forces when the rod drifted through the ether. Then the fictitious rod obtained by applying the dilation $x' = \gamma x$ to a longitudinally drifting rod must have the length that this rod would have if it were at rest. Consequently, the moving rod contracts by the amount γ^{-1}. The Lorentz contraction thus appears to result from a postulated similarity between molecular forces of cohesion and electrostatic forces.[16]

Fully explained in the *Versuch* of 1895, Lorentz's theory gained broad recognition before the end of the century. Two other physicists, Joseph Larmor of Cambridge and Emil Wiechert of Königsberg, proposed similar theories in the same period. In the three theories, the basic idea was to hybridize Maxwell's theory with the corpuscular concept of electricity and to reduce every optic and electromagnetic phenomenon to the interactions between electric particles through a stationary ether. Besides the optics of moving bodies, these theories explained a variety of magnetic and magneto-optic phenomena, and of course retrieved the confirmed predictions of Maxwell's theory. They benefited from the contemporary rise of an experimental microphysics, including the discoveries of x-rays (1895), radioactivity (1896), and the electron (1897). In 1896, the Dutch experimenter Pieter Zeeman revealed the magnetic splitting of spectral lines, which Lorentz immediately explained through the precession of the orbiting charged particles responsible for the lines. Being much lighter than hydrogen, these particles were soon identified to the corpuscle discovered in cathode rays by Emil Wiechert and Joseph John Thomson. Following Larmor's terminology, this corpuscle became known as the electron and replaced the ions in Lorentz's theory.[17]

4 Poincaré's criticism

In France, the mathematician Henri Poincaré had been teaching electrodynamics at the Sorbonne for several years. After reviewing the theories of Maxwell, Helmholtz, Hertz, Larmor, and Lorentz, he judged that the latter was the one that best accounted for the whole range of optic and electromagnetic phenomena. Yet he was not entirely satisfied with Lorentz's theory, because he believed it contradicted fundamental principles of physics. In general, Poincaré perceived an evolution of physics from the search of ultimate mechanisms to a "physique des principes" in which a few general principles served as guides in the formation of theories. Among these principles were three general principles of mechanics: the

[16]Lorentz, "De relative beweging van der aarde en den aether," Koninklijke Akademie van Wetenschappen, *Verslagen* (1892), transl. as "The relative motion of the earth and the ether" in *Collected papers* (ref. 11), vol. 4, 220-223.

[17]Lorentz, *Versuch einer Theorie der elektrischen un optischen Erscheinungen in bewegten Körpern* (Leiden, 1895), also in *Collected papers* (ref. 11), vol.5, 1-139; "Optische verschinitjnelsen die met de lading en de massa der ionen in verband staan," Koninklijke Akademie van Wetenschappen, Verslagen (1898), transl. as "Optical phenomena connected with the charge and mass of ions" in *Collected papers* (ref. 11) vol. 3, 17-39.

principle of relativity, the principle of reaction, and the principle of least action.[18]

For any believer in the mechanical nature of the electromagnetic ether, it was obvious that these three principles applied to electrodynamics, since ether and matter were together regarded as a complex mechanical system. In particular, it was clear that electromagnetic phenomena would be the same if the same uniform boost was applied to the ether and all material objects. It the boost was applied to matter only, effects of this boost were expected to occur. For instance, Maxwell believed that the force between two electric charges moving together uniformly on parallel lines had to vanish when their velocity reached the velocity of light. Poincaré thought differently. In his view, the ether only was a convenient convention suggested by the analogy between the propagation of sound and the propagation of light. In the foreword of his lectures of his lecture of 1887/8 on the mathematical theories of light, he wrote:[19]

> It matters little whether the ether really exists: that is the affair of the metaphysicians. The essential thing for us is that everything happens as if it existed, and that this hypothesis is convenient for us for the explanation of the phenomena. After all, have we any other reason to believe in the existence of material objects? That too, is only a convenient hypothesis; only this will never cease to do so, whereas, no doubt, some day the ether will be thrown aside as useless.

As we will see, Poincaré actually never abandoned the ether. But he refused to regard it as an ordinary kind of matter whose motion could affect observed phenomena. In his view, the principle of reaction and the principle of relativity had to apply to matter alone. In his lectures of 1899 on Lorentz's theory, he wrote: I consider it very probable that optical phenomena depend only on the relative motion of the material bodies present –light sources and apparatus– and this not only to first or second order but exactly.

It must be emphasized that at that time no other physicist believed in this acceptance of the relativity principle. Most physicists conceived the ether as a physical entity whose wind should have physical effects, even though the precision needed to test this consequence was not yet available. The few physicists, such as Paul Drude or Emil Cohn, who questioned the mechanical ether, felt free to violate principles of mechanics, including the relativity principle.[20]

Lorentz's theory satisfied Poincaré's relativity principle only approximately and did so through what Poincaré called two "coups de pouce": the local time and the Lorentz contraction. Moreover, it violated Poincaré's reaction principle, since Lorentz's equations implied that the net force acting on all the ions or electrons should be the space integral of $\partial(\mathbf{e} \times \mathbf{b})/c\partial t$, which does not vanish in general. In his contribution to Lorentz's jubilee of 1900, Poincaré iterated this criticism and further discussed the nature and impact of the violation of the reaction principle. In the course of this argument, about which more details will be given in

[18]H. Poincaré, *Electricité et optique. La Lumière et les théories électrodynamiques* [Sorbonne lectures of 1888, 1890 and 1899)], ed. J. Blondin and E. Néculcéa, (Paris, 1901).

[19]Poincaré, *Théorie mathématique de la lumière* (Sorbonne lectures, 1887-88), ed. J. Blondin (Paris, 1889), I.

[20]Poincaré, ref. 18, 536.

a moment, he relied on Lorentz's transformations (11) to compute the energy of a pulse of electromagnetic radiation from the standpoint of a moving observer. The transformed fields $\mathbf{e}'$ and $\mathbf{b}'$, he noted, are the fields measured by a moving observer. Indeed the force acting on a test unit charge moving with the velocity $\mathbf{u}$ is $\mathbf{e} + c^{-1}\mathbf{u} \times \mathbf{b} = \mathbf{e}'$ according to the Lorentz force formula. Poincaré went on noting that the local time $t' = t - ux/c^2$ was that measured by moving observers if they synchronized their clocks in the following manner:[21]

> I suppose that observers placed in different points set their watches by means of optical signals; that they try to correct these signals by the transmission time, but that, ignoring their translational motion and thus believing that the signals travel at the same speed in both directions, they content themselves with crossing the observations, by sending one signal from A to B, then another from B to A.

Poincaré only made this remark *en passant*, gave no proof, and did as if it had already been on Lorentz's mind. The proof goes as follows. When B receives the signal from A, he sets his watch to zero (for example), and immediately sends back a signal to A. When A receives the latter signal, he notes the time τ that has elapsed since he sent his own signal, and sets his watch to the time $\tau/2$. By doing so he commits an error $\tau/2 - t_-$, where t_- is the time that light really takes to travel from B to A. This time, and that of the reciprocal travel are given by $t_- = \mathrm{AB}/(c+u)$ and $t_+ = \mathrm{AB}/(c-u)$, since the velocity of light is c with respect to the ether (see fig. 4). The time τ is the sum of these two traveling times. Therefore, to first order in u/c the error committed in setting the watch A is $\tau/2 - t_- = (t_+ - t_-)/2 = u\mathrm{AB}/c^2$. At a given instant of the true time, the times indicated by the two clocks differ by $u\mathrm{AB}/c^2$, in conformity with Lorentz' expression of the local time.

Poincaré transposed this synchronization procedure from an earlier discussion on the measurement of time, published in 1898. There he noted that the dating of astronomical events was based on the implicit postulate "that light has a constant velocity, and in particular that its velocity is the same in all directions." He also explained the optical synchronization of clocks at rest, and mentioned its similarity with the telegraphic synchronization that was then being developed for the purpose of longitude measurement. As a member of the Bureau des Longitudes, Poincaré naturally sought an interpretation of Lorentz's local time in terms of cross-signaling. As a believer in the relativity principle, he understood that moving observers would never know their motion through the ether and therefore could only do as if these signals propagated isotropically.[22]

Poincaré thus provided a physical interpretation of the transformed time t' and the transformed fields $\mathbf{e}'$ and $\mathbf{b}'$, which only referred to a fictitious system for Lorentz. This interpretation greatly eased the use of this transformation, for it made the (first-order) invariance of optical phenomena a direct consequence of

[21] Poincaré, "La théorie de Lorentz et le principe de la réaction." In Recueil de travaux offerts par les auteurs à H.A. Lorentz à l'occasion du 25ème anniversaire de son doctorat le 11 décembre 1900, *Archives néerlandaises*, 5 (1900), 252-278, on 272.

[22] Poincaré, "La mesure du temps", *Revue de métaphysique et de morale*, 6 (1898), 371-384.

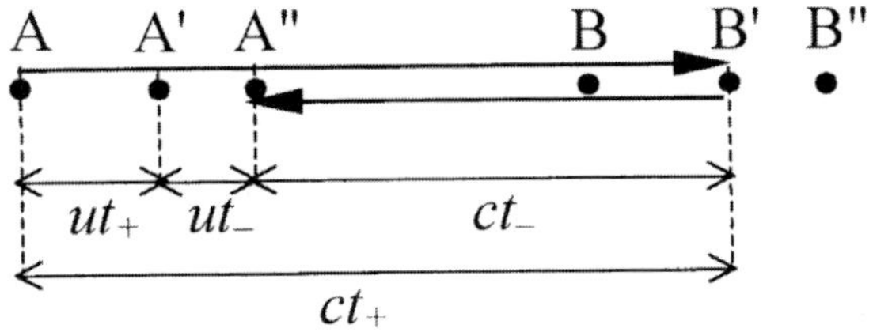

Figure 4: Cross-signaling between two observers moving at the velocity u through the ether. The points **A**, **A**$'$, **A**$''$, **B**, **B**$'$, **B**$''$ represent the successive positions of the observers in the ether when the first observer sends a light signal, when the second observer receives this signal and sends back another signal, and when the first observer receives the latter signal.

the formal invariance of the Maxwell-Lorentz equations. Yet it would be a mistake to believe that Poincaré thereby redefined the concepts of space and time. In his terms, the Lorentz-transformed quantities referred to the apparent states of the field for a moving observer. The true states remained those defined with respect to the ether. As we will see, Poincaré never gave up this view.

5 The Lorentz invariance

Strangely, Lorentz overlooked Poincaré's reinterpretation of his transformations, and kept reasoning in terms of a fictitious system brought to rest. So did other experts on the electrodynamics of moving bodies until at least 1904. Nevertheless, Lorentz took some of Poincaré's criticism seriously. In 1904, he offered a new version of his theory in which the invariance of optical phenomena held at every order in u/c, without the "coups de pouce" reproached by Poincaré. He knew since 1899 that the homogenous field equations for a system bound to the earth could be brought to the form they have for a system at rest in the ether through the transformations

$$x' = \gamma\epsilon x, \quad y' = \epsilon y, \quad z' = \epsilon z, \quad t' = \epsilon(\gamma^{-1}t - \gamma u x c^{-2})$$
$$\mathbf{e}' = \epsilon^{-2}(1,\gamma)(\mathbf{e} + c^{-1}\mathbf{u}\times\mathbf{b}), \quad \mathbf{b}' = \epsilon^{-2}(1,\gamma)(\mathbf{b} - c^{-1}\mathbf{u}\times\mathbf{e})\ , \tag{12}$$

where ϵ is an undetermined constant (for a given value of u) and the factor $(1,\gamma)$ means a multiplication by 1 of the component of the following vector parallel to **u** and a multiplication by γ of the component perpendicular to **u**. In 1904, he generalized this result to the coupling between electrons and field.[23]

[23] Lorentz, "Electromagnetic phenomena in a system moving with any velocity smaller than light," Royal Academy of Amsterdam, *Proceedings* (1904), also in *Collected papers* (ref. 11), vol. 5, 172-197.

Specifically, Lorentz realized that for a spherical electron subjected to the Lorentz contraction and carrying the electromagnetic momentum

$$\mathbf{p} = c^{-1} \int (\mathbf{e} \times \mathbf{b}) d\tau \ , \tag{13}$$

his transformations brought back the equation of motion

$$d\mathbf{p}/dt = e[\mathbf{e} + c^{-1}(\mathbf{u} + \mathbf{v}) \times \mathbf{b}] \tag{14}$$

of an electron of charge e to the form it has for a system at rest ($u = 0$), if and only if the constant ϵ had the value 1. On his way to this result, he derived the expression $\mathbf{p} = m_0 \gamma \mathbf{v}$ of the momentum, where $m_0 = e^2/6\pi R c^2$ is the electromagnetic mass of a spherical-shell electron of radius R. Lastly, Lorentz gave expressions of the transformed source terms of the field equations such that dipolar emission in the moving system transformed into dipolar emission in the system at rest. Combining all these results, he could assert that optical phenomena in a moving system were the same as in a system at rest.

This result only held in the dipolar approximation, because Lorentz's expression of the transformed source terms was not the one today regarded to be correct. Lorentz also neglected the spinning motion of the electrons, and overlooked the cohesive forces that the stability of his contractile electron required. His derivation of the invariance of optical phenomena was complex and indirect, for it involved a double-step transformation, the fictitious system at rest, and comparison between the states of this system and those of the real system. For other phenomena, there is no doubt that Lorentz still believed that motion with respect to the ether could in principle be detected.

Poincaré reacted enthusiastically to Lorentz memoir, because he saw in it an opportunity to satisfy the relativity principle in a complete and exact manner. He published the results of the ensuing reflections under the title "Sur la dynamique de l'électron," first as a short note of 5 June 1905 in the *Comptes rendus*, and as a bulky memoir in the *Rendiconti* of the Circolo matematico di Palermo for the following year. He first defined the "relativity postulate" as follows:

> It seems that the impossibility of experimentally detecting the absolute motion of the earth is a general law of nature; we naturally incline to assume this law, which we shall call the Postulate of Relativity, and to do so without any restriction.

Correcting Lorentz's expression of the transformed source terms, he then showed that "the Lorentz transformations"

$$\begin{aligned} &x' = \gamma(x - ut), \quad y' = y, \quad z' = z, \quad t' = \gamma(t - uxc^{-2}) \ , \\ &\mathbf{e}' = (1, \gamma)(\mathbf{e} + c^{-1}\mathbf{u} \times \mathbf{b}), \quad \mathbf{b}' = (1, \gamma)(\mathbf{b} - c^{-1}\mathbf{u} \times \mathbf{e}) \ , \end{aligned} \tag{15}$$

left the Maxwell-Lorentz equations invariant. These transformations are obtained by combining the transformations (12), which Lorentz used, with the Galilean transformation $x' = x - ut$. Poincaré showed that they formed a group, and used this property to determine the global scaling factor ϵ. He noted that the coordinate transformations left the quadratic form $x^2 + y^2 + z^2 - c^2t^2$ invariant and could thus be regarded as rotations in a four-dimensional space with an imaginary fourth coordinate. He obtained the relativistic law for the "addition" of velocities, for which the combined velocity always remains inferior to the limit c.[24]

Next, Poincaré showed that a model of the contractile electron could be conceived in which the cohesive forces (the so-called Poincaré tension) preserved the Lorentz invariance. He thus retrieved Lorentz's expression $\mathbf{p} = m_0\gamma\mathbf{v}$ for the momentum of the electron. Lastly, he argued that in order to be compatible with the postulate of relativity, gravitational interactions should propagate at the velocity of light; and he proposed modifications of Newton's law of gravitation that made it compatible with Lorentz invariance.

Thus, there is no doubt that Poincaré regarded Lorentz invariance as a general requirement for the laws of physics, and that he identified this formal condition with the principle of relativity. On the latter point, his only comment was:

> The reason why we can, without modifying any apparent phenomenon, confer to the whole system a common translation, is that the equations of an electromagnetic medium are not changed under certain transformations which I shall call the Lorentz transformations; two systems, one at rest, the other in translation, thus become exact images of one another.

The Palermo memoir, long and thorough as it was, said nothing on the interpretation to be given to the transformed coordinates and fields. Perhaps Poincaré believed this should not be the main point. Perhaps he had not yet been able to provide an operational understanding of Lorentz's local time at any order in u/c. There is no doubt, however, that he regarded the transformed fields and coordinates as the ones measured by moving observers. At the Saint-Louis conference of 1904, he repeated (and attributed to Lorentz!) his definition of the local time by optical cross-signaling. In his Sorbonne lectures of 1906, he proved that this definition remained valid at any order in u/c, and he characterized the Lorentz transformations as the ones giving the "apparent space and time coordinates."[25]

The same lectures and later talks on the "mécanique nouvelle" show that Poincaré nonetheless maintained the ether and the ordinary concepts of space and time. In his view, the clocks bound to the ether frame gave the true time, for it was only in this frame that the true velocity of light was c. The clocks of a moving frame only gave the apparent time. For those who would think that the difference with Einstein's theory of relativity is merely verbal, it is instructive to look at an argument Poincaré repeatedly gave to justify optical synchronization.[26]

---

[24]Poincaré, "Sur la dynamique de l'électron," Académie des Sciences, *Comptes-rendus*, 140, (1905), 1504-1508; "Sur la dynamique de l'électron," *Rendiconti del Circolo matematico di Palermo* (1906), also in Poincaré, *Oeuvres* (Paris, 1954), vol. 9, 494-550, on 495.

[25]Poincaré, ibid., 495; "L'état actuel et l'avenir de la physique mathématique" (Saint-Louis lecture), *Bulletin des sciences mathématiques*, 28 (1904), 302-324, transl. in Poincaré, *The foundations of science* (New York, 1929); "Les limites de la loi de Newton," Sorbonne lectures (1906-1907) ed. by H. Vergne in *Bulletin astronomique publié par l'observatoire de Paris*, 17 (1953), 121-365, chap. 11.

[26]Poincaré, ibid., 218-220; "La dynamique de l'électron," *Revue générale des sciences pures et appliquées* (1908), also in *Oeuvres* (ref. 24), vol. 9, 551-586.

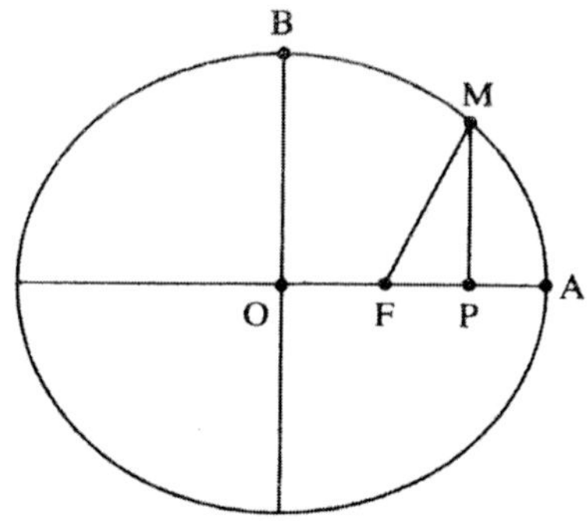

Figure 5: *Poincaré's light ellipsoid* ($a = OA,\ b = OB,\ f = OF$).

Simultaneity should be transitive, namely: if the clock A is synchronized with the clock B, and if the clock B is synchronized with the clock C, then the clock A should be synchronized with clock C for any given choice of the positions of the three clocks. Indeed, any breakdown of transitivity could be used to detect motion through the ether and thus to violate the relativity principle. Now consider an observer moving with the constant velocity $\mathbf{u}$ through the ether and emitting a flash of light at time zero. At the value t of the true time, this light is located on a sphere of radius ct centered at the emission point. Poincaré next considered the appearance of this light shell for a moving observer, the rulers of which are subjected to the Lorentz contraction. The result is an ellipsoid of revolution, the half-axes of which have the values $a = \gamma ct$ and $b = ct$ (see fig. 5). As the eccentricity is $e = \sqrt{1 - b^2/a^2} = u/c$, the focal distance $f = ea = \gamma ut$ is equal to the apparent distance traveled by the observer during the time t. Therefore, the Lorentz contraction is the contraction for which the position of the observer at time t coincides with the focus F of the light ellipsoid he has emitted.

Now consider a second observer traveling with the same velocity $\mathbf{u}$ and receiving the flash of light at the time t_+. The position M of this observer belongs to the ellipsoid $t = t_+$, and the distance FM represents the apparent distance between the two observers, which is invariable. According to a well-known property of ellipses, we have

$$FM + eFP = b^2/a \ , \tag{16}$$

where P denotes the projection of M on the larger axis. The length FP being equal to the difference x' of the apparent abscissas of the two observers, this implies

$$t_+ = \gamma FM/c + \gamma ux'/c^2 \ . \tag{17}$$

Suppose that the two observers synchronize their clocks by cross-signaling. The traveling time of the reverse signal is

$$t_- = \gamma FM/c - \gamma ux'/c^2 \tag{18}$$

Therefore, two events are judged simultaneous by these observers if and only if their true times differ by

$$(t_+ - t_-)/2 = \gamma u x'/c^2 \ . \tag{19}$$

This condition is obviously transitive.[27]

For any one familiar with Einstein's theory of relativity, this reasoning seems very odd. Indeed, the light ellipsoid corresponds to a fixed value of the time t in one reference frame and to space measured in another frame. In general, Poincaré's theory allows for the use of the "true time" in any reference system, whereas our relativity theory regards this sort of mixed reference as a mathematical fiction. This means that the conceptual basis of Poincaré's theory is not compatible with Einstein's, even though both theories are internally consistent and have the same empirical predictions (for the electrodynamics of moving bodies).[28]

Another oddity of Poincaré's theory is his naming the Lorentz contraction "a hypothesis." As we just saw, Poincaré showed that the contraction was necessary to the transitivity of optical synchronization, which itself derives from the relativity principle. He nonetheless spoke of a hypothesis, probably because he did not quite trust the implicit conventions made in this reasoning. In the Palermo memoir, he clearly indicated his dissatisfaction with the present state of the theory:

> We cannot content ourselves with simply juxtaposed formulas that would agree only by some happy coincidence; the formulas should, so to say, penetrate each other. Our mind will be satisfied only when we believe that we perceive the reason of this agreement, so that we may fancy that we have predicted it.

Poincaré meant that the Lorentz covariance of all forces in nature, including gravitation, could not be regarded as a mere consequence of the principle of relativity. He believed this symmetry also implied more arbitrary assumptions, such as the similarity between electromagnetic and other forces and the universality of the velocity of light as a propagation velocity.[29]

As Poincaré reminded his reader, one way to justify these assumptions was the electromagnetic view of nature, according to which electromagnetism should be the ultimate basis of all physics. More appealing to him was the following suggestion:

> The common part of all physical phenomena would only be an appearance, something that would pertain to our methods of measurement. How do we perform our measurements? By superposing objects that are regarded as rigid bodies, would be one first answer; but this is no longer true in

[27] Although Poincaré did not do so much, the expression of the local time t' can simply be obtained by requiring the apparent velocity of light to be equal to c. This condition implies $FM = ct'$, and $t = \gamma(t' + ux'/c^2)$. Calling x the true abscissa of the second observer at time t (with respect to the emission point of the flash), we also have $x' = \gamma(x - ut)$. Consequently, $t' = \gamma(t - ux/c^2)$, in conformity with the Lorentz transformations.

[28] The empirical equivalence of the two theories simply results from the fact that any valid reasoning of Einstein's theory can be translated into a valid reasoning of Poincaré's theory by arbitrarily calling the time, space, and fields measured in one given frame the true ones, and calling all other determinations apparent.

[29] Poincaré, ref. 24 (1906), 497.

the present theory, if one assumes the Lorentz contraction. In this theory, two equal lengths are, by definition, two lengths which light takes equal time to travel through. Perhaps it would be sufficient to renounce this definition so that Lorentz's theory would be as completely overturned [bouleversée] as Ptolemy's system was through Copernicus' intervention.

Lorentz's explanation of the null result of the Michelson-Morley experiment, Poincaré reasoned, implicitly rested on the convention that two lengths (sharing the same motion) are equal if and only if light takes the same (true) time to travel through them. What he meant by dropping this convention is not clear. Some commentators have speculated that he meant a revision of the concept of time, in Einstein's manner. This is not very likely, because the context of Poincaré's suggestion was length measurement instead of time measurement, and also because he ignored Einstein's point of view to the end of his life. More likely he was alluding to a suggestion he had earlier made at the Saint-Louis conference: "that the ether is modified when it moves relative to the medium which penetrates it."[30]

To sum up, in 1905/6 Poincaré obtained a version of the theory of relativity based on the principle of relativity and the Lorentz group. He believed this symmetry should apply to all forces in nature. He exploited it to derive the dynamics of the electron on a specific model and to suggest a modification of the law of gravitation. He nevertheless maintained the ether as the medium in which light truly propagated at the constant velocity c and clocks indicated the true time. He regarded the quantities measured in moving frames as only apparent, although the principle of relativity forbade any observational distinction between a moving frame and the ether frame. He understood the compatibility of the Lorentz transformations of coordinates with the optical synchronization of clocks and the invariance of the *apparent* velocity of light, but hesitated on the physical significance of the Lorentz contraction and never discussed the dilation of time.

6 Einstein's theory

Albert Einstein had an early interest in electrodynamics, if only because his family owned a small electrotechnical company. At age sixteen, he wrote a little essay on the state of the ether in an electromagnetic field. If we believe a late reminiscence, he also wondered about the appearance of a light wave for an observer traveling along with it. In 1896 he entered the Zürich Polytechnikum, where he learned electrodynamics in the standard continental style. Two years later he studied Maxwell's theory by himself from Drude's *Physik des Aethers*. Drude was a sympathizer of Ernst Mach's philosophy, and belonged to a tradition of German physics that favored phenomenological theories over mechanistic assumptions. In his rendering of Maxwell's theory, he avoided any picture of ether processes and propounded to redefine the ether as space endowed with special physical properties.[31]

[30] Poincaré, ref. 24 (1906), 498; ref. 25 (1904), 315 (Foundations).

[31] A. Einstein, "Über die Untersuchung des Aetherzustandes im magnetischen Felde," in John Stachel et al. (eds), *The collected papers of Albert Einstein*, vol. 1 (Princeton, 1987), 6-9; P.

In the summer of 1899, Einstein's reading of Hertz's Untersuchungen prompted the following comment, addressed to his lover Mileva Marić:

> I am more and more convinced that the electrodynamics of moving bodies, as it is presented today, does not agree with the truth, and that it should be possible to present it in a simpler way. The introduction of the name 'ether' into the electric theories has led to the notion of a medium of whose motion one could speak of without being able, I believe, to associate a physical meaning to this statement. I believe that electric forces can be directly defined only for empty space, [which is] also emphasized by Hertz. Further, electric currents will have to be regarded not as 'the vanishing of electric polarization in time' but as motion of true electric masses, whose physical reality seems to result from the electrochemical equivalents.... Electrodynamics would then be the science of the motions in empty space of moving electricities and magnetisms.

Einstein's criticism targeted Hertz's electrodynamics of moving bodies, which developed the Maxwellian idea of an ether fully dragged by matter. Einstein was also aware of the Maxwellian concept of electric current as "the vanishing of electric polarization in time," and suggested to replace it with the motion of ions moving through empty space. What he had in mind probably was a theory similar to Lorentz's and Wiechert's, in which electromagnetic phenomena are brought back to the interactions of ions through a stationary ether. Einstein added that "the radiation experiments" would decide between the two conceptions. He presumably meant to compare the intensities of light emitted from the same source in opposite directions.[32]

A month later, Einstein thought of another experiment concerning "the effect that the relative motion of bodies with respect to the luminiferous ether has on the velocity of propagation of light in transparent bodies." The physics professor, Heinrich Weber, to whom he explained this project and the motivating theory, told him to read a paper by Wilhelm Wien that contained a short account of Lorentz's theory and a description of many experiments on ether motion, including those of Fizeau and of Michelson-Morley. Einstein presumably welcomed Fizeau's result, which confirmed the stationary ether. But he may have doubted the import of the Michelson-Morley null result, because two years later, in the fall of 1901, he was still planning a new interferometric method "for the search of the relative motion of matter with respect to the luminiferous ether."[33]

Sometime after 1901, Einstein ceased to look for experimental tests of motion through the ether and adopted the relativity principle. Although the null-result of ether-drift experiments probably contributed to this move, Einstein's autobiographical remarks and the relativity paper of 1905 give an essential role to another kind of consideration. In the introduction to this paper, Einstein remarks that magneto-electric induction receives two very different interpretations in Lorentz's theory, according as it is the magnet or the electric conductor that is moving with respect to the ether. In the first case the motion of the magnet implies the exis-

Drude, *Physik des Aethers auf electromagnetischer Grundlage* (Stuttgart, 1894).

[32]Einstein to Marić [Aug 1899], ECP 1, 225-227.

[33]Einstein to Marić, 10 and 28? Sep 1899, ECP 1, 229-230, 233-235; W. Wien, "Über die Fragen, welche die translatorische Bewegung des Lichtethers betreffen," appendix to *Annalen der Physik*, 65 (1898): I-XVIII); Einstein to Grossmann, 6 Sep 1901, in Stachel (ref. 31), 315-316.

tence of an electric field e within the conductor (such that $\nabla \times \mathbf{e} = -\partial \mathbf{b}/\partial t$)). In the second case, there is no electric field within the conductor, and the Lorentz force ($\mathbf{v} \times \mathbf{b}$ per unit charge) is responsible for the motion of the electrons. Yet the induced current only depends on the relative motion of the coil and the magnet.[34]

This fairly obvious remark had already been made by several authors. Most of these, however, easily accepted the theoretical asymmetry and believed that finer details of electromagnetic induction or other phenomena would reveal effects of motion through the ether. Einstein thought differently. Following an epistemological trend expressed in Hertz's, Hume's, and Mach's writings, he rejected theoretical asymmetries that had no empirical counterpart. As Lorentz's stationary ether led to many asymmetries of this sort, it had to be rejected.

This way or reasoning explains why Einstein, unlike Poincaré, conflated the adoption of the relativity principle with the rejection of the ether. He thus found himself compelled to imagine a theory of electromagnetic propagation that would respect the relativity principle without disturbing the confirmed predictions of Lorentz's theory. The relativity principle implies that the measured velocity of light should be the same in any inertial frame. For one who has given up the ether, there is a simple way to satisfy this requirement: to make this velocity depend on the velocity of the emitter, as was the case in Newton's old corpuscular theory of light. According to later reminiscences, this was Einstein's first idea: he tried to modify the expression of the retarded interaction between two charged particles in such a way that it would depend on their relative motion only. Alas, difficulties soon came up. In particular, Einstein found that the light emitted by an accelerated source could sometime back up on itself, because the successive wave planes traveled at different velocities depending on the velocity of the source during their emission.[35]

Einstein gave up this theory, and long remained unable to conciliate Lorentz's theory with the relativity principle. According to the Kyoto lecture of 1922, he suddenly realized that a redefinition of the concept of time solved his problem during a conversation with his friend Michele Besso in the spring of 1905. In the same lecture, Einstein also indicated that he had earlier tried to assume the validity of the Maxwell-Lorentz equations in any inertial frame. This assumption of course implied that the velocity of light should be the same in any inertial frame, against the Galilean rule for the addition of velocities. The difficulty disappeared when Einstein realized that there was "an inseparable connection between time and signal velocity."[36]

---

[34] A. Einstein, "Zur Elektrodynamik bewegter Körper," *Annalen der Physik*, 17 (1905), 891-921, on 891. Einstein removed this asymmetry in 1905 (ibid. on 909-910) by making the separation between electric and magnetic field depend on the reference frame and by defining the electromotive force in a conductor as the electric field in a frame bound to this conductor.

[35] Einstein to Ehrenfest, 25 Apr 1912, quoted in John Stachel et al. (eds.) *The collected papers of Albert Einstein*, vol. 2 (Princeton, 1989), 263, and further ref. ibid.

[36] Einstein, "How I created the theory of relativity?" (from notes taken by Jun Ishiwara from Einstein's lecture in Kyoto on 14 Dec 1922), *Physics today*, 35: 8 (1982), 45-47.

This remark and the ensuing developments may have been eased by Einstein's readings. We surely know he had read Lorentz's *Versuch* of 1895, and was therefore aware of the local time and the role it served in preserving the form of the Maxwell-Lorentz equations to first order. He may also have known the exact form of the Lorentz transformations, for in 1904 several German theorists commented on them in journals that he regularly read. We also know that he read Poincaré's *La science et l'hypothèse*, which contained an eloquent plead for the relativity principle as well as a brief criticism of simultaneity:

> There is no absolute time. To say two durations are equal is an assertion which has by itself no meaning and which can acquire one only by convention. Not only have we no direct intuition of the equality of two durations, but we have not even direct intuition of the simultaneity of two events occurring in different places: this I have explained in an article entitled *La mesure du temps*.

The German version of this book, published in 1904 and perhaps the one read by Einstein, had a long citation of Poincaré's article of 1898, including:

> The simultaneity of two events or the order of their occurrence, and the equality of two time intervals must be defined so that the expression of the laws of physics should be the simplest possible; in other words, all those rules and definitions [conventions for time measurement] only are the fruits of an unconscious opportunism.

The German editor further mentioned the possibility that a new time coordinate may be a function of the older time *and space* coordinates. Lastly, Einstein may have read Poincaré's memoir of 1900, which contained the interpretation of Lorentz's local time in terms of optically synchronized clocks. Or he may have been aware of Emil Cohn's remark of 1904 that the local time was the time for which the propagation of light was isotropic.[37]

Whether Einstein borrowed this idea or rediscovered it by himself, he became aware of a relation between local time and optical synchronization. In his understanding of the relativity principle, there was no ether and all inertial frames were entirely equivalent. Therefore, the time and space coordinates defined in these frames all had the same status. The constancy of the velocity of light no longer resulted from the existence of the ether, and had to be postulated separately. According to the relativity principle, this property had to hold in any inertial system. The apparent absurdity of this consequence disappeared if time was *defined* in conformity with the light postulate. This definition turned out to imply the Lorentz transformations, without any recourse to the Maxwell-Lorentz equations.

Through reasoning of that kind, Einstein arrived at the "new kinematics" that formed the first part of his celebrated memoir of 1905 "On the electrodynamics of moving bodies." The following summary should suffice to exhibit the magnificent architecture of this memoir.[38]

---

[37] Poincaré, *La science et l'hypothèse* (Paris, 1902), 111; *Wissenschaft und Hypothese* (Leipzig, 1904), 286-289.

[38] Einstein, ref. 34.

Introduction: Einstein exposes the aforementioned asymmetry of electromagnetic induction in Lorentz's theory and uses it to plead for the strict validity of the relativity principle. He announces that this principle, together with the principle of the constancy of the velocity of light, leads to a new kinematics that solves the contradictions of the electrodynamics of moving body.

I. Kinematical part

§1. For two distant clocks A and B of identical constitution attached to a given reference frame, Einstein defines synchronicity by the condition $t_B - t_A = t'_A - t_B$, where t_A is the time of the clock A at which a light signal is sent from A, t_B the time of the clock B at which this signal reaches **B** and a replying signal is sent from B, and t'_A the time at which the latter signal arrives at A. This definition is arranged so that the propagation of light should be isotropic in the given frame. "In conformity with experience," Einstein further assumes that the ratio $2AB/(t'_A - t_A)$ is the universal constant c.

§2. Einstein states the two principles on which his new kinematics is built: the "relativity principle" according to which the laws of physics are the same in any inertial system, and the "principle of the constancy of the velocity of light" in a given inertial system. He then shows that simultaneity is a relative notion, because according to the above given criterion two clocks synchronized in a given reference frame are not in another.

§3. Einstein derives the Lorentz transformations by requiring the velocity of light to be the same constant in two different frames of reference in the two following cases: when the light path is parallel to the relative velocity of the two frames, and when it is normal to this velocity in one of the frames. His reasoning also relies on the group structure of the transformations in order to determine the global scaling factor of the transformations.

§4. Einstein gives the "physical consequences" of these transformations for the behavior of moving rigid bodies and clocks. The extremities of a rigid ruler moving edgewise at the velocity v in a given frame of reference coincide, at a given instant of this frame, with points of this frame whose distance is proportional to $\sqrt{1 - v^2/c^2}$. Einstein thus introduces the contraction of lengths as a perspectival effect. Most innovatively, he shows that a clock C traveling at the uniform speed v from the clock A to the clock B of a given reference frame appears to be slow compared to these clocks by a factor $1/\sqrt{1 - v^2/c^2}$. He predicts the same retardation if the clock C makes a U-turn at B and returns to A. He extends this result to any loop-wise trip of the clock C.

§5. Einstein gives the relativistic law for the composition of velocities.

II. Electrodynamic part

This part is devoted to the application of the new kinematics to electrodynamics.

§6. Einstein proves the covariance of the homogenous Maxwell-Lorentz equations, and uses the relevant field transformations to remove the theoretical asymmetry he condemned in his introduction: the force acting on a moving unit point charge at a given instant must now be regarded as the electric field acting on it in an inertial frame that has the same velocity as the charge does at this instant.

§7. Einstein uses the transformation of a plane monochromatic wave to derive the Doppler effect and stellar aberration.

§8. Einstein derives the transformation law for the energy of a light pulse, and uses this law to derive the work done by radiation pressure on a moving mirror.

§9. Einstein obtains the covariance of the inhomogeneous Maxwell-Lorentz equations, thus establishing "the conformity of the electrodynamic basis of Lorentz's theory... with the relativity principle." This remark implies that the new theory retrieves every consequence of Lorentz's theory for the electrodynamics and optics of moving bodies (including the Fresnel drag, for instance).

§10. Einstein obtains the relativistic equation of motion of an electron in an electromagnetic field by assuming the approximate validity of Newtonian mechanics in a quasi-tangent frame (in which the velocity of the electron remains small within a sufficiently small time interval) and transforming to the laboratory frame. He calls for experimental testing of the resulting velocity dependence of the mass of the electron. As he knew, the Göttingen experimenter Walther Kaufmann had performed several experiments of that kind in order to test the existence of an electromagnetic mass and to decide between competing models of the electron.

Most of the components of Einstein's paper appeared in others' anterior works on the electrodynamics of moving bodies. Poincaré and Alfred Bucherer had the relativity principle. Lorentz and Larmor had most of the Lorentz transformations, Poincaré had them all. Cohn and Bucherer rejected the ether. Poincaré, Cohn, and Abraham had a physical interpretation of Lorentz's local time. Larmor and Cohn alluded to the dilation of time. Lorentz and Poincaré had the relativistic dynamics of the electron. None of these authors, however, dared to reform the concepts of space and time. None of them imagined a new kinematics based on two postulates. None of them derived the Lorentz transformations on this basis. None of them fully understood the physical implications of these transformations. It all was Einstein's unique feat.

7 The inertia of energy

In the fall of 1905, Einstein wrote to his friend Conrad Habicht:

> Another consequence of the electrodynamics paper came to my mind. Together with Maxwell's fundamental equations, the relativity principle implies that mass is a measure of the energy content of bodies. Light transports mass. There should be a sensible diminution of mass in the case of radium. The line of thought is amusing and fascinating. But is not the dear Lord laughing about it? Is not he pulling me by the nose? This much I cannot know.

Although Einstein's extraordinary conclusion was entirely new, the paradoxes that led to it were not. In order to see that, we need to return to Poincaré's memoir of 1900 on the reaction principle in Lorentz's theory.[39]

Remember that Poincaré denounced the violation of this principle when applied to matter *alone*. Lorentz and other theorists were unshaken by this objection. They believed that the ether, stationary though it was, could well carry the missing momentum. In the name of the electromagnetic worldview, Max Abraham based his dynamics of the electron of 1902 on the concept of electromagnetic momentum. Ironically, he attributed the expression $c^{-1}\mathbf{e}\times\mathbf{b}$ of this momentum to Poincaré, who had only introduced it as an absurd contribution to the momentum balance.

Poincaré's abhorrence for this notion was not a mere consequence of his ghostly concept of the ether. It resulted from his Newtonian insight that any violation of the principle of reaction led, together with the relativity principle, to the possibility of perpetual motion. Suppose, with him and Newton, that two bodies, initially at rest and isolated from other bodies, act on each other in a non-balanced way by forces that depend only on their configuration. Connect the two bodies by a rigid bar. The resulting system begins to move. According to the principle of relativity, the net force acting on the system does not depend on the acquired velocity. Therefore, the system undergoes a forever accelerated motion.

This reasoning, which assumes direct action from matter to matter, does not immediately apply to electrodynamics. In this case, Poincaré examined the implications of Lorentz's theory for the motion of the center of mass of the matter-field system. Calling m the mass density of matter, and j the energy density $(\mathbf{e}^2+\mathbf{b}^2)/2$ of the field, the Maxwell-Lorentz equations imply the global momentum relation[40]

$$\int m\mathbf{v}d\tau + \int c^{-1}\mathbf{e}\times\mathbf{b}d\tau = \text{constant} , \tag{20}$$

and the local energy relation

$$\frac{\partial j}{\partial t} + \nabla\cdot(c\mathbf{e}\times\mathbf{b}) = -\rho\mathbf{v}\cdot\mathbf{e} . \tag{21}$$

In turn, these relations imply the balance

$$\frac{d}{dt}\int c^{-2}j\mathbf{r}d\tau + \int m\mathbf{v}d\tau + \int (c^{-2}\rho\mathbf{v}\cdot\mathbf{e})\mathbf{r}d\tau = \text{constant} . \tag{22}$$

[39] Einstein to Habicht, undated [Jun to Sep 1905], in The *collected papers of Albert Einstein*, vol. 3; Poincaré, ref. 21.

[40] In 1900, Poincaré had no reason to doubt the expression $m\mathbf{v}$ of the momentum density of matter. He assumed that eventual non-electromagnetic forces balanced each other.

When there is no energy transfer between matter and field, the third term vanishes and the theorem of the uniform motion of the center of mass of the system is saved by associating to the field the flow of a fictitious fluid of mass density j/c^2. In the general case, Poincaré added the fiction of a latent, ether-bound fluid. He set the local conversion rate between free and latent fluid to $\rho\mathbf{v} \cdot \mathbf{e}/c^2$, so that the center of mass of matter, free fluid, and latent fluid moved uniformly. It must be emphasized that he only introduced these fictitious entities to show more precisely how Lorentz's theory violated the theorem of the center of mass.

Poincaré went on to show that this violation, or the concomitant violation of the reaction principle, led to absurd consequences. For this purpose, he considered a Hertzian oscillator placed at the focus of a parabolic mirror and emitting radiation at a constant rate. This system moves with the absolute velocity $\mathbf{u}$ in the direction of emission, and is heavy enough so that the change of this velocity can be neglected for a given momentum change. For an observer at rest in the ether, the conservation of energy reads

$$S = J + (-J/c)u \ , \tag{23}$$

where S is the energy spent by the oscillator in a unit time, J the energy of the emitted wave train, and $-J/c$ the recoil momentum according to Lorentz's theory. For an observer moving at the velocity $\mathbf{u}$ of the emitter, the recoil force does not work, and the spent energy S is obviously the same. According to the Lorentz transformations for time and fields (to first order), this observer should ascribe the energy $J(1 - u/c)$ to the emitted radiation, and the value $(-J/c)(1 - u/c)$ to the recoil momentum. Hence the energy principle is satisfied for the moving observer, but the momentum law is not. Poincaré concluded that the violation of the principle of reaction in Lorentz's theory led to a first-order violation of the relativity principle for electromagnetic forces.[41]

In his Saint-Louis lecture of 1904, Poincaré acknowledged recent experimental confirmations of the radiation pressure, as well as Kaufmann's measurements of the velocity-dependence of the mass of the electron. The latter results, he now judged, "rather seemed to confirm... the consequences of the theory contrary to Newton's principle [of reaction]." In 1906, he explained how this violation could be conciliated with the relativity principle *at the electronic level*: the lack of invariance of the force acting on an electron under the Lorentz transformations is compensated by the velocity-dependence of its mass, so that the equations of motion written in different reference frames are equivalent. More generally, Poincaré argued that if all forces, including inertial ones, transformed like electromagnetic forces under a Lorentz transformation, then the balance of forces held in any reference frame.[42]

As everything worked fine at the microscopic level, Poincaré did not feel necessary to revisit the macroscopic radiation paradox of 1900. Had he done so, he would have seen that the discrepancy Ju/c^2 between the recoil force in the moving

[41]In 1898, Alfred Liénard had already pointed to the first-order modification of the Lorentz force through a Lorentz transformation.

[42]Poincaré, ref. 25 (1904), 310 (*Foundations*); ref. 24 (1906), 490, 503.

frame and in the ether frame could not be explained by any velocity dependence of the mass of the emitter, since the velocity change of the emitter is then negligible. With the benefit of hindsight, we may note that a decrease of the mass of the emitter by J/c^2 during the emission process solves the paradox. Indeed, in the ether frame this mass decrease implies a modification of the momentum variation of the emitter by $(J/c^2)u$, which exactly compensates the force Ju/c^2.

In his *Annalen* article of September 1905 (published in November), Einstein considered a radiation process from the point of view of two different observers, as Poincaré had done in 1900. The only difference is that he considered a symmetric process, which avoids any consideration of recoil momentum. For an observer bound to the emitter, the same energy $J/2$ is emitted by the light source in two opposite directions. For an observer moving at the velocity $\mathbf{u}$ with respect to the source on the emission line, Einstein's earlier transformation rule for the energy of light pulses gives $\gamma(1+u/c)J/2$ for the energy emitted in one direction and $\gamma(1-u/c)J/2$ for the energy emitted in the other. The sum of these energies exceeds the energy J by $J(\gamma-1)$. As the kinetic energy of the emitter is the product of its mass by $(\gamma-1)c^2$, a variation $-J/c^2$ of this mass during the emission restores the energy balance. From this remark, Einstein jumped to the general conclusion that "the mass of a body depends on its energy content."[43]

As appears from the letter to Habicht cited above, the inference was so novel that Einstein was not yet quite sure about it. The following year, he argued that mass-energy equivalence was the necessary and sufficient condition for an extension of the theorem of the center of mass to electromagnetic systems. At the beginning of this memoir, he noted that Poincaré's memoir of 1900 contained "the simple formal considerations on which the proof of this assertion is based." Indeed, Poincaré's equation

$$\frac{d}{dt}\int c^{-2}j\mathbf{r}d\tau + \int m\mathbf{v}d\tau + \int (c^{-2}\rho\mathbf{v}\cdot\mathbf{e})\mathbf{r}d\tau = \text{constant} \tag{22}$$

results from the equation

$$\frac{d}{dt}\left[\int c^{-2}j\mathbf{r}d\tau + \int m\mathbf{r}d\tau\right] = \text{constant} \tag{24}$$

for the uniform motion of the center of mass if the mass density of matter follows the increase $\rho\mathbf{v}\cdot\mathbf{e}$ of its energy content according to

$$Dm/Dt \equiv \partial m/\partial t + \nabla\cdot(m\mathbf{v}) = <\rho\mathbf{v}\cdot\mathbf{e}/c^2> , \tag{25}$$

wherein the symbols $< >$ indicate an average at the scale at which the mass density m is defined. As Einstein noted, the reasoning only makes sense if the expression of $m\mathbf{v}$ the momentum density is applicable and if the only relevant energies are the energy of the electromagnetic field and the internal energy of matter. This can be the case if the mass density m is defined at a macroscopic scale for which each

[43]Einstein, "Ist die Trägheit eines Körper von seinem Energieinhalt abhängig ?" *Annalen der Physik*, 18 (1905), 639-641.

volume element includes many molecules and if the macroscopic kinetic energy remains small.[44]

More concretely and more in the spirit of Poincaré's perpetual motion argument, Einstein considered an emitter and an absorber of radiation that faced each other and belonged to the same solid. He imagined the following cycle:

- The source emits a radiation pulse with the energy J in the direction of the absorber, which implies a recoil momentum for the solid.
- When the pulse reaches the absorber, the solid returns to rest.
- A massless carrier then brings the energy J back to the absorber.

At the end of this cycle, the solid has shifted by the amount $-(J/Mc)L/c$ (in a first approximation), where M is the mass of the solid and L the distance between emitter and absorber. In order to avoid the resulting sort of perpetual motion, Einstein assumed that the return of the energy J to the emitter involved a transfer of mass J/c^2. During this transfer, the center of mass of the global system does not move. Therefore, the solid moves by the amount $L(J/c^2)/M$ (in a first approximation), which compensates the shift in the first step of the cycle.

This was neither the last nor the most convincing of Einstein's derivation of the inertia of energy. Einstein nonetheless believed in this astonishing consequence of relativity theory.

Conclusions

The genesis of the theory of relativity was a long process that involved at least three key players and their critical reflections on the electrodynamics of moving bodies. Although Maxwell made optics a part of electrodynamics, he could not explain optical phenomena that depend on the molecular structure of matter. Toward the end of the nineteenth century, four circumstances favored investigations of this issue: Hertz's confirmation of Maxwell's theory; continental attempts to inject into this theory the molecular conception of electricity that belonged to the defeated German theories; the rise of an experimental microphysics of ions, x-rays, and electrons; and the multiplication of experiments on the optics of moving bodies. A number of theorists then improved the electrodynamics of moving bodies in competing approaches.

For the sake of simplicity, we may extract from this thriving physics the main events that contributed to the formation of relativity theory. A first essential step was taken by Lorentz, who replaced Maxwell's hybrid, macroscopic, ether-matter medium with a stationary ether in which electrons and other atomistic entities freely circulated. Exploiting the invariance properties of the fundamental equations for the interaction between electrons and fields, Lorentz accounted for the

[44] Einstein, "Das Prinzip der Erhaltung der Schwerpunktsbewegung und die Trägheit der Energie," *Annalen der Physik*, 20 (1906), 627-633. The reasoning can easily be extended to fast moving material elements: one only has to replace m with $m/\sqrt{1-v^2/c^2}$.

absence of effects of the motion of the earth through the ether, but only to a certain approximation. Poincaré made this absence of effects a general postulate. He gave a physical interpretation of the Lorentz transformations as those giving the space, time, and fields measured by moving observers. He obtained the exact form of these transformations. He used them to determine electron dynamics and to suggest a modification of Newton's theory of gravitation. Yet he maintained the ether as a privileged frame in which true time and space were defined. Einstein adopted the relativity principle, eliminated the ether, and placed the space and time determinations in any two inertial systems on exactly the same footing. Combining the relativity principle with that of the constancy of the velocity of light, he obtained the Lorentz transformations, the contraction of lengths, and the dilation of times. He showed how this symmetry permitted a consistent electrodynamics of moving bodies and determined the dynamics of the electron. He derived the inertia of energy.

The construction of the special theory of relativity did not end with Einstein's papers of 1905. Some features that today's physicists judge essential were added only later. For example, Hermann Minkowski and Arnold Sommerfeld developed the 4-dimensional notation and the relativistic tensor formulation of electromagnetism; Max Planck gave the relativistic definition of force and the Lagrangian formulation of relativistic dynamics; Max von Laue gave the kinematical interpretation of the Fresnel drag as a direct consequence of the relativistic combination of the velocity of the moving transparent body and the velocity of light within it.

Thus, Einstein was neither the first nor the last contributor to relativity theory. He learned much by reading the best authors of his time, and he partly duplicated results already obtained by Lorentz and Poincaré. Yet there is no doubt that his papers of 1905 marked a dramatic turn in our understanding of space, time, mass, and energy. His questioning of received ideas was most radical. His construction of alternative theories was most elegant, powerful, and durable. By rejecting the ether and propounding a new chronogeometry, he prepared the ground for further intellectual achievements, including general relativity and quantum theory.

Acknowledgments. I thank Thibault Damour and Bertrand Duplantier for useful suggestions.

Short bibliography

Albert Einstein, *The collected papers of Albert Einstein*, vols. 1-2. Ed. John Stachel et al. Princeton, 1987-1989.

Albert Einstein, *Oeuvres choisies*, ed. Françoise Balibar et al. 6 vols (Paris, 1985-1993).

Albert Einstein, "Autobiographisches," in P.A. Schilpp (éd.), *Albert Einstein: Philosopher-scientist* (Evanston, 1949), 1-94.

Jed Buchwald, *From Maxwell to microphysics: Aspects of electromagnetic theory in the last quarter of the nineteenth century* (Chicago, 1985).

Olivier Darrigol, *Electrodynamics from Ampère to Einstein* (Oxford, 2000).

Albrecht Fölsing, *Albert Einstein: Eine Biographie* (Frankfurt, 1993); transl. (New York, 1993).

Peter Galison, *Einstein's clocks, Poincaré's maps: Empires of time* (New York, 2003).

Christa Junknickel and Russel McCormmach, *Intellectual mastery of nature: Theoretical physics from Ohm to Einstein*, 2 vols. (Chicago, 1986).

Arthur, Miller, *Albert Einstein's special relativity: Emergence and early interpretation* (1905-1911) (Reading, 1981).

Abraham Pais, *"Subtle is the Lord…": The science and the life of Albert Einstein* (Oxford, 1982).

Michel Paty, *Einstein philosophe: La physique comme pratique philosophique* (Paris, 1993).

John Stachel, "History of relativity," in L. Brown, A. Pais, B. Pippard (eds.), *Twentieth century physics* (New York, 1995), 249-356.

Edmund Whittaker, *A history of the theories of aether and electricity*, 2 vols. (London, 1951).

Olivier Darrigol
CNRS : Rehseis
83, rue Broca
75013 Paris
France
email: darrigol@paris7.jussieu.fr

Einstein, 33 – 58
© Birkhäuser Verlag, Basel, 2005

Poincaré Seminar 2005

Special Relativity: A Centenary Perspective

Clifford M. Will

1 Introduction

A hundred years ago, Einstein laid the foundation for a revolution in our conception of time and space, matter and energy. In his remarkable 1905 paper "On the Electrodynamics of Moving Bodies" [1], and the follow-up note "Does the Inertia of a Body Depend upon its Energy-Content?" [2], he established what we now call special relativity as one of the two pillars on which virtually all of physics of the 20th century would be built (the other pillar being quantum mechanics). The first new theory to be built on this framework was general relativity [3], and the successful measurement of the predicted deflection of light in 1919 made both Einstein the person and relativity the theory internationally famous. The next great theory to incorporate relativity was the Dirac equation of quantum mechanics; later would come the stunningly successful relativistic theory of quantum electrodynamics.

Strangely, although general relativity had its crucial successes, such as the bending of starlight and the explanation of the advance of Mercury's perihelion, special relativity was not so fortunate. Indeed, many scholars believe that a lack of direct experimental support for special relativity in the years immediately following 1905 played a role in the decision to award Einstein's 1921 Nobel Prize, not for relativity, but for one of his other 1905 "miracle" papers, the photoelectric effect, which *did* have direct confirmation in the laboratory.

And although there were experimental tests, such as improved versions of the Michelson-Morley experiment, the Ives-Stilwell experiment, and others, they did not seem to have the same impact as the light-deflection experiment. Still, during the late 1920s and after, special relativity was inexorably accepted by mainstream physicists (apart from those who participated in the anti-Semitic, anti-relativity crusades that arose in Germany and elsewhere in the 1920s, coincident with the rise of Nazism), until it became part of the standard toolkit of every working physicist. Quite the opposite happened to general relativity, which for a time receded to the backwaters of physics, largely because of the perceived absence of further experimental tests or consequences. General relativity would not return to the mainstream until the 1960s.

On the 100th anniversary of special relativity, we see that the theory has been so thoroughly integrated into the fabric of modern physics that its validity is rarely challenged, except by cranks and crackpots. It is ironic then, that during the past several years, a vigorous theoretical and experimental effort has been launched, on an international scale, to find violations of special relativity. The motivation for this effort is not a desire to repudiate Einstein, but to look for evidence of new physics "beyond" Einstein, such as apparent violations of Lorentz invariance that might result from certain models of quantum gravity. So far, special relativity has passed all these new high-precision tests, but the possibility of detecting a signature of quantum gravity, stringiness, or extra dimensions will keep this effort alive for some time to come.

In this paper we endeavor to provide a centenary perspective of special relativity. In Section 2, we discuss special relativity from a historical and pedagogical viewpoint, describing the basic postulates and consequences of special relativity, at a level suitable for non-experts, or for experts who are called upon to teach special relativity to non-experts. In Section 3, we review some of the classic experiments, and discuss the famous "twin paradox" as an example of a frequently misunderstood "consistency" test of the theory. Section 4 discusses special relativity in the broader context of curved spacetime and general relativity, describes how long-range fields interacting with matter can produce "effective" violations of Lorentz invariance and discusses experiments to constrain such violations. In Section 5 we discuss whether gravity itself satisfies a version of Lorentz invariance, and describe the current experimental constraints. In Section 6 we briefly review the most recent extended theoretical frameworks that have been developed to discuss the possible ways of violating Lorentz invariance, as well as some of the ongoing and future experiments to look for such violations. Section 7 presents concluding remarks.

2 Fundamentals of special relativity

2.1 Einstein's postulates and insights

Special relativity is based on two postulates that are remarkable for their simplicity, yet whose consequences are far-reaching. They state [1]:

- The laws of physics are the same in any inertial reference frame.
- The speed of light in vacuum is the same as measured by any observer, regardless of the velocity of the inertial reference frame in which the measurement is made.

The first postulate merely adopts the wisdom, handed down from Galileo and Newton, that the laws of mechanics are the same in any inertial frame, and extends it to cover *all* the laws of physics, notably electrodynamics, but also laws yet to be discovered. There is nothing radical or unreasonable about this postulate. It is the second postulate, that the speed of light is the same to all observers, that is usually regarded as radical, yet it is also strangely conservative. Maxwell's equations stated that the speed of light was a fundamental constant, given by $c = 1/\sqrt{\epsilon_0 \mu_0}$, where ϵ_0 and μ_0 are the dielectric permittivity and magnetic permeability of vacuum, two constants that could be measured in the laboratory by performing experiments that had nothing obvious to do with light. That speed c, now defined to be *exactly* 299, 792, 458 m/sec, bore no relation to the state of motion of emitter or receiver. Furthermore, there existed a set of transformations, found by Lorentz, under which Maxwell's equations were invariant, with an invariant speed of light.

In addition, Einstein was presumably aware of the Michelson-Morley experiment (although he did not refer to it by name in his 1905 paper) which demonstrated no effect on the speed of light of our motion relative to the so-called "aether" [4]. While the great physicists of the day, such as Lorentz, Poincaré and others were struggling to bring all these facts together by proposing concepts such as "internal time", or postulating and then rejecting "aether drift", Einstein's attitude seems to have been similar to that expressed in the American idiom: "if it walks like a duck and quacks like a duck, it's a duck". If light's speed *seems* to be constant, then perhaps it really *is* a constant, no matter who measures it. Throughout his early career, Einstein demonstrated an extraordinary gift for taking a simple idea at face value and "running" with it; he did this with the speed of light; he did it with Planck's quantum hypothesis and the photoelectric effect, also in 1905.

2.2 Time out of joint

An immediate and deep consequence of the second postulate is that time loses its absolute character. First, the rate of time depends on the velocity of the clock. A very simple way to see this is to imagine a thought experiment involving three identical clocks. Each clock consists of a chamber of length h with a perfect mirror at each end. A light ray bounces back and forth between the mirrors, recording one "tick" each time it hits the bottom mirror. In the rest frame of each clock the speed of light is c (by the second postulate), so the duration of each "tick" is $2h/c$ according to observers on each clock. Two of the clocks are at rest in a laboratory, a distance d apart along the x-axis, arranged so that the light rays move in the y-direction. The two clocks have been synchronized using a light flash from a lamp midway between them. The third clock moves with velocity v in the x-direction (Fig. 1). As it passes each of the laboratory clocks in turn, its own reading and the reading on the adjacent laboratory clock are taken and later compared. The time difference between the readings on the two laboratory clocks is clearly d/v or $(d/v)/(2h/c)$ ticks. But from the point of view of the laboratory, the light ray on the moving clock moves in a saw-tooth manner as the mirrors move, with the distance along the hypotenuse of each tooth given by $l = \sqrt{h^2 + (vt)^2}$ where t is the time taken as seen from the lab. But at the speed of light, this time is given by l/c, so the duration of a "tick" on the moving clock from the lab viewpoint is given by $(2h/c)\gamma$, where $\gamma = 1/\sqrt{1 - v^2/c^2}$. Thus the number of ticks on the moving clock between its encounters with the lab clocks is $(d/v)/(2h/c) \times \sqrt{1 - v^2/c^2}$. If we define "proper time" $\Delta\tau$ as the time elapsed on a single clock between two events at its own location, and Δt as the time difference measured by the two

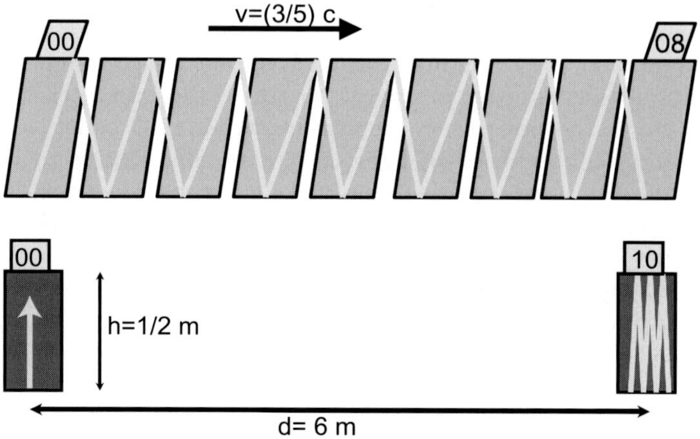

Figure 1: Time dilation of a clock moving at $v = 3/5c$ between two identical laboratory clocks a distance 6 m apart. The laboratory clocks each tick 10 times during the passage, while the moving clock ticks only 8 times because the light rays travel farther to complete each tick, as seen from the laboratory.

separated laboratory clocks, then

$$\Delta\tau = \Delta t\sqrt{1 - v^2/c^2}\,. \tag{1}$$

This is the time dilation: the time elapsed between two events along the path of a single moving clock is less than that measured by a *pair* of synchronized clocks located at the two events. The asymmetry is critical: A clock can only make time readings along its own world line, thus *two* synchronized clocks are required in the laboratory, in order to make comparisons with readings on the moving clock.

While this time dilation was already recognized at some level by Lorentz and others as a consequence of the Lorentz transformations, they were unable or unwilling to recognize its true meaning, because they remained wedded to the Newtonian view of an absolute time. Einstein, possibly because of his early contact with the machinery and equipment of his father's factories, was able to view time operationally: time is what clocks measure. If one thinks of a clock as any device that performs some precisely repetitive activity governed fundamentally by the laws of physics, then it becomes obvious that time in the moving frame really does tick more slowly than in the lab. And this is not some abstract, internal time, this is time measured by our mirror clock, by an atomic clock, by a biological clock, by a human heartbeat, all of which are governed by the laws of physics, which are the same in every inertial frame. From any conceivable observable viewpoint this *is* time.

In the thought experiment above we remarked that the laboratory clocks were synchronized. This seemingly obvious and innocuous statement also has deep consequences, because, as Einstein realized, if the speed of light is the same for all observers, then synchronization is relative. Consider two observers on the ground who synchronize their clocks by setting them to read the same when a light flash from a point midway between them is received. Now consider observers on a train moving by (who have previously synchronized their own clocks using the same method on the train). The light flash emitted by the lamp on the ground has speed c in both directions as seen from the train (second postulate), therefore the forward moving flash will encounter the forward ground clock (which is moving toward the lamp as seen from the train) *before* the backward moving flash encounters the rear ground clock (which is receding). The events of reception of the light flash by the two ground clocks are simultaneous in the ground frame, but are *not* simultaneous in the train's frame. Again, this was embodied mathematically in the Lorentz transformation, but it was Einstein who inferred this truth about time: events simultaneous in one frame, are not automatically simultaneous in a moving frame.

Much has been written about why Einstein was able to arrive at this new view of time, while his contemporaries, including great men like Lorentz and Poincaré, were not. Henri Poincaré is a case in point. By 1904 Poincaré understood almost everything there was to understand about relativity. In 1904 he journeyed to St. Louis to speak at the scientific congress associated with the World's Fair, on the newly relocated campus of my own institution, Washington University. In reading Poincaré's paper "The Principles of Mathematical Physics" [5], one senses that he is so close to having special relativity that he can almost taste it. Yet he could not take the final leap to the new understanding of time. This is ironic, because as Peter Galison has written [6], Poincaré was one of the world's leaders in the understanding of clock synchronization, having served on French and international agencies and committees charged with establishing the world-wide conventions for time-synchronization and time transfer that were needed for transportation, navigation and telegraphy. Surely Poincaré would have understood our example of the moving train, yet it seems that he could not go beyond viewing it as merely conventional. To Einstein, it reflected what clocks measure, and therefore reflected the true nature of time.

2.3 Spacetime and Lorentz invariance

If the speed of light is the same to all observers, then time and space can be put on a similar footing initially by measuring time in units of distance, so that t in meters stands for ct, and corresponds to the time it takes for light to travel one meter (3.336 nanoseconds). We will call this time in distance units the coordinate x^0. One can then describe space and time together on a spacetime diagram, with points representing "events", "worldlines" representing the trajectories of particles

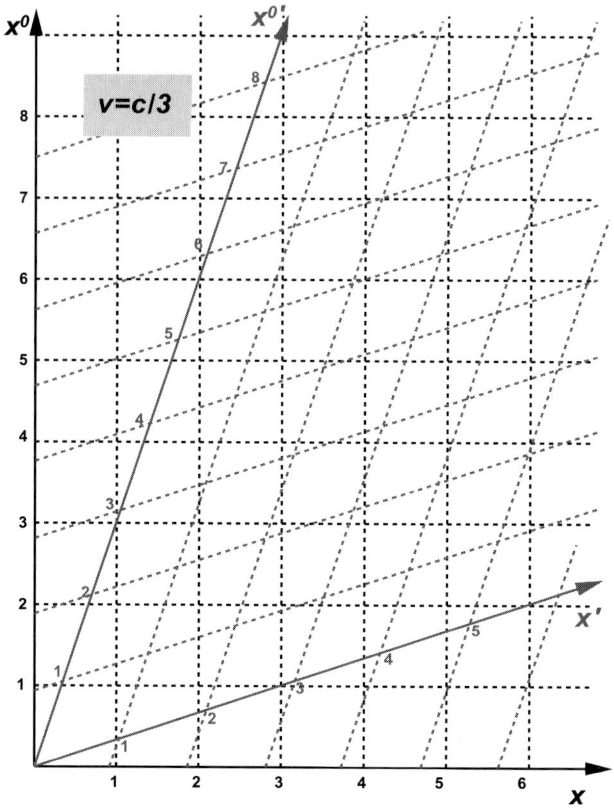

Figure 2: Spacetime diagram showing a laboratory frame and a frame moving at $v = c/3$.

through space and time and so on.

A train moving with speed v, with the caboose passing the origin at $x^0 = 0$ has the collection of world lines shown in Fig. 2 (one for each car in the train), each with slope $1/v$. The line passing through the origin is called the $x^{0\prime}$ axis, just as in Galilean relativity. By carefully considering how clocks on the train would be synchronized, either using a master lamp as in the example above, or by using round-trip signals (often called Einstein synchronization), it is easy to show that the collection of events on the train that are simultaneous with the origin lie along the x'-axis shown, with slope v. Later "lines of simultaneity" on the train are also shown. Figure 2 makes it clear how all observers can agree on the speed of light. A light ray emanating from the origin of Fig. 2 follows a 45° line, or a line that bisects the x and x^0 axes. But that line also bisects the x' and $x^{0\prime}$ axes, thus observers on the train will also find speed c for that ray.

These considerations establish only the slopes of the lines, however. They do not tell us where, for example, to mark 1 meter on the x'-axis. To resolve this, we return to our simple moving clock example, and notice that, while the time difference and spatial difference between the events describing one "tick" of the moving clock are given by $\Delta t' = 2h/c$ and $\Delta x' = 0$ in its own frame, and by the different values $\Delta t = \gamma(2h/c)$ and $\Delta x = v\Delta t = v\gamma(2h/c)$ in the lab frame, the quantity $\Delta s^2 \equiv -c^2\Delta t^2 + \Delta x^2$ is the same for the tick, whether calculated in the clock's frame or in the lab frame. This is the "invariant interval", given for general infinitesimal displacements by

$$\begin{aligned} ds^2 &= -c^2dt^2 + dx^2 + dy^2 + dz^2 \\ &= -(dx^0)^2 + dx^2 + dy^2 + dz^2 \\ &= \eta_{\mu\nu}dx^\mu dx^\nu \,, \end{aligned} \tag{2}$$

where $\eta_{\mu\nu} = \mathrm{diag}(-1, 1, 1, 1)$ is the Minkowski metric, Greek indices run over four spacetime values, and we use the Einstein convention of summing over repeated indices. If one then asks, what linear transformations from one inertial frame to a moving inertial frame will leave this interval invariant in form, or equivalently will leave the Minkowski metric invariant, the answer is the Lorentz transformations: for a boost in the x-direction, they are given by

$$\begin{aligned} (x^0)' &= \gamma(x^0 - vx)\,, \\ x' &= \gamma(x - vx^0)\,. \end{aligned} \tag{3}$$

For a general boost with velocity v^i, they are given by $x^{\alpha'} = \Lambda^{\alpha'}_{\beta} x^\beta$, where

$$\Lambda^{0'}_0 = \gamma\,, \quad \Lambda^{0'}_i = \Lambda^{i'}_0 = -\gamma v^i\,, \quad \Lambda^{i'}_j = \delta^i_j + (\gamma - 1)v^i v^j/v^2\,. \tag{4}$$

This is called Lorentz invariance of the interval (or metric). The form of the interval is also invariant under ordinary rotations, and under displacements such as $x^\alpha \to x^\alpha + a^\alpha$. Collectively this larger 10 parameter invariance is called Poincaré invariance. The Lorentz transformations then allow one to establish the scale of the axes of the moving frame, as shown in Fig. 2 for the case $v = c/3$.

These are the same transformations, of course, as those found to leave Maxwell's equations invariant. Einstein's first postulate, that the laws of physics should be the same in every inertial frame, therefore places a stringent constraint on the design of any future fundamental laws, namely that they should be Lorentz invariant, at least when viewed from an inertial frame. This constraint has guided the great advances in fundamental theory of the 20th century, such as relativistic quantum mechanics and the Dirac equations, quantum electrodynamics, quantum chromodynamics, superstring theory, not to mention general relativity.

2.4 Special relativistic dynamics

By considering the acceleration of a charged particle in an electromagnetic field and imposing the principle of relativity [1], Einstein concluded that the equations of dynamics would have to be modified. Further, in another characteristic example of his ability to use a simple thought experiment to derive profound consequences, Einstein established the equivalence between mass and energy [2]. He considered the simple situation of a particle emitting an equal amount of electromagnetic radiation in opposite directions. He then considered the same situation from the viewpoint of a moving inertial frame. By imposing conservation of energy in both frames, and using the transformation laws for electromagnetic radiation, he concluded, working in the low-velocity limit, that the difference in kinetic energy of the particle before and after the emission, as seen in the moving frame, had to be given by $\frac{1}{2}Ev^2/c^2$, where E is the energy of the emitted light. But since kinetic energy in this limit is given by $\frac{1}{2}mv^2$, then the mass of the particle must have changed by E/c^2 during the emission of energy E.

What emerged from these considerations was a new relativistic dynamics. One must replace the Newtonian formulation of $\mathbf{F} = m\mathbf{a}$ with a relativistically correct formulation $\vec{F} = d\vec{p}/d\tau$, where the force $\vec{F}$ is now a four-vector, $\vec{p}$ is the four-momentum, given for a particle of rest mass m_0 by $\vec{p} = m_0\vec{u}$, where the four-velocity $\vec{u}$ has components $u^\alpha = dx^\alpha/d\tau$, and where $d\tau = ds/c$ denotes proper time along the particle's worldline. If the force is provided by electromagnetic fields, then $F^\nu = (e/c)u_\mu F^{\mu\nu}$, where e is the charge of the particle, and $F^{\mu\nu}$ is the antisymmetric Faraday tensor, whose components in a given inertial frame may be identified as $F_{i0} = E_i$, $F_{ij} = \epsilon_{ijk}B_k$, where E^i and B^i are the normal electric and magnetic fields. This dynamics, along with Maxwell's equations, can be derived from the action

$$
\begin{aligned}
I &= -\sum_a m_{0a}c\int(-\eta_{\mu\nu}u_a^\mu u_a^\nu)^{1/2}d\tau + \sum_a \frac{e_a}{c}\int A_\mu(x_a^\nu)dx_a^\mu \\
&\quad -\frac{1}{16\pi}\int\sqrt{-\eta}\,\eta^{\mu\alpha}\eta^{\nu\beta}F_{\mu\nu}F_{\alpha\beta}d^4x\,, \qquad (5)
\end{aligned}
$$

where u_a^μ is the four-velocity of the particle, $A_\mu(x^\nu)$ is the electromagnetic four-vector potential, and $F_{\mu\nu} \equiv \partial A_\nu/\partial x^\mu - \partial A_\mu/\partial x^\nu$. In ordinary variables, in a given inertial frame, the action takes the form

$$
\begin{aligned}
I &= -\sum_a m_{0a}c^2\int(1 - v_a^2/c^2)^{1/2}dt + \sum_a e_a\int(-\Phi + \mathbf{A}\cdot\mathbf{v_a}/c)dt \\
&\quad +\frac{1}{8\pi}\int(E^2 - c^2B^2)d^3x dt\,, \qquad (6)
\end{aligned}
$$

where $\Phi = -A_0$, $\mathbf{E} = -\nabla\Phi - \dot{\mathbf{A}}/c$, and $\mathbf{B} = \nabla\times\mathbf{A}$.

3 Classic tests of special relativity

3.1 The Michelson-Morley experiment

From today's perspective the null result of the 1887 Michelson-Morley aether-drift experiment marked the beginning of the end for the Newtonian notions of absolute space and time. Yet it took almost 20 years for the new view of spacetime to be realized. The experiment was beautiful in its simplicity, and should have been a "slam dunk" for conventional 19th century physics. If the speed of light is a fundamental constant, then it must take this value in some preferred frame, presumably that of a luminiferous aether, which would be at rest with respect to the universe, and which would provide the medium that every one thought was necessary for the propagation of light. For any observer moving relative to the aether, the speed of light would be formed by subtracting the velocity vector of the observer from that of the light ray. In one of the interferometers that Michelson had pioneered for measuring the speed of light itself, the speed of light up and down an arm that was parallel to our motion through the aether would be $c+v$ and $c-v$, while the speed along an arm perpendicular to our motion would be $\sqrt{c^2+v^2}$. For an equal-arm interferometer of length h, the difference in round trip travel time along the two arms would then be, to first order in $(v/c)^2$, $\Delta T = (h/c)(v/c)^2$. This would be reflected in a change in the interference pattern of the recombined beams, that would shift as the apparatus was rotated, thereby interchanging the roles of the two arms.

But instead of the predicted shift, Michelson and Morley found no effect, and placed an upper limit on a shift 40 times smaller than the shift predicted [4], and later experiments only improved the bounds (see [7] for a review up to 1955). Attempts to explain this by arguing that the aether was "dragged" by the Earth proved to be untenable. Lorentz wrote to Lord Rayleigh in 1892, "I am totally at a loss to clear away this contradiction ... Can there be some point in the theory of Mr. Michelson's experiment which has been overlooked?" [7]. Lorentz and FitzGerald attempted to resolve the problem by proposing that the interferometer arms parallel to the motion through the aether were shortened by the factor $\sqrt{1-v^2/c^2}$, but could not suggest what this meant [8, 9].

Special relativity resolved the Michelson-Morley experiment instantly. In the rest frame of the experiment, the speed of light is the same, irrespective of the instrument's motion relative to the universe, so the experiment should automatically give a null result. Indeed, the aether now becomes completely irrelevant. Alternatively, from the point of view of a frame at rest relative to the universe, careful consideration of how length is measured in special relativity showed that the interferometer arm moving parallel to its length must be shortened by the precise Lorentz-FitzGerald factor. The null experimental result could be derived from either frame of reference.

In placing the Michelson-Morley (MM) experiment in a modern context, it is useful to view it not as an interferometer experiment, but as a clock anisotropy experiment. Each arm of the interferometer can be thought of as a clock just like the clocks used in Sec. 2.2 above. The fundamental question then becomes, is the rate of a clock independent of its orientation relative to its motion through the universe? Most modern incarnations of the MM experiment are clock anisotropy experiments. For example, MM experiments using lasers [10, 11] compare two laser resonant cavities by beating their frequencies against each other as one or both rotate relative to the universe.

One can invent a way to parametrize the MM experiment so as to quantify how the null result could be violated, that turns out to be useful in more general contexts. Suppose that, working in the rest frame of the universe (we may discard the aether, but the rest-frame of the universe, as reflected by the rest frame of the cosmic background radiation, has a well defined meaning), the speed of light is c. But suppose that the Lorentz-FitzGerald contraction of the parallel arm is given by the factor $\sqrt{1-v^2/c_0^2}$, where c_0 is a different speed (measured in the universe rest frame), that is connected with whatever dynamics determines the structure of the walls of the cavity that forms our clock. Then it is easy to show that, while the time for one tick of the clock perpendicular to the motion is given by $(2h/c)(1/\sqrt{1-v^2/c^2})$, the time for one tick of the parallel clock is $(2h/c)[\sqrt{1-v^2/c_0^2}/(1-v^2/c^2)]$. To first order in $(v/c)^2$, the differential clock time is given by $(h/c)(v/c_0)^2\delta$, where $\delta = (c_0/c)^2 - 1$.

If Lorentz invariance holds, then the electrodynamics that governs the solids that form the cavity must involve the same c as that which governs the propagation of light, hence $c_0 = c$, $\delta = 0$ and we recover the null prediction for the MM experiment. Below we will discuss classes of theories that involve curved spacetime plus certain kinds of long-range fields, in which this no longer holds. Figure 4 shows selected bounds on δ that were achieved in the original MM experiment, and in later experiments of the MM type by Joos and a 1979 test using laser technology by Brillet and Hall [11]. In that Figure, units are chosen so that $c_0 = 1$.

3.2 Invariance of c

Several classic experiments have been performed to verify that the speed of light is independent of the speed of the emitter. If the speed of light were given by $\mathbf{c} + k\mathbf{v}$, where $\mathbf{v}$ is the velocity of the emitter, and k is a parameter to be measured or bounded, then orbits of binary star systems would appear to have an anomalous eccentricity unexplainable by normal Newtonian gravity. However, at optical wavelengths, this test is not unambiguous because light is absorbed and reemitted by the intervening interstellar medium, thus losing the memory of the speed of the source, a phenomenon known as extinction. But at X-ray wavelengths, the path length of extinction is tens of kiloparsecs, so nearby X-ray binary sources in our galaxy may be used to test the velocity dependence of light. Using data on pulsed 70 keV X-ray binary systems, Her S-1, Cen X-3 and SMC X-1, Brecher [12] obtained a bound $|k| < 2 \times 10^{-9}$, for typical orbital velocities $v/c \sim 10^{-3}$.

At the other extreme, a 1964 experiment at CERN used ultrarelativistic particles as the source of light. Neutral pions were produced by the collisions of 20 GeV protons on stationary nucleons in the proton synchrotron. With energies larger than 6 GeV, the pions had $v/c \geq 0.99975$. Photons produced by the decay $\pi^0 \to \gamma_1 + \gamma_2$ were collimated and timed over a 30 meter long flight path. Because the protons in the synchrotron were pulsed, the speed of the photons could be measured by measuring the arrival times of their pulses as a function of the varying location of the detector along the flight path. The result for the speed was $2.9977 \pm 0.0004 \times 10^8$ m/sec, in agreement with the laboratory value [13]. This experiment thus set a bound $|k| < 10^{-4}$ for $v \approx c$.

3.3 Time dilation

The observational evidence for time dilation is overwhelming. Ives and Stilwell [14] measured the frequency shifts of radiation emitted in the forward and backward direction by moving ions of H_2 and H_3 molecules. The first-order Doppler shift cancels from the sum of the forward and backward shifts, leaving only the second-order time-dilation effect, which was found to agree with theory. (Ironically, Ives was a die-hard opponent of special relativity.)

A classic experiment performed by Rossi and Hall [15] showed that the lifetime of μ-mesons was prolonged by the standard factor $\gamma = 1/\sqrt{1 - v^2/c^2}$. Muons are created in the upper atmosphere when cosmic ray protons collide with nuclei of air, producing pions, which decay to muons. With a rest half-life of 2.2×10^{-6} s, a muon travelling near the speed of light should travel only 2/3 of a kilometer on average before decaying to a harmless electron or positron and two neutrinos. Yet muons are the primary component of cosmic radiation detected at sea level. But with time dilation and a typical speed of $v/c \sim 0.994$, their lives as seen from Earth are prolonged by a factor of nine, easily enough for them to reach sea level. Rossi and Hall measured the distribution of muons as a function of altitude and also measured their energies, and confirmed the time dilation formula. In fact, since collisions between cosmic ray muons and DNA molecules are a non-negligible source of natural genetic mutations, one could even argue that special relativity plays a role in evolution!

In an experiment performed in 1966 at CERN, muons produced by collisions at one of the targets in the accelerator were deflected by magnets so that they would move on circular paths in a "storage ring". Their speeds were 99.7 percent of the velocity of light, and the observed twelve-fold increase in their lifetimes agreed with the prediction with 2 percent accuracy [16].

3.4 Lorentz invariance and quantum mechanics

The integration of Lorentz invariance into quantum mechanics has provided a string of successes for special relativity. The first was the discovery of the Dirac equation, the relativistic generalization of Schrödinger quantum mechanics, with its prediction of anti-particles and elementary particle spin. Another was the development of relativistic quantum field theory. QFT naturally embodies the Pauli exclusion principle, by requiring that the creation and annihilation operators of spinor fields satisfy anticommutation relations in order to obey Lorentz invariance. Since the Pauli exclusion principle explains the occupation of atomic energy levels by electrons, one could argue, with but a hint of chauvinism, that special relativity explains Chemistry! The modern incarnations of QFT, such as

Quantum Electrodynamics, Electroweak Theory, Quantum Chromodynamics all have Lorentz invariance as foundations. However, until recently, the experimental successes of such theories have not been used to attempt to quantify how well Lorentz invariance holds. We will return to this subject in Sec. 6.

3.5 Consistency tests of special relativity

Over the years, special relativity has been subjected to a series of tests, not of its experimental predictions, but of its very logic. Many of its predictions, such as the slowing of time on moving clocks, were deemed to be so strange, so beyond normal experience, that there had to be something wrong with the theory. The idea was to find "paradoxes", simple situations where the theory could be shown to be logically inconsistent.

Of course, there are no paradoxes! To be sure, the idea of time dilation may be hard to understand or to swallow, but there is absolutely nothing paradoxical about it.

The most popular of these is, of course, the twin paradox. In his 1905 paper, Einstein himself presents the situation clearly [1]: "If one of two synchronous clocks at A is moved in a closed curve with constant velocity until it returns to A, the journey lasting t seconds, then by the clock which has remained at rest the travelled clock on its arrival at A will be $\frac{1}{2}tv^2/c^2$ seconds slow."

The more modern versions of the story go something like this: On New Year's Day 3000, an astronaut (A) sets out from Earth at speed $0.6\,c$ and travels to the nearest interstellar Space Station, Clinton-1, which is 3 light-years away as measured in the Earth frame of reference (Fig. 3). Having reached Clinton-1, she immediately turns around and returns to Earth at the same speed, arriving home on New Year's Day 3010, by Earth time. The astronaut has a twin brother (B), who remains on Earth.

From the point of view of Earth's inertial frame, astronaut A's clock runs slow, with her proper time elapsed on the outbound journey being given by Eq. (1), amounting to 4 years, compared with 5 years on Earth. The times elapsed on the return journey are the same (the total proper time elapsed during the accelerated motion needed for the turnaround can be made as small as one likes by applying large accelerations for a short time). Astronaut A returns having aged 8 years, compared to the 10 years aging of her twin brother.

The "paradox" is then stated as follows: from the astronaut A's point of view, Earth's clocks run slow, so A should return *older* than her brother, not *younger*. Since this is a logical contradiction, relativity is untenable.

The flaw in the "paradox" is the failure to comprehend what is meant by "A sees Earth's clock run slow". A cannot compare her clock with Earth's clock because she is nowhere near Earth except at the start of the journey. Instead, an inertial frame moving outbound with A's velocity must be created, with a set of observers carrying clocks synchronized with hers. The readings on Earth's clock can only be read by one of these observers who happens to be passing the Earth

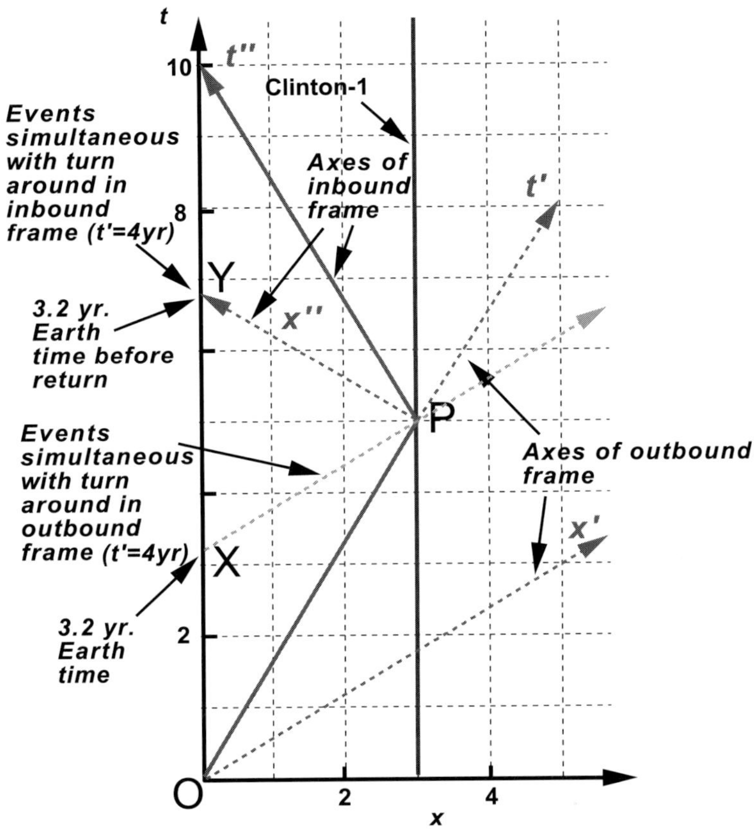

Figure 3: Twin Paradox as seen from traveller's viewpoint

at that moment of time. But because of the relativity of simultaneity, the event in this outbound frame that is simultaneous with A's turnaround event P is *not* the 5-year mark on Earth, but is event X on Fig. 3, which is at Earth year 3003.2. So observers in A's outbound frame do agree that Earth's clock has run slow compared to hers, 3.2 years compared to 4 years. But while A decelerates and accelerates for the return journey, that outbound inertial frame continues flying off at $0.6\,c$ forever, and A must pick up a new inertial frame inbound at $0.6\,c$. In *that* frame, the event that is simultaneous with the turnaround is at event Y, Earth year 3006.8, 3.2 years before the return. Again, observers in the inbound inertial frame agree that Earth's clock runs slow during the return journey, 3.2 years, compared to A's 4 years. But the analysis using the two inertial frames has failed to account for the 3.6 years between events X and Y.

This is not a paradox, it's merely sloppy accounting (perhaps the twin paradox should be renamed the Enron of Relativity). With a knowledge of the relativity of simultaneity, astronaut A could easily conclude that the gap between the two lines of simultaneity corresponding to her turnaround is 3.6 years; alternatively she could consult observers in an infinite sequence of inertial frames corresponding to all the velocities of her spacecraft from v to $-v$ and add up all the infinitesimal increments of Earth's clock as read by these observers, and account for the 3.6 missing years. Either way, she reaches the unambiguous conclusion that she ages a total of 8 years, while her twin ages 10 years.

It is sometimes claimed that the resolution of the twin paradox must ultimately involve general relativity, because the traveller accelerates, and acceleration is equivalent to gravitation. As the discussion above shows, acceleration plays no role in the analysis, other than to provide the asymmetry whereby the traveller must occupy more than one inertial frame, while the home-bound twin occupies a single inertial frame throughout. The relativity of simultaneity is the key, not gravity.

In fact, the relativity of simultaneity is the key to resolving essentially all of the "paradoxes" that have been devised to test the logical structure of special relativity, such as the "pole in the barn" paradox (a rapidly moving pole is short enough to fit inside a barn, at least momentarily, from the barn's point of view, but can't possibly fit from the pole's point of view), the "space-war paradox", "the jumping frog paradox" and others. For discussion of these and many other paradoxes, see [17].

4 Special relativity and curved spacetime

Special relativity and general relativity are often viewed as being independent. One reason for this apparent division is that Einstein presented special relativity 100 years ago in 1905, while general relativity was not published in its final form until 1916. Another reason is that the two parts of the theory have very different realms of applicability: special relativity mainly in the world of microscopic physics, and general relativity in the world of astrophysics and cosmology.

But in fact, the theory of relativity is a single, all-encompassing theory of space-time, gravity and mechanics. Special relativity is actually an approximation to curved space-time that is valid in sufficiently small regions of space-time (called "local freely falling frames"), much as small regions on the surface of an apple are approximately flat, even though the overall surface is curved. Special relativity can therefore be used whenever the scale of the phenomena being studied is small compared with the scale on which the curvature of space-time (i.e. gravity) begins to be noticed. For most applications in atomic or nuclear physics, this approximation is so accurate that special relativity can be assumed to be exact.

Historically, however, Einstein's journey from special to general relativity was tortuous and difficult. It began in 1907 with what he has called "the happiest thought" of his life. According to numerous experiments, all laboratory-sized bodies fall with the same acceleration, regardless of their mass, composition or structure, in a given external gravitational field. Einstein was probably aware of experiments performed by Eötvös around the turn of the 20th century [18], that demonstrated this "universality of free fall" to parts in 10^9. The modern bounds are at the level of parts in 10^{13} [19].

From this simple fact, Einstein noticed that if an observer were to ride in an elevator falling freely in a gravitational field, then all bodies inside the elevator would move uniformly in straight lines as if gravity had vanished. Conversely, in an accelerated elevator in free space, where there is no gravity, the bodies would fall with the same acceleration because of their inertia, just as if there were a gravitational field.

Einstein's great insight was to postulate that this "vanishing" of gravity in free fall or its "presence" in an accelerating frame applied not only to mechanical motion but to all the laws of physics, such as electromagnetism. Thus, in an accelerating frame, a light ray moving horizontally would be seen to be deflected downward, and a ray moving upward or downward would have its frequency shifted [20, 21].

For the next 8 years, Einstein looked for a theory that would embody this principle of equivalence, be compatible with Lorentz invariance in the absence of gravity, and reflect his goals of elegance and simplicity, succeeding finally in the fall of 1915 [3].

4.1 Einstein's equivalence principle

Our modern viewpoint of the foundations of general relativity is based on an extension and embellishment of Einstein's principle of equivalence. Much of this viewpoint can be traced back to Robert Dicke, who contributed crucial ideas about the foundations of gravitation theory between 1960 and 1965. These ideas were summarized in his influential Les Houches lectures of 1964 [22] and resulted in what has come to be called the Einstein equivalence principle (EEP), which states that

- test bodies fall with the same acceleration independently of their internal structure or composition (universality of free fall, also called the weak equivalence principle, or WEP);
- the outcome of any local non-gravitational experiment is independent of the velocity of the freely-falling reference frame in which it is performed (local Lorentz invariance, or LLI)
- the outcome of any local non-gravitational experiment is independent of where and when in the universe it is performed (local position invariance, or LPI).

The Einstein equivalence principle is the heart of gravitational theory, for it is possible to argue convincingly that if EEP is valid, then gravitation must be described by "metric theories of gravity", which state that (i) spacetime is endowed with a symmetric metric, (ii) the trajectories of freely falling bodies are geodesics of that metric, and (iii) in local freely falling reference frames, the non-gravitational laws of physics are those written in the language of special relativity. For further discussion, see [23].

One way to see that spacetime cannot be flat is the following. Consider two freely-falling frames on opposite sides of the Earth. According to the Einstein equivalence principle, space-time is Minkowkian in each frame, but because the frames are accelerating toward each other, the two space-times cannot be extended and meshed into a single Minkowskian space-time. In the presence of gravity, space-time is flat locally but curved globally.

4.2 Metric theories of gravity

The simplest way to incorporate the Einstein equivalence principle mathematically into the special relativistic dynamics of particles and fields is to replace the Minkowski metric in the action of Eq. (5) with the curved-spacetime metric $g_{\mu\nu}$, and to replace ordinary derivatives with covariant derivatives, yielding the action

$$\begin{aligned} I &= -\sum_a m_{0a} c \int (-g_{\mu\nu} u_a^\mu u_a^\nu)^{1/2} d\tau + \sum_a \frac{e_a}{c} \int A_\mu(x_a^\nu) dx_a^\mu \\ &\quad -\frac{1}{16\pi} \int \sqrt{-g}\, g^{\mu\alpha} g^{\nu\beta} F_{\mu\nu} F_{\alpha\beta} d^4x \,, \end{aligned} \tag{7}$$

where $d\tau = ds/c$, with $ds^2 = g_{\mu\nu} dx^\mu dx^\nu$. The only way that "gravity" enters is via the metric $g_{\mu\nu}$. Any theory whose equations for matter can be cast into this form is called a metric theory.

As a result, the non-gravitational interactions couple *only* to the spacetime metric $g_{\mu\nu}$, which locally has the Minkowski form $\eta_{\mu\nu}$ of special relativity. Because this local interaction is only with $\eta_{\mu\nu}$, local non-gravitational physics is immune from the influence of distant matter, apart from tidal effects. Local physics is Lorentz invariant (because $\eta_{\mu\nu}$ is) and position invariant (because $\eta_{\mu\nu}$ is constant in space and time).

General relativity is a metric theory of gravity, but so are many others, including the Brans-Dicke theory. In this sense, superstring theory is not metric, because there is a residual coupling of external, gravitation-like fields, to matter. Theories in which varying non-gravitational constants are associated with dynamical fields that couple to matter directly are also not metric theories.

4.3 Effective violations of local Lorentz invariance

How could violations of LLI arise? From the viewpoint of field theory, violations would generically be caused by other long-range fields in addition to $g_{\mu\nu}$ which also couple to matter, such as scalar, vector and tensor fields. Theories that have this property are called non-metric theories. A simple example of such a theory is one in which the matter action is given by

$$\begin{aligned} I &= -\sum_a m_{0a}c \int (-g_{\mu\nu}u_a^\mu u_a^\nu)^{1/2} d\tau + \sum_a \frac{e_a}{c} \int A_\mu(x_a^\nu) dx_a^\mu \\ &\quad -\frac{1}{16\pi} \int \sqrt{-h}\, h^{\mu\alpha} h^{\nu\beta} F_{\mu\nu} F_{\alpha\beta} d^4x \,, \end{aligned} \tag{8}$$

where $h_{\mu\nu}$ is a second, second-rank tensor field. Locally, one can always find coordinates (local freely-falling frame) in which $g_{\mu\nu} \to \eta_{\mu\nu}$, but in general $h_{\mu\nu} \not\to \eta_{\mu\nu}$; instead $h_{\mu\nu} \to (h_0)_{\mu\nu}$, where $(h_0)_{\mu\nu}$ is a tensor whose values are determined by the cosmology or nearby mass distribution. In the rest frame of the distant matter distribution, $(h_0)_{\mu\nu}$ will have specific values, and there is no reason *a priori* why those should correspond to the Minkowski metric (unless $h_{\mu\nu}$ were identical to $g_{\mu\nu}$ in the first place, in which case one would have a metric theory). Also, in a frame moving with respect to the distant sources of $h_{\mu\nu}$, the local values of $(h_0)_{\mu\nu}$ will depend on the velocity of the frame, thereby producing *effective* violations of Lorentz invariance in electrodynamics.

A number of explicit theoretical frameworks were developed between 1973 and 1990 to treat non-metric theories of this general type. They include the "$TH\epsilon\mu$" framework of Lightman and Lee [24], the $\chi - g$ framework of Ni [25], the c^2 framework of Haugan and coworkers [26, 27], and the extended $TH\epsilon\mu$ framework of Vucetich and colleagues [28].

In the c^2 framework, one assumes a class of non-metric theories in which the particle and interaction parts of the action Eq. (8) can be put into the local special relativistic form, using units in which the limiting speed of neutral test particles is unity, and in which the sole effect of any non-metric field coupling to electrodynamics is to alter the effective speed of light. The result is the action

$$\begin{aligned} I &= -\sum_a m_{0a} \int (1 - v_a^2)^{1/2} dt + \sum_a e_a \int (-\Phi + \mathbf{A} \cdot \mathbf{v_a}) dt \\ &\quad +\frac{1}{8\pi} \int (E^2 - c^2 B^2) d^3x dt \,. \end{aligned} \tag{9}$$

Because the action is explicitly non-Lorentz invariant if $c \neq 1$, it must be defined in a preferred universal rest frame, presumably that of the 3K microwave background. In this frame, the value of c^2 is determined by the cosmological values of the non-metric field. Even if the non-metric field coupling to electrodynamics is a tensor field, the homogeneity and isotropy of the background cosmology in the preferred frame is likely to collapse its effects to that of the single parameter c^2. Detailed calculations of a variety of experimental situations show that those "preferred-frame" effects depend on the magnitude of the velocity through the preferred frame (~ 350 km/sec), and on the parameter $\delta \equiv c^{-2} - 1$. In any metric theory or theory with local Lorentz invariance, $\delta = 0$.

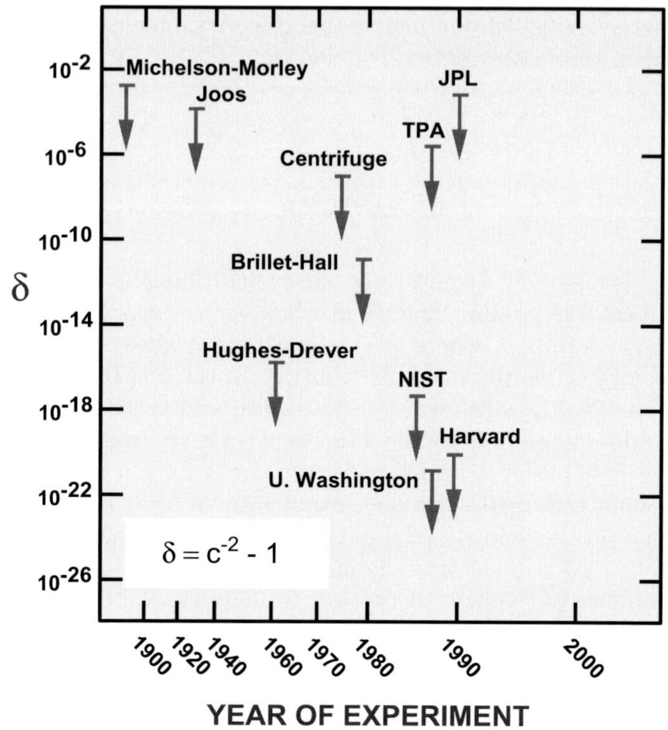

Figure 4: Bounds on violations of local Lorentz invariance

One can then set observable upper bounds on δ using a variety of experiments. In the Michelson-Morley experiment, by considering the behavior of amorphous solids in the dynamics above, one can show that the length of the "parallel" clock is shortened by the factor $\sqrt{1-v^2}$; in our units, the speed c_0 of Sec. 3.1 is unity. Thus the MM experiment sets the bound $\delta < 10^{-3}$.

Better bounds on δ have be set by other "standard" tests of special relativity, such as descendents of the Michelson-Morley experiment [4, 7, 11], a test of time-dilation using radionuclides on centrifuges [29], tests of the relativistic Doppler shift formula using two-photon absorption (TPA) [30], and a test of the isotropy of the speed of light using one-way propagation of light between hydrogen maser atomic clocks at the Jet Propulsion Laboratory (JPL) [31].

Very stringent bounds $|\delta| < 10^{-21}$ have been set by "mass isotropy" experiments of a kind pioneered by Hughes and Drever [32, 33]. The idea is simple: in a frame moving relative to the preferred frame, the non-Lorentz-invariant electromagnetic action of Eq. (9) becomes anisotropic, dependent on the direction of the velocity $\mathbf{V}$. Those anisotropies then are reflected in the energy levels of electromagnetically bound atoms and nuclei (for nuclei, we consider only the electromagnetic contributions). For example, the three sublevels of an $l = 1$ atomic or nuclear wavefunction in an otherwise spherically symmetric atom can be split in energy, because the anisotropic perturbations arising from the electromagnetic action affect the energy of each substate differently. One can study such energy anisotropies by first splitting the sublevels slightly using a magnetic field, and then monitoring the resulting Zeeman splitting as the rotation of the Earth causes the laboratory **B**-field (and hence the quantization axis) to rotate relative to **V**, causing the relative energies of the sublevels to vary among themselves diurnally. Using nuclear magnetic resonance techniques, the original Hughes-Drever experiments placed a bound of about 10^{-16} eV on such variations. This is about 10^{-22} of the electromagnetic energy of the nuclei used. Since the magnitude of the predicted effect depends on the product $V^2\delta$, and $V^2 \approx 10^{-6}$, one obtains the bound $|\delta| < 10^{-16}$. Energy anisotropy experiments were improved dramatically in the 1980s using laser-cooled trapped atoms and ions [34, 35, 36]. This technique made it possible to reduce the broading of resonance lines caused by collisions, leading to improved bounds on δ shown in Figure 4 (experiments labelled NIST, U. Washington and Harvard, respectively).

5 Is gravity Lorentz invariant?

The strong equivalence principle (SEP) is a generalization of EEP which states that in local "freely-falling" frames that are large enough to include gravitating systems (such as planets, stars, a Cavendish experiment, a binary system, *etc.*), yet that are small enough to ignore tidal gravitational effects from surrounding matter, local *gravitational* physics should be independent of the velocity of the frame and of its location in space and time. Also *all* bodies, including those bound by their own self-gravity, should fall with the same acceleration. General relativity satisfies SEP, whereas most other metric theories do not (*eg.* the Brans-Dicke theory).

It is straightforward to see how a gravitational theory could violate SEP [37]. Most alternative metric theories of gravity introduce auxiliary fields which couple to the metric (in a metric theory they can't couple to matter), and the boundary values of these auxiliary fields determined either by cosmology or by distant matter can act back on the local gravitational dynamics. The effects can include variations in time and space of the locally measured effective Newtonian gravitational constant G (preferred-location effects), as well as effects resulting from the motion of the frame relative to a preferred cosmic reference frame (preferred-frame effects). Theories with auxiliary scalar fields, such as the Brans-Dicke theory and its generalizations, generically cause temporal and spatial variations in G, but respect the "Lorentz invariance" of gravity, *i.e.* produce no preferred-frame effects. The reason is that a scalar field is invariant under boosts. On the other hand, theories with auxiliary vector or tensor fields can cause preferred-frame effects, in addition to temporal and spatial variations in local gravitational physics. For example, a timelike, long-range vector field singles out a preferred universal rest frame, one in which the field has no spatial components; if this field is generated by a cosmic distribution of matter, it is natural to assume that this special frame is the mean rest frame of that matter. A number of such "vector-tensor" metric theories of gravity have been devised [37, 38, 39]; see [23] for a review.

General relativity embodies SEP because it contains only one gravitational field $g_{\mu\nu}$. Far from a local gravitating system, this metric can always be transformed to the Minkowski form $\eta_{\mu\nu}$ (modulo tidal effects of distant matter and $1/r$ contributions from the far field of the local system), a form that is constant and Lorentz invariant, and thus that does not lead to preferred-frame or preferred-location effects.

The theoretical framework most convenient for discussing SEP effects is the parametrized post-Newtonian (PPN) formalism [40, 41, 23], which treats the weak-field, slow-motion limit of metric theories of gravity. This limit is appropriate for discussing the dynamics of the solar system and for many stellar systems, except for those containing compact objects such as neutron stars. If one focuses attention on theories of gravity whose field equations are derivable from an invariant action principle (Lagrangian-based theories), the generic post-Newtonian limit is characterized by the values of five PPN parameters, γ, β, ξ, α_1 and α_2. Two in particular, α_1 and α_2, measure the existence of preferred-frame effects. If SEP is valid, $\alpha_1 = \alpha_2 = \xi = 4\beta - \gamma - 3 = 0$, as in general relativity. In scalar-tensor theories, $\alpha_1 = \alpha_2 = \xi = 0$, but $4\beta - \gamma - 3 = 1/(2 + \omega)$, where ω is the "coupling parameter" of the scalar-tensor theory. In Rosen's bimetric theory, $\alpha_2 = c_0/c_1 - 1$, $\alpha_1 = \xi = 4\beta - \gamma - 3 = 0$, where c_0 and c_1 are the cosmologically induced values of the temporal and spatial diagonal components of a flat background tensor field, evaluated in a cosmic rest frame in which the physical metric has the Minkowski form far from the local system.

Within the PPN formalism the variations in the locally measured Newtonian gravitational constant G_{local} can be calculated explicitly: viewed as the coupling constant in the gravitational force between two point masses at a given separation, it is given by

$$G_{\text{local}} = 1 - (4\beta - \gamma - 3 - 3\xi)U_{\text{ext}} - \frac{1}{2}(\alpha_1 - \alpha_2)V^2 - \frac{1}{2}\alpha_2(\mathbf{V}\cdot\mathbf{e})^2 + \xi U_{\text{ext}}(\mathbf{N}\cdot\mathbf{e})^2 , \tag{10}$$

where U_{ext} is the potential of an external mass in the direction $\mathbf{N}$, $\mathbf{V}$ is the velocity of the experiment relative to the preferred frame, $\mathbf{e}$ is the orientation of the two masses and units have been chosen so that $G_{\text{local}} = 1$ in the preferred frame far from local matter sources. Thus G_{local} can vary in magnitude with variations in U_{ext} and V^2, and can also be anisotropic, that is can vary with the orientation of the two bodies. Other SEP-violating effects include planetary orbital perturbations and precessions of planetary and solar spin axes. A variety of observations have placed the bounds

$$|\alpha_1| < 10^{-4} , \quad |\alpha_2| < 4 \times 10^{-7} . \tag{11}$$

See [23, 42] for further details about tests of preferred-frame effects in gravity.

6 Tests of local Lorentz invariance at the centenary

6.1 Frameworks for Lorentz symmetry violations

During the past decade there has been a major renewal of interest in developing new ways to test Lorentz symmetry, using laboratory experiments and astrophysical observations. Part of the motivation for this comes from quantum gravity. Quantum gravity asserts that there is a fundamental length scale given by the Planck length, $L_p = (\hbar G/c^3)^{1/2} = 1.6 \times 10^{-33}$ cm, but since length is not an invariant quantity (Lorentz-FitzGerald contraction), then there could be a violation of Lorentz invariance at some level in quantum gravity. In brane world scenarios, while physics may be locally Lorentz invariant in the higher dimensional world, the confinement of the interactions of normal physics to our four-dimensional "brane" could induce apparent Lorentz violating effects. And in models such as string theory, the presence of additional scalar, vector and tensor long-range fields that couple to matter of the standard model could induce effective violations of Lorentz symmetry, as we discussed in Sec. 4.3. These and other ideas have motivated a serious reconsideration of how to test Lorentz invariance with better precision and in new ways.

Kostalecky and collaborators developed a useful and elegant framework for discussing violations of Lorentz symmetry in the context of the standard model of particle physics [43, 44, 45]. Called the Standard Model Extension (SME), it takes the standard SU(3) $\times$ SU(2) $\times$ U(1) field theory of particle physics, and modifies the terms in the action by inserting a variety of tensorial quantities in the quark, lepton, Higgs, and gauge boson sectors that could explicitly violate LLI. SME extends the earlier classical frameworks ($TH\epsilon\mu$, c^2, $\chi - g$) to quantum field theory and particle physics. The modified terms split naturally into those that are odd under CPT (i.e. that violate CPT) and terms that are even under CPT. The result is a rich and complex framework, with many parameters to be analysed and tested by experiment. Such details are beyond the scope of this paper; for a review of SME and other frameworks, the reader is referred to the recent article by Mattingly [46].

Here we confine our attention to the electromagnetic sector, in order to link the SME with the c^2 framework discussed above. In the SME, the Lagrangian for a scalar particle ϕ with charge e interacting with electrodynamics takes the form

$$\begin{aligned}\mathcal{L} &= [\eta^{\mu\nu} + (k_\phi)^{\mu\nu}](D_\mu\phi)^\dagger D_\nu\phi - m^2\phi^\dagger\phi \\ &\quad -\frac{1}{4}[\eta^{\mu\alpha}\eta^{\nu\beta} + (k_F)^{\mu\nu\alpha\beta}]F_{\mu\nu}F_{\alpha\beta}\,, \end{aligned} \tag{12}$$

where $D_\mu\phi = \partial_\mu\phi + ieA_\mu\phi$, and where $(k_\phi)^{\mu\nu}$ is a real symmetric trace-free tensor, and $(k_F)^{\mu\nu\alpha\beta}$ is a tensor with the symmetries of the Riemann tensor, and with vanishing double trace. It has 19 independent components. There could also be a CPT-odd term in $\mathcal{L}$ of the form $(k_A)^\mu\epsilon_{\mu\nu\alpha\beta}A^\nu F^{\alpha\beta}$, but because of a variety of pre-existing theoretical and experimental constraints, it is generally set to zero.

The tensor $(k_F)^{\mu\alpha\nu\beta}$ can be decomposed into "electric", "magnetic" and "odd-parity" components, by defining

$$\begin{aligned}(\kappa_{DE})^{jk} &= -2(k_F)^{0j0k}\,,\\ (\kappa_{HB})^{jk} &= \frac{1}{2}\epsilon^{jpq}\epsilon^{krs}(k_F)^{pqrs}\,,\\ (\kappa_{DB})^{kj} &= -(k_{HE})^{jk} = \epsilon^{jpq}(k_F)^{0kpq}\,.\end{aligned} \tag{13}$$

In many applications it is useful to use the further decomposition

$$\begin{aligned}\tilde\kappa_{tr} &= \frac{1}{3}(\kappa_{DE})^{jj}\,,\\ (\tilde\kappa_{e+})^{jk} &= \frac{1}{2}(\kappa_{DE}+\kappa_{HB})^{jk}\,,\\ (\tilde\kappa_{e-})^{jk} &= \frac{1}{2}(\kappa_{DE}-\kappa_{HB})^{jk} - \frac{1}{3}\delta^{jk}(\kappa_{DE})^{ii}\,,\\ (\tilde\kappa_{o+})^{jk} &= \frac{1}{2}(\kappa_{DB}+\kappa_{HE})^{jk}\,,\\ (\tilde\kappa_{o-})^{jk} &= \frac{1}{2}(\kappa_{DB}-\kappa_{HE})^{jk}\,.\end{aligned} \tag{14}$$

The first expression is a single number, the next three are symmetric trace-free matrices, and the final is an antisymmetric matrix, accounting thereby for the 19 components of the original tensor $(k_F)^{\mu\alpha\nu\beta}$.

In the rest frame of the universe, these tensors have some form that is established by the global nature of the solutions of the overarching theory being used. In a frame that is moving relative to the universe, the tensors will have components that depend on the velocity of the frame, and on the orientation of the frame relative to that velocity.

In the case where the theory is rotationally symmetric in the preferred frame, the tensors $(k_\phi)^{\mu\nu}$ and $(k_F)^{\mu\nu\alpha\beta}$ can be expressed in the form

$$\begin{aligned}(k_\phi)^{\mu\nu} &= \tilde\kappa_\phi(u^\mu u^\nu + \frac{1}{4}\eta^{\mu\nu})\,,\\ (k_F)^{\mu\nu\alpha\beta} &= \tilde\kappa_{tr}(4u^{[\mu}\eta^{\nu][\alpha}u^{\beta]} - \eta^{\mu[\alpha}\eta^{\beta]\nu})\,,\end{aligned} \tag{15}$$

where [] around indices denote antisymmetrization, and where u^μ is the four-velocity of an observer at rest in the preferred frame. With this assumption, all the tensorial quantities in Eq. (14) vanish in the preferred frame, and, after suitable rescalings of coordinates and fields, the action (12) can be put into the form of the c^2 framework, with

$$c = \left(\frac{1-\frac{3}{4}\tilde\kappa_\phi}{1+\frac{1}{4}\tilde\kappa_\phi}\right)^{1/2}\left(\frac{1-\tilde\kappa_{tr}}{1+\tilde\kappa_{tr}}\right)^{1/2}\,. \tag{16}$$

Another class of frameworks for considering Lorentz invariance violations is kinematical. They involve modifying the relationship between energy E and momentum p for each particle species. Assuming that rotational symmetry in the preferred frame is maintained, then one adopts a parametrized dispersion relation of the form

$$E^2 = m^2 + p^2 + E_{Pl} f^{(1)} |p| + f^{(2)} p^2 + \frac{f^{(3)}}{E_{Pl}} |p|^3 + \dots , \tag{17}$$

where E_{Pl} is the Planck energy. Frameworks like these are useful for discussing effects that might be relics of quantum gravity, and for discussing particle physics and high-energy astrophysics experiments.

6.2 Modern searches for Lorentz symmetry violation

A variety of modern "clock isotropy" experiments have been carried out to bound the electromagnetic parameters of the SME framework. For example, comparing the frequency of electromagnetic cavity oscillators of various configurations with atomic clocks as a function of the orientation of the laboratory has placed bounds on the coefficients of the tensors $\tilde{\kappa}_{e-}$ and $\tilde{\kappa}_{o+}$ at the levels of 10^{-15} and 10^{-11}, respectively [46]. Direct comparisons between atomic clocks based on different nuclear species place bounds on SME parameters in the neutron and proton sectors, depending on the nature of the transitions involved. The bounds achieved range from 10^{-27} to $10^{-32}\, GeV$ [46].

Astrophysical observations have also been used to bound Lorentz violations. For example, if photons satisfy the Lorentz violating dispersion relation (17), then the speed of light $v_\gamma = \partial E / \partial p$ would be given by

$$v_\gamma = 1 + \frac{(n-1) f_\gamma^{(n)} E^{n-2}}{2 E_{Pl}^{n-2}} \,. \tag{18}$$

By bounding the difference in arrival time of high-energy photons from a burst source at large distances, one could bound contributions to the dispersion for $n > 2$. The best limit, $|f^{(3)}| < 128$ comes from observations of 1 and 2 TeV gamma rays from the blazar Markarian 421 [47].

Other testable effects of Lorentz invariance violation include threshold effects in particle reactions, birefringence in photon propagation through empty space, gravitational Cerenkov radiation, and neutrino oscillations. Mattingly [46] gives a thorough and up-to-date review of both the theoretical frameworks and the experimental results.

7 Concluding remarks

At the centenary of special relativity, I can think of no better tribute to the impact and influence of Einstein's relativistic contributions than to cite how they now affect daily life. This unique confluence of abstract theory, high precision technology and everyday applications involves the Global Positioning System (GPS). This navigation system, based on a constellation of 24 satellites carrying atomic clocks, uses precise time transfer to provide accurate absolute positioning anywhere on Earth to 15 meters, differential or relative positioning to the level of centimeters, and time transfer to a precision of 50 nanoseconds. It relies on clocks that are stable, run at the same or well calibrated rates, and are synchronized. However, the difference in rate between GPS satellite clocks and ground clocks caused by the special relativistic time dilation is around -7,000 ns per day, while the difference caused by the gravitational redshift is around 46,000 ns per day. The net effect is that the satellite clocks tick faster than ground clocks by around 39,000 ns per day. Consequently, general relativity *must* be taken into account in order to achieve the 50 ns time transfer accuracy required for 15 m navigation. In addition, the satellite clocks must be synchronized with respect to a fictitious clock on the Earth's rotation axis, in order to avoid the inevitable inconsistency in synchronizing clocks around a closed path in a rotating frame (called the Sagnac effect). For a detailed discussion of relativity in GPS, see [48]; for a popular essay on the subject, see [49]. GPS is a spectacular example of the unexpected and unintended benefits of basic research. While Einstein often used trains to illustrate principles and consequences of relativity, one can now find practical, everyday consequences of relativity in trains, planes and automobiles.

Acknowledgments

This work was supported in part by the National Science Foundation under Grant No. PHY 03-53180.

References

[1] A. Einstein, *Ann. d. Physik* **17**, 891 (1905).

[2] A. Einstein, *Ann. d. Physik* **17**, 639 (1905).

[3] A. Einstein, *Ann. d. Physik* **49**, 769 (1916).

[4] A. A. Michelson and E. W. Morley, *Am. J. Sci.* **34**, 333 (1887).

[5] J. H. Poincaré, in *Physics for a New Century: Papers Presented at the 1904 St. Louis Congress* (Tomash Publishers/American Institute of Physics, Washington, 1986), p. 281.

[6] P. Galison, *Einstein's Clocks, Poincaré's Maps: Empires of Time*, (W. W. Norton, New York, 2003).

[7] R. . Shankland, S.W. McCuskey, F.C. Leone and G. Kuerti, *Rev. Mod. Phys.* **27**, 167 (1955).

[8] G.F. FitzGerald, *Science* **13**, 390 (1889).

[9] H.A. Lorentz, *Versl. K. Ak. Amsterdam* **1**, 74 (1892).

[10] T.S. Jaseja, A. Javan, J. Murray and C.H. Townes, *Phys. Rev.* **133**, A1221 (1964).

[11] A. Brillet and J.L. Hall, *Phys. Rev. Lett.* **42**, 549 (1979).

[12] K. Brecher, *Phys. Rev. Lett.* **39**, 1051 (1977).

[13] T. Alväger, F.J.M. Farley, J. Kjellman and I. Wallin, *Phys. Lett.* **12**, 260 (1977).

[14] H.E. Ives and G.R. Stilwell, *J. Opt. Soc. Am.* **28**, 215 (1938).

[15] B. Rossi and D.B. Hall, *Phys. Rev.* **59**, 223 (1941).

[16] F.J. M.Farley, J. Bailey, R.C.A. Brown, M. Giesch, H. Jöstlein, S. van der Meer, E. Picasso and M. Tannenbaum, *Nuovo Cimento* **45**, 281 (1966).

[17] E.F. Taylor and J.A. Wheeler, *Spacetime Physics: Introduction to Special Relativity* (W. H. Freeman, New York, 1992).

[18] R.V. Eötvös, V. Pekár and E. Fekete, *Ann. d. Physik* **68**, 11 (1922).

[19] S. Baessler, B.R. Heckel, E.G. Adelberger, J.H. Gundlach, U. Schmidt and H.E. Swanson, *Phys. Rev. Lett.* **83**, 3585 (1999).

[20] A. Einstein, *Jahrb. d. Radioactivität und Elektronik* **4**, 411 (1907).

[21] A. Einstein, *Ann. d. Physik* **35**, 898 (1911).

[22] R.H. Dicke, in *Relativity, Groups and Topology*, edited by C. DeWitt and B. DeWitt (Gordon and Breach, New York, 1964), p. 165.

[23] C.M. Will, *Theory and Experiment in Gravitational Physics* (Cambridge University Press, Cambridge, 1993).

[24] A.P. Lightman and D.L. Lee, *Phys. Rev. D* **8**, 364 (1973).

[25] W.-T. Ni, *Phys. Rev. Lett.* **38**, 301 (1977).

[26] M.P. Haugan and C.M. Will, *Physics Today* **40**, 69 (May) (1987).

[27] M.D. Gabriel and M.P. Haugan, *Phys. Rev. D* **41**, 2943 (1990).

[28] J.E. Horvath, E.A. Logiudice, C. Reveros and H. Vucetich, *Phys. Rev. D* **38**, 1754 (1988).

[29] D.C. Champeney, G.R. Isaak and A.M. Khan, *Phys. Lett.* **7**, 241 (1963).

[30] E. Riis, L.-U. A. Anderson, N. Bjerre, O. Poulson, S.A. Lee and J.L. Hall, *Phys. Rev. Lett.* **60**, 81 (1988); *ibid* **62**, 842 (1989).

[31] T.P. Krisher, L. Maleki, G.F. Lutes, L.E. Primas, R.T. Logan, J.D. Anderson and C.M. Will, *Phys. Rev. D* **42**, 731 (1990).

[32] V.W. Hughes, H.G. Robinson and V. Beltran-Lopez, *Phys. Rev. Lett.* **4**, 342 (1960).

[33] R.W.P. Drever, *Phil. Mag.* **6**, 683 (1961).

[34] J.D. Prestage, J.J. Bollinger, W.M. Itano and D.J. Wineland, *Phys. Rev. Lett.* **54**, 2387 (1985).

[35] S.K. Lamoreaux, J.P. Jacobs, B.R. Heckel, F.J. Raab and E.N. Fortson, *Phys. Rev. Lett.* **57**, 3125 (1986).

[36] T.E. Chupp, R.J. Hoare, R.A. Loveman, E.R. Oteiza, J.M. Richardson, M.E. Wagshul and A.K. Thompson, *Phys. Rev. Lett.* **63**, 1541 (1989).

[37] C. M. Will and K. Nordtvedt, Jr., Astrophys. J. **177**, 757 (1972).

[38] R.W. Hellings and K. Nordtvedt, Jr., *Phys. Rev. D* **7**, 3593 (1973).

[39] C. Eling and T. Jacobson, *Phys. Rev. D* **69**, 064005 (2004).

[40] K. Nordtvedt, Jr., *Phys. Rev.* **169**, 1017 (1968).

[41] C.M. Will, *Astrophys. J.* **163**, 611 (1971).

[42] C.M. Will, *Living Rev. Relativ.* **4**, 4 (2001) [Online article]: cited on 28 Feb 2005, http://www.livingreviews.org/lrr-2001-4.

[43] D. Colladay and V.A. Kostalecky, *Phys. Rev. D* **55**, 6760 (1997).

[44] D. Colladay and V.A. Kostalecky, *Phys. Rev. D* **58**, 116002 (1998).

[45] V.A. Kostalecky and M. Mewes, *Phys. Rev. D* **66**, 056005 (2002).

[46] D. Mattingly, *Living Rev. Relativ.* **8**, 5, (2005). [Online article]: cited on 1 October 2005, http://www.livingreviews.org/lrr-2005-5.

[47] S.D. Biller, *et al.*, *Phys. Rev. Lett.* **83**, 2108 (1999).

[48] N. Ashby, *Living Rev. Relativ.* **6**, 1 (2003) [Online article]: cited on 28 Feb 2005, http://www.livingreviews.org/lrr-2003-1.

[49] C.M. Will, *Einstein's Relativity and Everyday Life*, [Online Essay]: http://www.physicscentral.com/writers/writers-00-2.html

Clifford M. Will
McDonnell Center for the Space Sciences and Department of Physics
Washington University, St. Louis, MO 63130
USA
email: `cmw@wuphys.wustl.edu`

Einstein, 59 – 133

Poincaré Seminar 2005

The Geometry of Relativistic Spacetime

Jacques Bros and Ugo Moschella

This paper aims to be a pedagogical excursion across the land of relativistic spacetime. By diving into the past, two thousand and three hundred years ago, the first part traces back to the good old Euclid's geometry for constructing without harm the flat Minkowski's spacetime of special relativity. Then by crossing the year 1905 on the way back, it does not forget to cheer Albert Einstein's papers on the subject! But the future lusts for the wide horizons of curved spacetimes and larger dimensions, born with general relativity... The second part proposes a sightseeing tour in the most accessible ones. Have fun in the de Sitter and anti-de Sitter spacetimes !

From Euclid's Geometry to Minkowski's Spacetime

Jacques Bros

> "...the word *relativity-postulate* for the requirement of the invariance under the group G_c seems to me very feeble. Since the postulate comes to mean that only the four-dimensional world in space and time is given by phenomena, but that the projection in space and in time may still be undertaken with a certain degree of freedom, I prefer to call it the *postulate of the absolute world* (or briefly the world-postulate)."

H. MINKOWSKI

Cologne Conference, September 1908

Introduction and general survey

From a variety of viewpoints, the theory of relativity appears as one of the major conceptual events that have ever happened in the adventure of knowledge. It is therefore highly pertinent that the scientific community celebrates the "century commemoration" of the revelation of special relativity by two of the four fundamental papers that were published by Einstein in the year 1905 [1] [2]. Since then, the historians of science have been able to accumulate a crop of information about the complex genesis and the multiple and intricate aspects of that extraordinary intellectual adventure. However, strangely enough an important pedagogical work still remains to be done, if one retains from that adventure one of its most striking aspects, namely the existence of a united geometrical representation of space and time, called *spacetime*, and the logical necessity of its introduction on the basis of the special properties of the velocity of light. In fact, we think it worthwhile and possible to communicate this geometrical representation not only to learned scientists, but also to any scientifically-curious and/or philosophically-minded student. Let us explain why we think that it is 1) worthwhile and 2) possible.

1) A wide communication of it is worthwhile, because we have here to deal with a genuine "jewel of human knowledge", in which Physics, Mathematics (at a rather elementary level, see 2) below) and Philosophy are intimately related. Physics at first: one century after its discovery, one can say that in our present knowledge of the universe, the validity of this joint representation of space and time extends from the spacetime scales of microphysics to those of cosmology, which represents a scaling factor of more than 10^{40}. Then Philosophy and Mathematics: we have to deal with an overwhelming "ontological fusion" of the categories of space and time, through a mental representation which belongs to the platonician

world of geometrical concepts. Here is what can be felt as a real shock for the human mind ! With respect to our usual separate perceptions of space and time, the new geometrical conception of spacetime is as much revolutionary as was the idea of the sphericity of the earth and the computation of its circumference by Eratosthenes with respect to the primitive conception of a flat earth. In the latter case, it is only the development of long-distance travels that have made this idea more and more acceptable for the "common sense" throughout the centuries. In the former case, only motions whose velocity is substantial compared with the velocity of light provide an evidence that the new spacetime framework gives a correct representation of the physical reality. This is indeed attested as well by the motions of particles which are the ultimate components of matter as by the motions of astronomical objects observed by telescopes. It is only the fact (basic in our social existence!) that all of us are "slowly moving travelers with respect to one another" which comforts us every day in our feeling that the flow of physical time is the same for all of us and therefore perceived as *absolute* (our watches run at the same rhythm!); but this viewpoint, which is encoded in the usual "Galilean kinematics" is only the low-velocity approximation of the physically relevant representation of spacetime. The basic character of the physical spacetime is that the lapse of time measured by an experimentalist between two successive events A and B depends on the particular motion which has been adopted by this experimentalist for proceeding from A to B. But this fact becomes conceivable to us if we compare it with the following one which is familiar to our perception: the distance which is measured by an experimentalist between two given points A and B of space depends on the particular path which has been adopted by this experimentalist for going from A to B. As a matter of fact, what may seem here as purely metaphoric turns out to be a deep structural analogy in geometrical terms.

2) A wide communication of it is possible, once one has realized that these purely geometrical aspects of relativity theory can actually be transmitted in the old Greek spirit of Euclid's geometry. In fact, let us recall (if forgotten) that this so-called "elementary geometry", revived in a second golden age by the European geometers (from seventeenth to nineteenth centuries), was given to the pupils of secondary schools of the old Europe as the most secure guide for training the faculties of logics and rational thought ! Here we would like to make the point that (at the age of computers…) this framework might also be the most secure one for transmitting to everyone who is interested a simple, but sound idea of *what is the spacetime of relativity theory* ! The simplest the argument, the strongest the impact for the mind !

From the viewpoint of the historian of science, the adventure of relativistic theory can be seen as the unexpected, although unavoidable issue of the major crisis of nineteenth-century physics, in which the concept of a fixed reference medium in the universe, called the *ether*, was in open conflict with the recently discovered laws of electromagnetism. Among a lot of experimental as well as theoretical results, crucial experiments had been proposed and performed as soon as 1887: these were the famous Michelson and Morley experiments about the constancy of the velocity of light. Then almost twenty years of maturation were still necessary for the conceptual elaboration of the theory of special relativity to be performed. Although it was revealed to the scientific community in the year 1905 by Einstein's revolutionary paper entitled "On the electrodynamics of moving bodies" [1], the theory made a basic use of formulae established previously by Lorentz; moreover its further formulation greatly benefitted from the group-theoretical analysis of Poincaré also delivered in 1905 [4], while it found its achievement in 1908 through Minkowski's illuminating geometrical work [3]. It is indeed the latter which has to be granted for introducing the appropriate new concept of *absolute spacetime*, a concept whose fate was to go far beyond the theory of special relativity, since it played an essential role in the further discovery and formulation of the theory of general relativity by Einstein in 1916.

It will be precisely our purpose to focus on the concept of spacetime and at first on its logical introduction, which may be presented in a spirit that parallels the axiomatization of Euclid's geometry, thanks to an appropriate axiom about the "universality" of the velocity of light. This spacetime, which can be regarded after Minkowski as an *absolute* framework for describing the kinematics of special relativity, is a representation space whose points are interpreted as the "physical events". Any motion which is physically possible between two given events A and B is represented by a certain *world-line* with end-points A and B. There is an absolute orientation of such world-line, which can be called its "time-arrow": its physical meaning is that one of the end-point events, e.g. B, is in the future of the other one A. The pair of events (A, B) is also said to be *causally separated*; it is not the case for all pairs of events. The limits of causality are determined by the world-lines of *light-rays* passing by each event: the Minkowski spacetime is thus basically equipped with a *light-webbed structure*. In that geometrical representation, one is thus led to distinguish radically the "absolute properties", also called "relativistic invariant properties" from the properties which are "relative to a reference frame" and thereby comparable with the effects of spatial perspective in the usual Euclidean geometry. The basic absolute property of Minkowski spacetime is the fact that it is a mathematical space equipped with a *pseudo-distance*, which is closely linked with the existence of the light-webbed structure of the universe: along the world-lines of light-rays, this pseudo-distance vanishes ! The most striking feature of this absolute pseudo-distance is the *inverse triangular inequality*, which is responsible for the overwhelming phenomenon of "Langevin twins": The "length" of one side (e.g. the aging of the twin at rest) is *longer* than the sums of the "lengths" of the other two sides of the triangle (namely the aging of the traveling twin). As a matter of fact, eventhough the full spacetime is (in mathematical terms) an abstract four-dimensional manifold, such an overwhelming property as the aging difference for twins with different motions can be visualized in terms of planar geometry. It is in fact sufficient to consider two-dimensional sections of spacetime in which a single dimension of space is involved for having a fully correct and intuitive geometrical picture of the Minkowskian triangular inequality. Similarly, one can easily visualize in such a planar section of spacetime the phenomenon of relativistic perspective called "the contraction of lengths". Of course, the last important step for our understanding of spacetime concerns the way in which the usual three-dimensional Euclidean geometry is embedded in the Minkowskian four-dimensional spacetime. The fact that different embeddings hold for observers in relative uniform motion is implied by the notion of Lorentz frame; there appears the relevance of the group of Poincaré transformations. All these aspects of elementary Minkowskian geometry following from an axiomatic Euclid-type construction will be covered in our second part (Sec.2); a short preliminary part (Sec.1) is devoted to the use of geometry in mathematical physics, as an introduction to the concept of spacetime.

At that point, one might have the feeling that nothing more has to be added about the kinematics of special relativity, but this is not so. In fact, the conceptual revolution that it represents is so rich that after the basic articles of 1905 and 1908 in which it was delivered, several aspects of it deserved to be deepened and clarified: this was performed around 1960 in two directions.

a) If the parallel between the Euclidean geometry of our usual three-dimensional space and the Minkowskian geometry of four-dimensional spacetime is actually complete in the physical world, this parallel has to be checked not only for the geometry of straight world-lines, namely for uniform motions, but for arbitrary (smooth) curved world-lines, namely for accelerated motions. The interpretation of Minkowskian pseudo-length as a *proper time* measured by a clock along the world-line of the motion and the geometrical property asserting that such a pseudo-length is always smaller than that of the corresponding uniform motion originating and terminating at the same events had to be tested experimentally. This basic property of Minkowskian geometry, which can be nicely summarized by saying that "In proper-time distances, the straight-line is the *longest* distance between two points (namely two events)", was already present in Einstein's article [1] under the physical terminology of "clock slowing-down phenomenon". However, it remained to be checked experimentally that clocks submitted to accelerated motions were as insensitive to the accelerations as graduated ribbons were insensitive to curvature for measuring Euclidean curvilinear distances. What was in question in such investigations had to do with the physical nature of the clocks, considered as trustful measuring instruments, whose robustness with regard to the accelerations had to be quantitatively estimated. Thanks to the progress of physics during the twentieth century, the set of traditional clocks (called "dynamical") was enriched by a new class of clocks, based on microphysics phenomena and called *"atomic clocks"*, whose precision degree and robustness were far higher. Around 1960 (see in particular Sherwin's paper [5]), this property of insensitivity to accelerations has been established (and confirmed since then with higher and higher precision) for various types of atomic clocks. These results then exclude radically the last objections of the opponents to the "twin paradox" (see [5]). In particular, they allow one to present a completely acceptable version of the twin phenomenon in uniformly accelerated motions, namely a version which is *biologically bearable* by human experimentalists, even though for technical reasons it remains presently a "gedanken-experiment". Moreover, these manifestations of the Minkowskian geometrical structure in accelerated motions give an opportunity to state clearly that they must not be confused with possible effects of general relativity. In fact, the latter occur substantially when the accelerations are caused by the presence of large masses of matter, which produces an additional curvature effect on the Minkowskian geometry of spacetime.

b) Since 1959 with the articles of Terrell [6] and of V. Weisskopf [7], problems of relativistic perspective have been reconsidered. Progresses have been made on the problem of what should be the real optical appearance of a fast-moving *extended* object with respect to an observer linked to a given Lorentz frame. The understanding of the phenomenon of "contraction of lengths" was thus revisited and corrected for the case of extended objects. Much more recently, impressive visualizations of moving objects with relativistic velocities have been given thanks to the help of computer technique (see [8] and references therein).

An account of the previous developments a) and b) will be given below respectively in Sec.3 and Sec.4. Sec.5 and the companion paper by Ugo Moschella will illustrate the fundamental role played by the conceptual framework of Minkowski spacetime in two domains of physics whose orders of magnitude of spacetime distances differ by 10^{40}; we mean respectively particle physics and cosmology. A short final part (Sec.6) will serve as a bridge between the two papers.

It is at the scale of particle physics phenomena that the validity of special relativity and of its expression in the Minkowski spacetime framework appears with its full strength. In fact, the second revolutionary discovery which can be found in the second Einstein's paper [2] on special relativity in 1905, namely the equivalence relation of mass and energy $E = mc^2$, provides the relevant kinematical framework for understanding the energy-balance of all the nuclear and electromagnetic reactions. In geometrical terms, this framework corresponds to supplement Minkowski's spacetime with the introduction of another identical Minkowskian space, interpreted as the space of energy-momentum vectors of material points. This framework gives a remarkably good description of the kinematics of high-energy particle physics. In the Minkowskian energy-momentum space, Einstein's relation $E = mc^2$ is visualized under the form of the mass hyperboloid, called the *mass shell* of the particles: it is the surface which represents the set of all possible states of a free relativistic particle with mass m. This description includes the case of photons: for these "massless particles", the mass shell coincides with the "light-cone". In the energy-momentum space, the law of conservation of total energy-momentum admits a simple geometrical formulation. In that space, the Minkowskian triangular inequality accounts for the production of any number of particles in high-energy collisions of two particles (including the massless case of photons). All that constitutes the basic background for the formulation of the theory of high-energy particle collisions in the general framework of relativistic quantum field theory. In particular, the world-line representation of free particles and of their multiple collisions in Minkowski's spacetime obeying the rules of relativistic kinematics plays a basic role in the corresponding quantum field-theoretical treatment of particle physics: it explains the so-called *Landau singularities* of the multiparticle scattering functions.

At cosmological scales, the concept of spacetime introduced by Minkowski is still valid, provided one includes as a new revolutionary ingredient the notion of *curvature*: here is the geometrical content of *general relativity*. There are two reasons for this curvature phenomenon: while the first one is the local density of matter (or "gravific mass") which is present near each event in the universe, the other one is linked with the expansion of the universe; it is encoded in the so-called *cosmological constant* in the equations of tentative geometrical models of the universe, whose simplest one (with zero mass density) is the de Sitter universe (1917) presented in the companion paper. Under this respect, the role of Minkowskian geometry for the *local* description of the universe throughout its evolution parallels the role of planar Euclidean geometry for the *local* description of the surface of the earth. In mathematical terms, the latter is a two-dimensional Euclidean manifold: the straight-line distance of planar geometry is replaced by the *geodesical distance* between two points of the surface of the earth, *which is the shortest one* with respect to all possible paths joining these two points on the surface. Similarly, the universe (considered throughout its evolution) appears as a four-dimensional Minkowskian (one also says "Lorentzian") manifold: between two causally-separated events, there is a *geodesical time-like distance, which is the longest one* with respect to all possible world-lines joining these two events. For instance, when one estimates the age of the universe to be of the order of 14 or 15 billions of years, one has in mind the value of such a geodesical time-like distance between an event that can be called "the big bang" (in the most currently accepted cosmological models) and the event called "here and now" by the inhabitants of the earth in the year 2005. However, it is philosophically questioning to remain conscious of the following: according to the structure of Minkowskian manifold of the universe, any other world-line that relates those two events is covered in a *shorter* time-like distance. According to the motion which is associated with that world-line, it can be ... one century, one year, one day, one second ... or even zero, if one considers a light-ray trajectory, namely a world-line which is composed of pieces of light-like geodesics ....

1 On the use of geometry in mathematical physics and the concept of spacetime

1.1 Geometry of description and geometry of representation

As we all know it, Euclidean geometry (in two or three dimensions) corresponds to an idealized description of the space which surrounds us, as it is felt by our visual and tactile perceptions. The etymology of the word "geometer" (and for instance in France its standard meaning as a profession...) is still reminding us of the fact that, since very ancient times, this branch of mathematics was progressively elaborated from the consideration of practical physical problems, such as the measurement and sharing of ground pieces; the description of the trajectories of celestial bodies also provided another powerful motivation for the development of geometry. It is not a triviality, but a subject of wondering and of philosophical questioning that the idealized notions of "elementary geometry" (points, lines etc. ...) equipped with logical relations called axioms or postulates, allow us to construct "rigorous proofs" of nontrivial properties of the geometrical pictures. While their experimental checking in physical space is fully satisfactory, these properties also appear to us with the strength of evidence as elements of an "absolute reality of the mind", namely of a very special "world of Platonician ideas": the world of geometrical concepts. One can then say that, *as a geometry of description*, Euclidean geometry appears as the oldest manifestation of the spirit of mathematical physics.

Another considerable achievement in the history of mathematics is the fundamental correspondence between numbers and geometrical concepts which started from the length measurement procedure and resulted in the elaboration of Cartesian coordinates and of the so-called "analytic geometry". As it may be already familiar to pupils at the terminal level of high-school, this implies a relationship between algebra and geometry whose interest is two-fold. On the one-hand, the properties of geometrical curves can be equivalently represented by algebraic equations relating the coordinates of their points. This representation is unique, once the choice of a system of coordinates has been specified. For example in orthogonal coordinates, the equation of the unit circle $x^2 + y^2 - 1 = 0$ makes use of the standard Pythagore theorem for characterizing the points $M = (x, y)$ of that curve. On the other hand, any numerical relation between two quantities x and y (always representable by an equation of the form $f(x, y) = 0$) admits a pictorial representation by a curve in a plane equipped with given coordinate axes; this pictorial representation is specially interesting when x and y denote physical quantities related by a physical law. In fact, the curve which one thus constructs represents all the "states" of the observed phenomenon, each state being characterized by a pair of values of the quantities x and y which are simultaneously observed and thus associated with a particular point $M = (x, y)$ of the curve. The geometrical constructions which may be associated with the pictorial representation of a physical phenomenon in a plane or in a three-dimensional space equipped with coordinates pertain to what we shall call a *geometry of representation.* By using such a terminology, we adopt typically a viewpoint of mathematical physicist: while geometry presents all its mathematical characteristics, in particular the fact that its logical arguments are immediately perceived by a special type of global visual intuition, all its elements are here given a physical interpretation in terms of a certain category of phenomena; in other words, these phenomena are actually represented in terms of geometrical concepts.

1.2 The use of geometry in more than three dimensions

From a purely mathematical viewpoint, the correspondence between numbers and geometrical concepts can be extended to n–dimensional abstract spaces $\mathbf{R}^n$, with n larger than three. The concept of "point in $\mathbf{R}^n$" is now introduced as a n–tuplet of coordinates $M = (x_1, \ldots, x_n)$. The concept of "surface of dimension p" with $2 \leq p \leq n-1$ (called "curve" for $p = 1$ and "hypersurface" for $p = n-1$) is then introduced as a subset of points of $\mathbf{R}^n$ whose coordinates satisfy $n-p$ independent equations; correspondingly, these coordinates can also be expressed by parametric equations involving p independent parameters. If one wishes, one can equip the space $\mathbf{R}^n$ with a Euclidean distance, which is obtained by an obvious extrapolation from the usual one, two and three-dimensional cases. By definition, the squared length of a linear segment $MN = (a_1, a_2, \ldots, a_n)$ (or squared distance between the two points M and N) is $d(M,N)^2 = a_1^2 + a_2^2 + \cdots + a_n^2$, which implies the usual triangular inequality $d(O,N) \leq d(O,M) + d(M,N)$. The equation $x_1^2 + x_2^2 + \cdots + x_n^2 = 1$ is represented geometrically by the "unit hypersphere". In any two-dimensional or three-dimensional section of $\mathbf{R}^n$ defined by linear equations in terms of the coordinates, one recovers respectively a plane or a three-dimensional space equipped with the usual Euclidean distance. So one can develop a set of geometrical concepts, relations and constructions which generalize those of the usual geometry; this can be done at will either in terms of equations or in a purely geometrical language.

From the viewpoint of mathematical physics, the use of geometry in more than three dimensions turns out to be necessary, if one wishes to represent phenomena whose description necessitates more than three independent quantities. A typical example is the six dimensional space $\mathbf{R}^6_{\mathbf{x}_1,\mathbf{x}_2} = \mathbf{R}^3_{\mathbf{x}_1} \times \mathbf{R}^3_{\mathbf{x}_2}$ of the positions $(\mathbf{x}_1, \mathbf{x}_2)$ of pairs of material points (or pointlike particles) in mutual interaction. Trajectories of such pairs are represented by curves in $\mathbf{R}^6$, described in terms of a time parameter t by equations of the form $\mathbf{x}_1 = \mathbf{x}_1(t)$, $\mathbf{x}_2 = \mathbf{x}_2(t)$. Another type of geometrical representation which is also often used in physics with strong motivations is *complex geometry*: for example the extension of functions of the real *frequency* variable to (analytic) functions of the corresponding complex variable in a domain of the complex plane $\mathbf{C}$ is of current use. It is in fact a basic property of structural functions describing linear response phenomena, which provides a convenient visual representation of *resonance* phenomena by real or complex *poles*. In particle physics a similar use of complex geometry in spaces $\mathbf{C} \times \cdots \times \mathbf{C} = \mathbf{C}^n$

of various variables (positions, times, momenta, energies) plays an important conceptual role.

In the following, we shall be concerned with a very special type of geometry of representation, called *spacetime*, whose purpose is to provide a visualization of the motion phenomena *throughout their whole history.* If we consider motions in the Euclidean space $\mathbf{R}^3$, providing as usual a geometry of description of the world which surrounds us, we need an additional time-coordinate and therefore an affine space $\mathbf{R}^4$ for representing geometrically all the events of the world. Such a map is intended to picture in an idealistic way the whole history of the world: the motion of any material point (or of any observer) will be represented as a curve, called a *world-line*, which describes all its history from the remote past to the far future. The usual notion of trajectory will then appear as the *projection* of the world-line onto the Euclidean space $\mathbf{R}^3$. The world-line is a geometrical concept which contains all the information on the motion, which is not the case for the trajectory: two different world-lines (i.e. motions) may project onto the same trajectory.

1.3 Galilean spacetime as a geometry of representation of motion phenomena

In its simplest form, which we shall call *Galilean spacetime*, the concept of spacetime appears as a geometry of representation for the phenomena of motion, *as they are perceived by a privileged observer called* $\mathcal{O}_0$, submitted to the following prejudice: *the time interval that elapses between two events A and B is an absolute quantity*; its value is the same for observers moving in an arbitrary way between A and B, provided they are equipped with identical clocks.

Keeping the previous notations, $x = \mathbf{x}$ now denotes a point, or equivalently three coordinates called *space coordinates*, in the usual Euclidean space $\mathbf{R}^3$ in which we are living, while $y \equiv t$ denotes a time coordinate. A point $X = (\mathbf{x}, t)$ in $\mathbf{R}^4$ represents the *event* which takes place at time t at the point $\mathbf{x}$ of Euclidean space $\mathbf{R}^3$. In particular, the origin O represents the event called "here and now" (at a certain instant...) by the observer $\mathcal{O}_0$, who stands "at rest" at $\mathbf{x} = 0$; by definition, this means that the observer's world-line is the time-axis with equation $\mathbf{x} = 0$. For $\mathcal{O}_0$, the coordinate hyperplane with equation $t = 0$ represents the set of all *simultaneous events* which constitute the "present". Similarly, for every fixed value t_0 of t, the hyperplane with equation $t = t_0$ is a complete set of simultaneous events, which we call *set of simultaneity* and which belongs to the future or to the past according to whether t is positive or negative. The whole *future* and the whole *past* are represented respectively by the open half-spaces $t > 0$ and $t < 0$ of $\mathbf{R}^4$. In such a representation of the events, one says that the time-axis associated with the Euclidean space $\mathbf{R}^3$ of "present events" constitute the *reference frame* of the observer $\mathcal{O}_0$ (the choice of the "present time" $t = 0$ is of course a matter of convention for $\mathcal{O}_0$).

Let $\mathcal{O}_{\mathbf{v}_0}$ be an observer in uniform motion with vector velocity $\mathbf{v}_0$ with respect to $\mathcal{O}_0$ and passing by O: this means that the two observers $\mathcal{O}_0$ and $\mathcal{O}_{\mathbf{v}_0}$ share the same and unique event O that we called "here and now". The time-axis $\Delta_{\mathbf{v}_0}$ for this observer is defined by the corresponding world-line, namely the straight line with (vector) equation $\mathbf{x} = \mathbf{v}_0 t$ (see fig. 1).

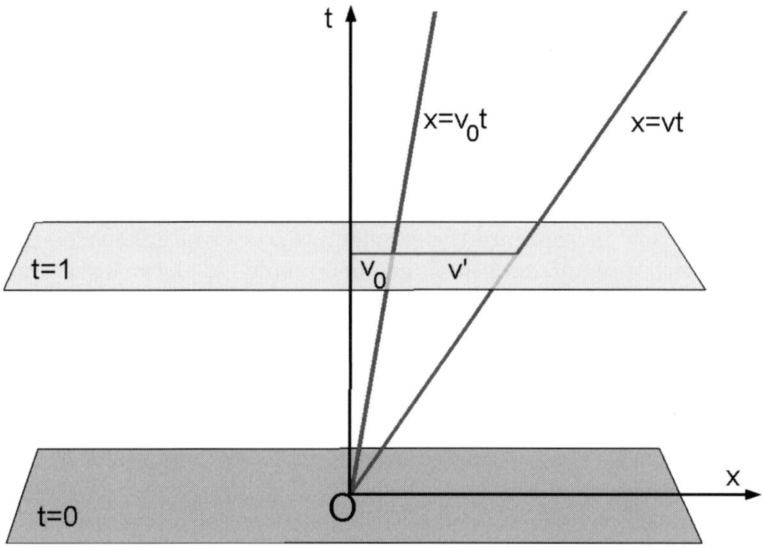

Figure 1: The Galilean spacetime

For any such observer, the sets of simultaneity $t = t_0$ are the same as for the observer $\mathcal{O}_0$. More precisely, every event $M = (\mathbf{x}, t)$ of spacetime is perceived by the observer $\mathcal{O}_{\mathbf{v}_0}$ as having coordinates $(\mathbf{x}', t')$ such that $\mathbf{x}' = \mathbf{x} - \mathbf{v}_0 t$ and $t' = t$. This change of coordinates from $\mathcal{O}_0$ to $\mathcal{O}_{\mathbf{v}_0}$ is also called a *Galilean transformation*; it implies the basic property of *additivity of velocities*: a uniform motion with world-line $\mathbf{x} = \mathbf{v}t$ is seen by $\mathcal{O}_{\mathbf{v}_0}$ as a uniform motion with equation $\mathbf{x}' = \mathbf{v}'t$, with velocity vector $\mathbf{v}' = \mathbf{v} - \mathbf{v}_0$. For example, in a train whose velocity is v_0 =100 kmh, a passenger walking longitudinally with velocity v' =5 kmh has a velocity with respect to the earth which is v =105 kmh or 95 kmh according to whether the forward or backward direction of the train has been chosen by that passenger...

We note that the Galilean changes of coordinates do not preserve the notion of orthogonality in $\mathbf{R}^4$. If for convenience we choose to represent the simultaneity sets as "horizontal spaces" (the dimension of space being unfortunately reduced to two in our visual perception...) and the time-axis of the observer at rest $\mathcal{O}_0$ by a vertical line, the reference frame for $\mathcal{O}_{\mathbf{v}_0}$ will associate the *oblique* time-axis $\Delta_{\mathbf{v}_0}$ with the horizontal space. But the observer at rest enjoys no special physical properties with respect to any other observer in uniform motion (that's the *"Galilean principle of relativity"* which follows from the law of inertia). So the verticality of the time-axis could have been chosen for representing the world-line of any given uniform motion: there is nothing deep in that choice. One can also say that the

Galilean spacetime is defined for $\mathcal{O}_0$ *up to the arbitrariness in the choice of the time-axis* or in mathematical terms *up to a Galilean transformation*: it is the equivalence class of all these representations. But the *same* representation of spacetime is then also acceptable by any observer $\mathcal{O}_{\mathbf{v}_0}$ in uniform motion, which expresses precisely in geometrical terms the content of the Galilean relativity principle.

Here it is also worthwhile to point out that, in contrast with the "horizontal" Euclidean subspaces $\mathbf{R}^3$, the Galilean spacetime $\mathbf{R}^4$ is only an *affine* space; it is not equipped with any physically sensible global notion of orthogonality and distance. But this is consistent with our standard perception: why would space and time strangely mix each other in some supergeometry? Galilean spacetime is just a geometry of representation in a very poor sense: it has no global geometrical structure. But let us now incorporate the strange properties of light velocity and then discover that such a phantasmic supergeometry holds in the realistic spacetime of physics, namely in the four-dimensional world called *Minkowski's spacetime* !!

2 Postulates and construction of Minkowski's spacetime

Preliminary Remark. The postulates and the construction which we propose do not pretend to be the most economical ones from the viewpoint of formal logics. In particular, we must draw the attention of the reader to the important mathematical article by E.C. Zeeman entitled *"Causality Implies the Lorentz Group"* [9]. We shall briefly indicate at the end of Sec.2-1 how the latter can be interpreted in our approach, which is much more pedestrian since making use of the basic physical concept of uniform motion and of the familiar representations of Euclid's geometry.

We shall introduce five postulates for our construction of the *spacetime of special relativity*. The first two postulates introduce a representation of spacetime conceived by the observers at rest, while the third and fourth postulates express minimal properties to be shared by all the observers in uniform motion. The contents of the first and third postulates are easily accepted as being already satisfied in the Galilean spacetime, but the second and fourth postulates introduce the *world-lines of light* as playing a fundamental role in spacetime. In fact, these postulates express in a geometrical way the revolutionary result obtained at first by the experiments of Michelson and Morley: *For all observers, either at rest or in uniform motion, the velocity of light in the vacuum is a universal constant c; neither it depends from the motion and from the nature of the light-emitter, nor from the direction of emission and the various changes of direction of the light beams considered (e.g. obtained by the interposition of mirrors), nor from the wave-length of the light.* Renewed experiments which make use of a variety of experimental devices and whose range extends to electromagnetic waves outside the spectrum of visible light (including in particular the propagation of radiowaves) have been repeatedly performed throughout the twentieth century. They all have confirmed the universality property of c, even if its precise value ($c = 299,776\ldots$km/sec as measured in 1940 by Anderson) is now thought to be possibly fluctuating with time at astronomical scales and also depending on the type of clocks (atomic or dynamical) for time measurements. The overwhelming fact about the universality property of c is that light does not satisfy the usual (Galilean) property of additivity of velocities: by switching on a lamp on a train, it is impossible to make its light travel at the velocity c plus the velocity of the train !!

Finally, it is pertinent and (as we shall see) useful for our construction to add a fifth postulate: the latter requires that, *in the limit of very low velocities* (those which we perceive in our life), Galilean spacetime has to be an excellent approximation of the new spacetime. Here lies the wisdom of all revolutions in the domain of science: the old theory is not thrown away as completely perverse, it is honestly recognized as a good first-order approximation of the new theory when the order of magnitude of certain variables lies within certain limits.

Notation. In all the following, the symbol $A \doteq B$ will be used when this equality serves as a definition either of A in terms of B or of B in terms of A. Examples: a vector $\mathbf{x} \doteq (x_1, x_2, x_3)$; $x_1^2 + x_2^2 + x_3^2 \doteq \mathbf{x}^2$, the squared norm of $\mathbf{x}$; the norm $|\mathbf{x}| \doteq (\mathbf{x}^2)^{\frac{1}{2}}$.

2.1 The postulates and the light-cone structure of spacetime

First postulate: the spacetime representation

All the observers at rest in the Euclidean space $\mathbf{R}^3_\mathbf{x}$ (where $\mathbf{x} = (x_1, x_2, x_3)$) agree on the existence of a geometrical representation of all "events" of the universe by points in a space $\mathbf{R}^4_{\mathbf{x},t} = \mathbf{R}^3_\mathbf{x} \times \mathbf{R}_t$, with the same notions of simultaneity sets $t = t_0$ as in the Galilean spacetime. The time-axis is the world-line of the observer $\mathcal{O}_0$; the time-axis together with the "present hyperplane" $t = 0$ constitute the reference frame of observers at rest, its origin O being the "present event" ("here and now") of the observer $\mathcal{O}_0$.

This postulate calls for three remarks:

i) The *events*, and thereby their representation by points in $\mathbf{R}^4$ are conceived as *"absolute elements of reality"*; however, the given system of coordinates $(\mathbf{x}, t)$ privileges the class of observers at rest, whose world-lines are all the parallels to the time-axis. The basic problem of our construction will be to determine the corresponding systems of coordinates for any observer in uniform motion. As in the Galilean spacetime, the world-line of any observer in uniform motion is a straight line. For example $\Delta_\mathbf{v}$ is the world-line of the observer $\mathcal{O}_\mathbf{v}$ whose motion is defined as in the Galilean case.

ii) All the observers at rest are supposed to be equipped with identical devices for measuring lengths (i.e. *graduated rods*) and for measuring time-intervals (i.e. *clocks*). The fact that all observers at rest agree on their Euclidean representation of *space* is trivial for us (after more than 2000 years of cartographical techniques...). The fact that they agree on the simultaneity of two events requires a procedure of *"synchronization of clocks"* through the emission of light-signals, and we shall see in Sec.2-2 that the full analysis of the notion of simultaneousness relies physically on the use of light-signals. For the moment, we only introduce this notion for the observers at rest.

iii) In all our pictorial representations, the time-axis will be represented as vertical and upward-oriented; the ascending arrow indicates the future. The Euclidean space $\mathbf{R}^3_{\mathbf{x}}$ with equation $t = 0$ is then considered as horizontal. In many arguments, it will be sufficient to consider a *single* space variable $x = x_1$, namely the planar section (Ox_1, Ot) of spacetime, with the axis Ox horizontal and rightward-oriented as usual.

Second postulate: the light-cone

All the world-lines of light rays emitted from the event O by any *(moving or at rest) light-emitter are represented in $\mathbf{R}^4_{\mathbf{x},t}$ by the linear generatrices of the cone C^+ with equation $|\mathbf{x}| = ct$, $t > o$, which is called the "future light-cone of O". Similarly, all the light rays emitted in the past of O by any (moving or at rest) light-emitter and which are detected at O have world-lines which are carried by the generatrices of the cone C^- with equation $|\mathbf{x}| = -ct$, $t < o$, which is called the "past light-cone of O". The whole set of light world-lines passing at O is the set of generatrices of the "light-cone C of O" (see fig. 2), with quadratic equation*

$$c^2t^2 - (x_1^2 + x_2^2 + x_3^2) = 0.$$

Similarly, with each event $X = (\mathbf{x}, t)$ of $\mathbf{R}^4_{\mathbf{x},t}$, one can associate the "light-cone $C(X)$ of X", which is obtained from C by the action of the translation with vector $[OX]$ in $\mathbf{R}^4_{\mathbf{x},t}$.

It is worthwhile to emphasize that the *absolute localization on the cone C* of the world-lines of light rays passing at O did not hold in the usual Galilean spacetime representation, since light was treated there as any other motion and therefore obeyed the principle of additivity of velocities. To be more illustrative, let us consider light-propagation along a single direction of space Ox represented as our horizontal axis, but with the two possible orientations of light rays emitted from O, namely the rightward light ray (towards positive x's) and the leftward light ray (towards negative x's). The world-lines of these two light rays in the planar section (Ox, Ot) of spacetime are respectively the half-lines C_R and C_L with equations $x = ct$, $t > 0$ and $x = -ct$, $t > 0$ (see fig. 3): they are the traces of the future light-cone C^+ in the planar section (Ox, Ot). If the light rays emitted from O are emitted from a train with velocity v in the direction Ox, its propagation is still observed by an observer at rest as having the velocity c and not $c+v$ or $c-v$, which would have been the case according to the Galilean viewpoint. In the planar section (Ox, Ot) of Galilean spacetime, the world-lines of the light rays emitted from O would have had equations of the form $x = (\pm c + v)t$ (resp. $x = (\pm c - v)t$), depending on the velocity v (resp. $-v$) along Ox of the light-emitter at O. Therefore, the Galilean world-lines of these light rays (considered for all possible values of v) would cover the whole half-plane of positive t's, namely "the absolute Galilean future". It is thus crucial to understand that in the new

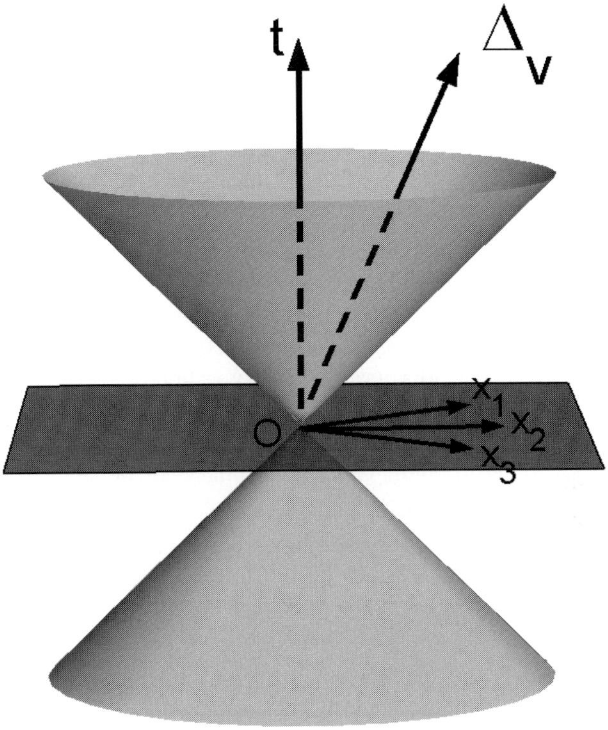

Figure 2: The light-cone

"relativistic spacetime" that we construct, the half-lines C_R and C_L and more generally the cone C^+ are *new absolute data.*

Remark on the choice of units. Instead of using the very large value of c expressed in km/sec which would make unpracticable the geometrical representation of spacetime, we can choose time and space units in such a way that $c = 1$. For example, we can adopt the choice of *year* and *light-year* which is standard in astronomy. The light-cone C is then well-represented as the cone with equation $t^2 - \mathbf{x}^2 \doteq t^2 - (x_1^2 + x_2^2 + x_3^2) = 0$ and the light world-lines C_R and C_L are then well-pictured along the diagonals of the axes (Ox, Ot) (fig. 3). Another possible convention whose advantage is also to keep the same geometrical representation but without fixing the value of c consists in considering that one plots the variable ct instead of t. Here it is relevant to notice that the variable ct has the "physical dimension" of a distance, which prepares us to understand why it can be treated on the same footing as the space coordinates $\mathbf{x}$ in the following.

Third postulate: isochronousness of all uniform motions
For every observer $\mathcal{O}$ in uniform motion, let $t_{\mathcal{O}}$ be its time-variable , measured by a clock which is identical with that of $\mathcal{O}_0$. Its world-line is a straight line denoted by Δ which carries the time-axis of $\mathcal{O}$. Let then X_1, X_2, X_3 be three events in Δ. We postulate that it is equivalent that their time-coordinates t_1, t_2, t_3 satisfy the equality $t_2 - t_1 = t_3 - t_2$, namely that X_2 be the middle of the segment X_1X_3, or that the corresponding times $(t_{\mathcal{O}})_1, (t_{\mathcal{O}})_2, (t_{\mathcal{O}})_3$ measured by $\mathcal{O}$ satisfy the equality $(t_{\mathcal{O}})_2 - (t_{\mathcal{O}})_1 = (t_{\mathcal{O}})_3 - (t_{\mathcal{O}})_2$.

This postulate is of course trivially satisfied in the *absolute time* viewpoint of Galilean spacetime. Here one only requires that the flow of time measured via a regular sequence of events by an observer $\mathcal{O}$ is also perceived as regular *up to a change in the scale*, when the same successive events linked to $\mathcal{O}$ are detected (with an identical clock) by $\mathcal{O}_0$, and thereby by any other observer in uniform motion with respect to $\mathcal{O}$. (Note that from a more realistic viewpoint such a regular detection is obtained by $\mathcal{O}_0$ at the reception of light beams emitted by $\mathcal{O}$, but this is easily seen to be equivalent to the regularity of the sequence of coordinates (t_1, t_2, t_3) by the Thales property).

Fourth postulate: "Physical" uniform motions and the universality of c
a) The only uniform motions considered as having a physical meaning are those whose velocity v is smaller than c. For such motions whose world-line $\Delta_{\mathbf{v}}$ contains the event O, $\Delta_{\mathbf{v}} \setminus O$ is made up of two half-lines $\Delta_{\mathbf{v}}^+$ and $\Delta_{\mathbf{v}}^-$ which are respectively contained in the convex conical volumes V^+ and V^-:

V^+ is the set of all events $(\mathbf{x}, t)$ such that $|\mathbf{x}| < ct,\ t > 0$, called "the absolute future of O";

V^- is the set of all events $(\mathbf{x}, t)$ such that $|\mathbf{x}| < -ct,\ t < 0$, called "the absolute past of O".

Similarly, for each event X one can introduce the convex conical volumes $V^+(X)$ and $V^-(X)$, namely respectively the absolute future and past of X, whose union contains all the world-lines Δ of the uniform motions passing at X. The future light-cone $C^+(X)$ (resp. past light-cone $C^-(X)$) thus appears as the boundary of the corresponding future cone $V^+(X)$ (resp. past cone $V^-(X)$).

b) For every observer $\mathcal{O}_{\mathbf{v}}$ with world-line $\Delta_{\mathbf{v}}$ graduated by the time-variable $t_{\mathbf{v}}$, there exist coordinates $\mathbf{x}_{\mathbf{v}}$ of the space perceived at rest *by $\mathcal{O}_{\mathbf{v}}$, such that any event $X = (\mathbf{x}, t)$ of the light-cone C is detected by $\mathcal{O}_{\mathbf{v}}$ as having coordinates $(\mathbf{x}_{\mathbf{v}}, t_{\mathbf{v}})$ satisfying the relation $|\mathbf{x}_{\mathbf{v}}| = c|t_{\mathbf{v}}|$.*

Part a) of the postulate, which requires that the light-velocity is an absolute limit to the propagation velocity of any physical system to which an observer can be linked, will appear as deeper than a pure physical requirement. It will in fact be seen below that the lines of spacetime which could be interpreted as world-lines of motions with velocity larger than c (or *"superluminal motions"*) are necessarily given another interpretation, which is of purely *spatial* nature. So the requirement a) is deeply involved in the self-consistency of the relativistic spacetime representation.

Part b) again pertains to the basic statement about the constancy of the velocity of light. It can also be seen as contained in the principle of relativity which claims that all the physical laws, and therefore in particular the velocity of light, are the same for all observers in uniform motion: no rest frame is physically privileged as it was presupposed in the old concept of ether.

Fifth postulate: validity of the Galilean approximation
For every observer $\mathcal{O}$ in uniform motion or at rest, there is a Galilean representation of spacetime which is an excellent approximation of the exact spacetime for the description of motions whose relative velocity with respect to $\mathcal{O}$ is very small with respect to c.

The precise mathematical formulation of this postulate will appear clearly in the following.

Remark. In the present approach, the interpretation of the basic result of [9] seems to be the following. Let us assume that the light-cone structure of the spacetime $\mathbf{R}^4$ holds for the observer at rest $\mathcal{O}_0$ as in our first and second postulates. Let us now consider observers in *unspecified* motion, for which the spacetime $\mathbf{R}^4$ is also perceived with a lightcone structure (implying the same universal velocity of light c). Let us assume that for such observers the causality order of events X, Y (denoted $X < Y$) is defined by the fact that Y belongs to the future cone $V^+(X)$ of X, and that *this order coincides with the one perceived by the observer at rest.* Then it is proven that such observers are *necessarily in uniform motion* and that *their scales of time and length are linear functions of those of the observer at rest* so that the whole structure of Minkowski's spacetime follows. In particular, our postulate three concerning the "isochronousness property" of uniform motions would then be redundant. However, as it has been pointed out in [9], the result does not hold in two-dimensional spacetime; a nontrivial use of the dimension larger than two has been made in that work. Our approach is rather opposite: in view of its pedagogical nature, it aims to exhibit already in two-dimensional spacetime (which is much simpler to describe) how the construction of Minkowski's spacetime can be worked out. In fact, this will be made in detail from Sec.2-2 to Sec.2-6. It is only in Sec.2-7 that we shall be ready to tackle the four-dimensional spacetime equipped with the group of general Lorentz transformations. For the sake of completeness, we have been led to include in that subsection some technical details which may be skipped in a first reading: the main geometrical result to be understood (with the help of fig. 9 and of the remark at the end of Sec.2-7) concerns the pairs "(time-axis, space hyperplane)" which are associated with every uniform motion.

2.2 Simultaneousness revisited

The notion of *absolute simultaneousness*, namely the identity of every simultaneity space $t = t_0$ for all the observers (at rest or in motion) is encoded in the Galilean spacetime representation. However this viewpoint is purely idealistic, because for each observer the property of simultaneity of two events is a physical property

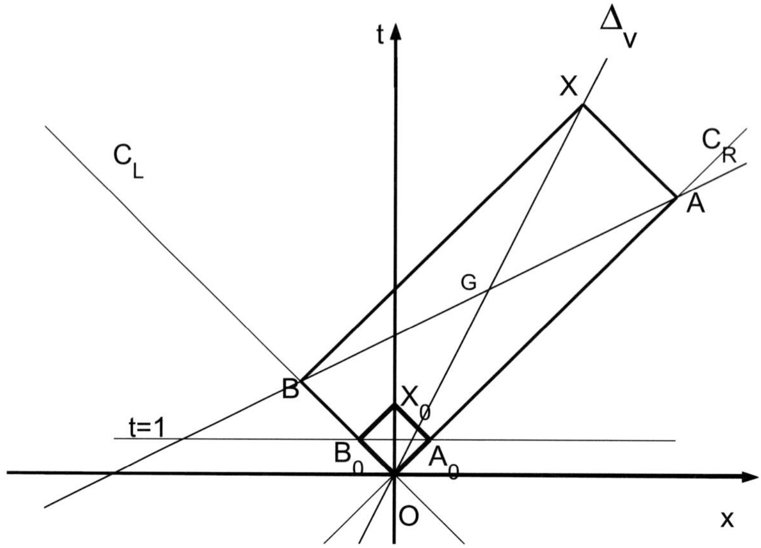

Figure 3: Simultaneous events

which has to be checked via some procedure implying the use of lengths and time measurements. Now in view of the universality of the velocity of light, the use of light-signals will be particularly helpful for clarifying the notion of *simultaneousness relatively to each observer at rest or in uniform motion.*

We shall describe a physical procedure for characterizing simultaneous events whose geometrical representation in spacetime is quite simple. It only requires observers and light-signals moving in a single space dimension Ox, which allows one to represent phenomena in the two-dimensional section (Ox, Ot) of spacetime. We are led to use the geometrical representation of light world-lines as being all parallel either to C_R or to C_L (according to our first and second postulates). For simplicity, chosen units are years and lightyears so that $c = 1$.

For the observer at rest $\mathcal{O}_0$, the procedure must of course confirm that (for instance) the events $A_0 \doteq (x = 1, t = 1)$ and $B_0 \doteq (x = -1, t = 1)$ are simultaneous. To that purpose, one considers rightward and leftward light rays emitted from O and reflected (by mirrors) at the respective points $x = 1$ and $x = -1$. The world-lines of these reflected light rays are respectively parallel to C_L and C_R and therefore converge at the event $X_0 \doteq (x = 0, t = 2)$ of the world-line of $\mathcal{O}_0$, which allows the latter to conclude that the "mirror events" A_0 and B_0 are simultaneous: since the velocity of light is the same in right and left directions, the mirror events have been simultaneously produced at half of the time of X_0 (namely $t = 1$). As seen on fig. 3, the geometrical representation of the previous light-signal procedure exhibits that the quadrilateral $(OA_0X_0B_0)$ is a parallelogram. We also notice that this procedure is useful for allowing all the observers at rest to synchronize their clocks with respect to $\mathcal{O}_0$'s clock and therefore to agree on the same representation of spacetime. For instance the observer situated at $x = 1$ (i.e. whose world-line has the equation $x = 1$) will be warned by $\mathcal{O}_0$ that he should assign the time $t = 1$ to the event A_0, at which he receives the light signal coming from O.

Now we can repeat the same construction for any given observer $\mathcal{O}_v$ in uniform motion, with $|v| < 1$. We use again two rightward and leftward light rays emitted from O and therefore represented along C_R and C_L, but we now set the mirrors (at some points $x_A > 0$ and $x_B < 0$) in such a way that the world-lines of the two reflected light rays intersect at an event X which belongs to the world-line Δ_v of $\mathcal{O}_v$. Here again the two mirror events A and B are such that the quadrilateral $(OAXB)$ formed by the four light world-lines is a parallelogram, and it then follows that, except when $v = 0$, the linear segment AB is *not* parallel to the axis Ox (fig. 3).

Now in view of b) of the fourth postulate, the forward and backward travels of light corresponding to the world-line segments OA and AX (resp. OB and BX) are performed during the same time for $\mathcal{O}_v$, since performed at the same universal velocity. Therefore if $t_v(X)$ denotes the time of the event X measured by $\mathcal{O}_v$, the times of the mirror events A and B measured by $\mathcal{O}_v$ will be both equal to $\frac{t_v(X)}{2}$: these two events are therefore to be considered as *simultaneous* by $\mathcal{O}_v$. Moreover (in view of the same postulate), the events A and B will be produced at spatial coordinates $x_v = \pm\frac{t_v(X)}{2}$. We shall now use our third postulate for proving the following property.

All the points of the straight line (AB) represent the events which appear to be simultaneous to A and B for the observer $\mathcal{O}_v$.

We consider at first the event G at the intersection of OX and AB. Since (in the parallelogram $(OAXB)$) one has $OG = GX$, the observer $\mathcal{O}_0$ perceives the event G at the time $t(G) = \frac{t(X)}{2}$. Then in view of the third postulate, the event G is also perceived by the observer $\mathcal{O}_v$ at the time $t_v(G) = \frac{t_v(X)}{2}$, which shows that G is simultaneous to A and B for $\mathcal{O}_v$.

Let now P be any point on the half-line with origin G and containing A, and let E and F be the intersections of the straight line (OX) respectively with the parallels to C_R and C_L by P. Thales property then yields (fig. 4):

$$\frac{GF}{GX} = \frac{GP}{GA} = \frac{EG}{OG}, \quad \text{and therefore} \quad EG = GF.$$

By introducing the point Q, symmetric of P with respect to G one then gets a parallelogram $(EPFQ)$. Therefore the same argument as above applies to the light rays emitted at E, reflected at P and Q and converging at F: it shows that P,Q and G are simultaneous with respect to $\mathcal{O}_v$. Since the symmetric pair (P, Q) may vary arbitrarily on the straight line (AB), this line is a line of simultaneity for $\mathcal{O}_v$ (corresponding to the time $\frac{t_v}{2}$).

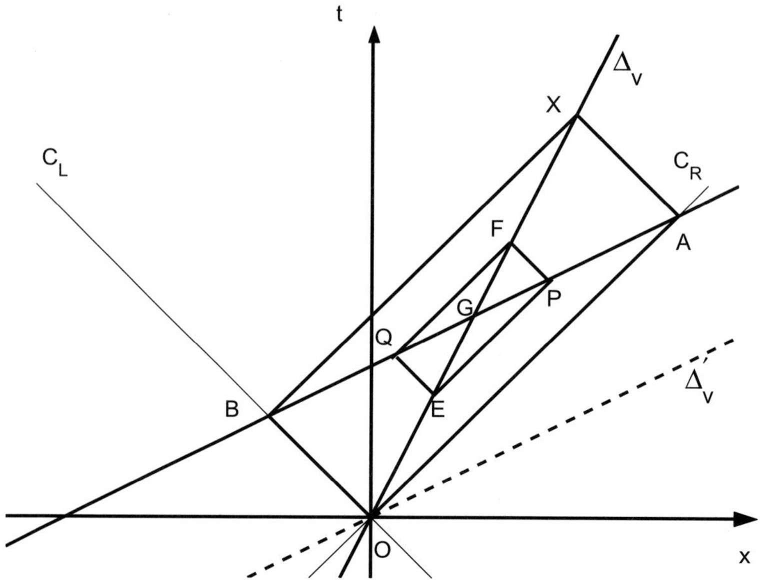

Figure 4: Conjugate axes

Since the choice of t_v was arbitrary in the previous argument, one concludes that the lines of simultaneity for the observer $\mathcal{O}_v$ in the plane (Ox, Ot) are all the parallels to (AB); *in particular the straight line Δ'_v parallel to (AB) and containing O represents the "present events" ($t_v = 0$) for $\mathcal{O}_v$* . As seen on fig. 4, half of the line Δ'_v (on the right of O for the choice $v > 0$) contains events at $t > 0$, which are therefore perceived as belonging to the future by $\mathcal{O}_0$ together with all the observers at rest, while the other half (on the left of O) contains events at $t < 0$, perceived as belonging to the past by the same observers.

The direction Δ'_v, obtained from Δ_v by the previously described *parallelogram construction*, is said to be *conjugate of Δ_v with respect to the light world-lines C_R and C_L*. Points $X = (x, t)$ and $X' = (x', t')$ of Δ_v and Δ'_v satisfy the equations

$$x = vt, \quad x' = \frac{1}{v}t', \quad \text{and therefore} \quad tt' - xx' = 0.$$

This calls for two comments:

i) conjugacy or pseudo-orthogonality relation

The relation $tt' - xx' = 0$ (or in unit-independent form $(ct)(ct') - xx' = 0$) can be called a pseudo-orthogonality relation between the vectors $[OX]$ and $[OX']$, by analogy with the orthogonality relation $xx' + yy' = 0$ in a Euclidean plane. Such a relation, which expresses the geometrical property of *conjugacy* of the pair (Δ_v, Δ'_v) with respect to the pair (C_R, C_L), introduces *a joint geometrical structure of space and time*, which will appear still stronger in the analysis of Sec. 2-3.

For the moment, we can simply notice the following properties of conjugate pairs (Δ_v, Δ'_v):

a) when v varies, Δ_v and Δ'_v are turning in opposite ways (one clockwise and one anticlockwise) in the plane (Ox, Ot).

b) when v tends to 1 (resp. -1), both lines tend together to C_R (resp. C_L).

c) there is a single conjugate pair which is orthogonal, namely the supports of the axes of coordinates Ox, Ot.

Here, however, one must stress that the choice of orthogonal space and time axes Ox, Ot for the observers at rest is a pure convention, as it was already the case for the Galilean spacetime representation. A more general, but equivalent choice *which does not ascribe a special role to observers at rest* would be the following. One first gives oneself the pair of light world-lines (C_R, C_L) and one chooses for (Ox, Ot) any pair of straight lines which are conjugate with respect to (C_R, C_L) (defined intrinsically through the parallelogram construction). The analysis above would have given the same result, namely that the time and space axes for any observer $\mathcal{O}_v$ are carried by conjugate pairs (Δ_v, Δ'_v) with respect to (C_R, C_L). Among them, the special pair which is orthogonal (namely the bisectors of (C_R, C_L)) would then be associated with *a certain* uniform motion having no special physical properties: in fact, it was one of the primary ideas of special relativity theory that systems in uniform motions are physically indistinguishable . So, as in the Galilean case, we keep the idea that the orthogonality of the rest system is only a convenient convention, but there is a whole class of equivalent representations of the planar relativistic spacetime in which the following notions have an *absolute* meaning: *i) the light lines (C_R, C_L) and ii) the systems of conjugate pairs (Δ_v, Δ'_v) for the coordinate axes of uniform motions, including the rest system.*

ii) "superluminal motions"

For $\mathcal{O}_0$, the line Δ'_v might be interpreted as the world-line of a superluminal motion with velocity $\frac{1}{v}(=\frac{c^2}{v})$.... But this would be very strange, since all the events of that line are perceived as simultaneous by $\mathcal{O}_v$: for the latter, a hypothetic observer $\mathcal{O}'_v$ with world-line Δ'_v would then have the "ubiquity property" ($t_v = 0, x_v$ arbitrary)! The interpretation of this motion would become even more paradoxical for an observer $\mathcal{O}_w$ with velocity w such that $v < w < c$. In fact, one can easily check geometrically (by using the property a) of conjugate pairs in the previous remark) that for $\mathcal{O}_w$ the line Δ'_v is parametrized by a time-coordinate t_w which is negative decreasing, while t is positive increasing: for $\mathcal{O}_w$, the hypothetic observer $\mathcal{O}'_v$ would be traveling back to the past !

The latter remark strengthens the meaning of part a) of our fourth postulate and justifies that the cones V^+ and V^- be considered respectively as the *absolute future and past* of the event O. It can now be fully understood that all events represented by points X *outside the union of* V^+ and V^- (like the points of any line Δ'_v) are in *"acausal" relation* with the event O: no physical signal can propagate either from O to X or from X to O.

2.3 Space-ships' flight: the anniversary curve

So far, we have discovered the conjugate directions of the space and time coordinate axes of all observers in uniform motions, but what remains unknown are the *scales* of time and length along these axes. As a matter of fact, we already see that only the scale of time remains a problem, since once it is known, the scale of length immediately follows from the knowledge of the velocity of light (universal for all uniform motions).

To set this problem of time scaling in an illustrative way, we consider a set of space-ships flying away simultaneously from the same place, let us say at the event O, along the unique horizontal direction Ox, but with various velocities v_i either rightwards ($0 < v_i < 1$) or leftwards ($-1 < v_i < 0$) (with units such that $c = 1$); one of them remains at rest ($v_0 = 0$). All space-ships contain observers $\mathcal{O}_{v_i}$ equipped with identical clocks, and all these observers are invited to celebrate the anniversary of their common departure by representing these events (each anniversary event in the corresponding space-ship) by points correctly situated in spacetime. *On what curve H of the plane (Ox, Ot) will all these points be situated?*

In the case of Galilean spacetime where time is absolute, the answer to that question is trivial, namely the straight line with equation $t = 1$. In the present framework of spacetime, governed by the five postulates stated in Sec. 2-1, one determines the curve H by showing that it must satisfy the following property.

Theorem. For each point X of H, there exists a tangent to H at X whose direction is conjugate of (OX) with respect to the pair (C_R, C_L).

Proof. This result follows directly from the conjugacy property of space and time axes established in Sec. 2-2 together with our *fifth* postulate. In fact, we know that for a given observer $\mathcal{O}_v$ whose world-line $\Delta_v = (OX)$ contains the anniversary event X ($x_v = 0, t_v = 1$), the straight line of simultaneous events ($t_v = 1$) is the parallel by X to the conjugate direction of Δ_v; in view of the parallelogram construction, this parallel intersects C_R and C_L in two points M and N such that X is the middle of MN. Now the fifth postulate asserts that for observers $\mathcal{O}'_v$ with velocity v' very close to v (this is what means "with very small relative velocities with respect to $\mathcal{O}_v$") the corresponding anniversary event $X_{v'}$ should be represented *with an excellent approximation* by the Galilean representation of $\mathcal{O}_v$, namely by the point at the intersection of the world-line $\Delta_{v'}$ and of the straight line with equation $t_v = 1$, i.e. (MN). This means that, in mathematical language, the straight line (MN) has to be the *tangent* to the unknown curve H at the point X (see fig. 5).

Now it is well-known in elementary geometry that *every curve H, whose tangent at each point X intersects two given (nonparallel) straight lines C_R, C_L at two points M, N such that X is the middle of MN is a branch of hyperbola with asymptotes C_R and C_L.*

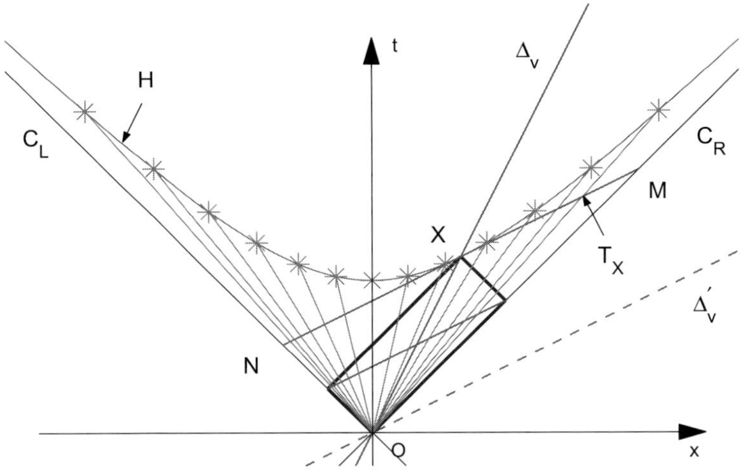

Figure 5: The anniversary curve

Since it must contain the anniversary event at rest $X_0 = (x = 0, t = 1)$, the anniversary curve H is therefore uniquely determined as the branch of hyperbola whose equation is $t^2 - x^2 = 1, \quad t > 0$ (fig. 5). The anniversary point $X = X_v$ of any observer $\mathcal{O}_v$ is thus given by the formulae

$$t(v) = \frac{1}{\sqrt{(1-v^2)}}, \quad x(v) = \frac{v}{\sqrt{(1-v^2)}} \quad \text{(where } |v| < c = 1\text{)}.$$

It is convenient to introduce instead of the velocity v the parameter χ called the *rapidity* which is defined by $v = \tanh\chi$; χ is a *"hyperbolic angle"* which takes all possible values from $-\infty$ to $+\infty$. The previous formulae can then be rewritten equivalently in the following form, which is similar to the angular parametrisation of the circle:

$$t(v) = \cosh\chi, \quad x(v) = \sinh\chi.$$

2.4 Minkowskian (pseudo-)distance and the inverse triangular inequality: the twin "paradox"

From the algebraic viewpoint, the hyperbola with equation $t^2 - x^2 = a^2$ present strong similarities with the circles with center O and radius R, whose equations are $x^2 + y^2 = R^2$ in orthonormal coordinates. They are *"level curves" of a certain "quadratic form"* $X \to Q(X)$ (with $X = (x, t)$ or $X = (x, y)$) specified by a second-degree homogeneous polynomial ($Q(X) = t^2 - x^2$ or $Q(X) = x^2 + y^2$)

This mathematical analogy between the hyperbola and the circle admits here a physical counterpart which is very striking. In fact, after the analysis of Sec. 2-3 we naturally come to the idea that our problem of space-ship travelers and its solution are quite comparable to the following very elementary situation in Euclid's planar geometry. Consider walkers equipped with identical graduated rods who start from the same point O along various straight lines and cover the same distance R: they all have reached the circle with center O and radius R. While the latter statement appears trivial to us because of our visual perception of geometry, the former result concerning the "anniversary curve" H tells us that individual time-measurements made by observers in uniform motion or, as one says, *"proper-time measurements"* inform us about the existence of a certain kind of "time-like distance" in spacetime between events related by physical causality. For that "time-like distance" which we shall also call *"Minkowskian distance"*, H appears as a unit level-curve with starting point O and *in the future of* O. Of course all the level-curves of that Minkowskian distance will appear as homothetic hyperbolae centered at O with equations $t^2 - x^2 = a^2$; they are obtained from H by either a dilatation or a contraction scale factor and completion by the "past branches". In fact, each of these hyperbolae contains *two branches* which are distinguished by the sign of t: the branch on which t remains positive (as the anniversary curve H) is contained in the (absolute) future V^+ of O, while the branch on which t remains negative is in the (absolute) past V^- of O: this is the case for the "negative anniversary curve" which is the set of all past events X from which a one-year travel until O is possible via a uniform motion.

In Euclidean space the notion of distance $d(A, B)$ between two points is characterized by the validity of the triangular inequality: $d(A, B) \leq d(A, C) + d(B, C)$, the equality being obtained if and only if the points A, B, C are on the same straight line. This fact is illustrated geometrically by constructing such triangles (ABC) with given side-lengths a, b and c: one just has to check the intersection property of circles with centers A and B, whose sum of radii b and a is larger than $d(A, B) = c$ (fig. 6).

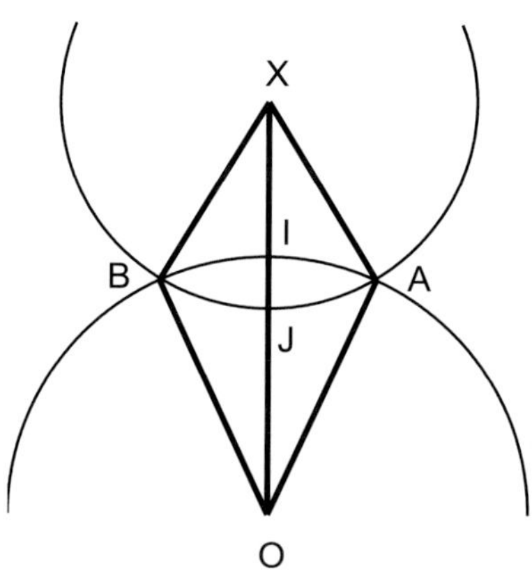

Figure 6: $d(O, X) \leq d(O, A) + d(A, X)$

In the spacetime plane (Ox, Ot), which we shall now properly call *the Minkowskian plane*, a similar geometrical construction shows that there exists again a triangular inequality for the Minkowskian distance d_M, but *with the inverse sign*, namely we have:

Minkowskian triangular inequality: *Let three points O, A, X be such that A and X be in V^+, with X in the future of A ($X \in V^+(A)$), then the corresponding Minkowskian distances satisfy the inequality:*

$$d_M(O, X) \geq d_M(O, A) + d_M(A, X),$$

the equality being obtained if and only if the points O, A, X belong to the same straight line.

The fact that $d_M(O, X) = d_M(O, A) + d_M(A, X)$ when O, A and X are aligned just expresses the additivity of the corresponding proper time intervals measured by an observer whose world-line is (OAX). Let us now consider the general case when O, A and X form a (non-flattened) triangle. We then consider two branches of hyperbola containing the point A: the first one, called H_O^+ is centered at O and lies in the future cone of O, while the other one, called H_X^- is centered at X and lies in the past cone of X (see fig. 7).

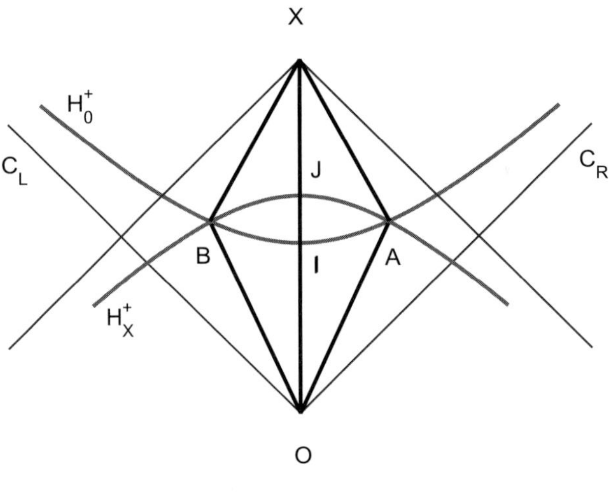

Figure 7: $d_M(O, X) \geq d_M(O, A) + d_M(A, X)$

H_O^+ and H_X^- intersect each other at A and at another point B (such that the straight lines (AB) and (OX) have conjugate directions with respect to (C_R, C_L)). Now the straight line (OX) intersects H_O^+ and H_X^- respectively in two points I and J such that the order of increasing times for the events along (OX) is: O, I, J, X. We therefore have

$$d_M(O, X) = d_M(O, I) + d_M(I, J) + d_M(J, X) \geq d_M(O, I) + d_M(J, X).$$

But since H_O^+ and H_X^- are level-curves for Minkowskian distances we have:

$$d_M(O,I) = d_M(O,A) = d_M(O,B) \quad \text{and} \quad d_M(J,X) = d_M(A,X) = d_M(B,X),$$

which implies the Minkowskian triangular inequality.

We notice that what makes the difference between the Euclidean and the Minkowskian cases is the *concavity* of the region between one branch of hyperbola and its asymptotes, to be compared with the *convexity* of the region inside a circle.

The "twin paradox"

The physical interpretation of this inverse triangular inequality is the famous *"twin paradox"* of special relativity, which actually no longer appears as a paradox if one gets rid of the concept of absolute time, since it expresses in a very illustrative way the content of the Minkowskian geometrical structure of spacetime.

One compares the aging of two persons between two events such as O and X at which they meet together. X can be chosen on the time-axis Ot and one of these persons is supposed to stay on the earth, namely on the world-line (OX). During that time, the other person (which we can imagine in O as the twin of the former) is submitted to a *one-year* travel in uniform motion (with a velocity v which is not small with respect to c) until the event A is reached; then this traveler comes back to the earth with the opposite uniform motion, namely with the opposite velocity $-v$. So two years have past between O and X for the traveller, while the aging of the twin at rest was two years *plus* the time represented by the Minkowskian distance $d_M(I,J)$.

Exercise: Compute $d_M(I,J)$ in terms of $\frac{v}{c}$. In terms of the rapidity χ, one finds that

$$d_M(I,J) = 2(\cosh \chi - 1).$$

What should the value of $\frac{v}{c}$ be equal to in order to produce a shift of one year between the ages of the twins ?

2.5 Spatial equidistance and the "Lorentz contraction" of lengths

In order to complete the coordinatization of spacetime associated with an observer $\mathcal{O}_v$, we reconsider the anniversary event $X = X_v$ of such an observer, situated at the intersection of the curve H and of the world-line Δ_v. Since the points M and N of the tangent to H at X belong respectively to the light world-lines C_R and C_L and represent events which are simultaneous for $\mathcal{O}_v$ with the time $t_v = 1$, they also define the spatial-distance unit for $\mathcal{O}_v$ in view of our fourth postulate (part b)). One can thus write (with a standard choice of orientation) $M = (x_v = 1, t_v = 1)$, $N = (x_v = -1, t_v = 1)$. This defines the spatial unit vector $[OX'_v]$ of $\mathcal{O}_v$ to be such that the quadrilateral $(OX'_v M X_v)$ is a parallelogram (fig. 8). OX'_v is thus the unit vector of the space-axis Δ'_v of $\mathcal{O}_v$, conjugate to Δ_v with respect to (C_R, C_L).

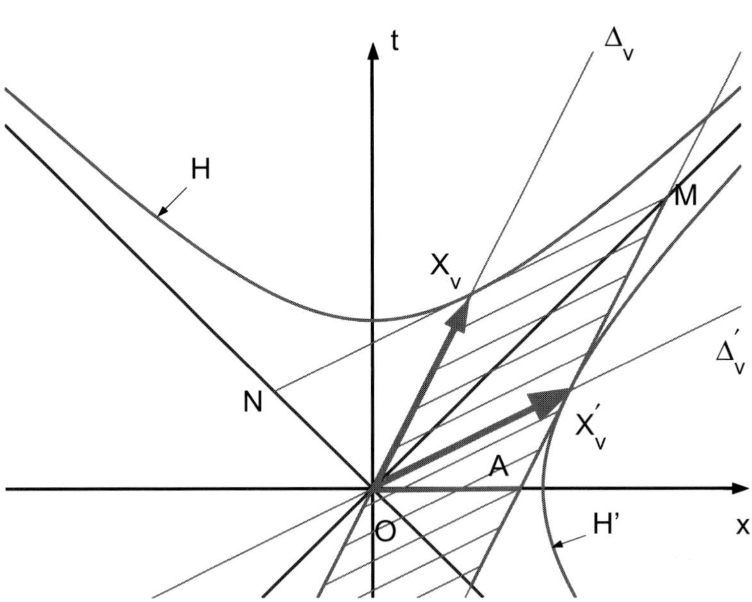

Figure 8: Equidistance curve and "contraction of lengths"

The curve of spatial equidistance H': It is clear that the point X'_v is the transform of X_v by the symmetry S_R with axis C_R which exchanges the rest-frame coordinate axes Ox and Ot. This means that if one puts $X_v = (x,t)$ and $X'_v = (x',t')$, then $x' = t, t' = x$. Therefore X'_v belongs to the curve H' with equation

$$t'^2 - x'^2 = -1, \quad x' > 0,$$

which is a branch of hyperbola with asymptotes C_R and C_L, obtained from H by applying the symmetry S_R.

As we shall see in Sec.3-2, the curve H' can be physically interpreted as *the world-line of a uniformly accelerated motion.* What is remarkable is the fact that an observer submitted to that motion always remains spatially equidistant from the fixed event O, since the latter is the center of the hyperbola H'. It even remains *perpetually contemporaneous* of the event O (as this will be fully explained in Sec.3).

Let us now consider the hyperbola composed of the curve H' and of the opposite branch (from the side $x' < 0$), together with all the homothetic hyperbolae $H'(a)$ with equations $t^2 - x^2 = -a^2$ (taken for all values of a). These are level curves of the Minkowskian quadratic form $Q(X) = t^2 - x^2$ which cover the two regions of spacetime defined by $|t| < |x|$, and respectively $x > 0$ and $x < 0$. These two regions in which $Q(X)$ remains *negative* are called *space-like regions.* Any point X in either one of these regions represents an event which is in "acausal" relation with respect to O.

The spatial triangular inequality

The previous construction shows that for any spacelike event X' in a hyperbola $H'(a)$, the usual Euclidean spatial distance $d(O, X')$ between O and X (measured in the system (Δ, Δ') such that $\Delta' = (OX)$) is given by

$$d(O, X)^2 = -Q(X).$$

Let now $(OA'X')$ be a triangle whose three sides have spacelike directions. Then the corresponding (spatial) lengths of these sides satisfy the following Minkowskian triangular inequality

$$d(O, X') \geq d(O, A') + d(A', X').$$

The proof of the latter is immediate by noticing that the symmetric of the triangle $(OA'X')$ with respect to the axis C_R (or C_L) is a triangle (OAX) whose all sides have time-like directions; moreover by construction, the spatial lengths of the sides of the triangle $(OA'X')$ are equal to the Minkowskian (proper-time) distances of the corresponding sides of (OAX). Therefore the triangular inequality for (OAX) (see Sec.2-4) can be transported for $(OA'X')$.

The contraction of lengths

Another surprising property which results from the Minkowskian geometry of spacetime is the famous apparent contraction of lengths. Here is the argument, which can easily be understood geometrically with the help of fig. 8. Consider a one-dimensional rigid body in uniform motion linked with the observer $\mathcal{O}_v$; at the time $t_v = 0$, it can be represented for example as the linear segment $[OX'_v]$ (with unit length for $\mathcal{O}_v$). Then the set of world-lines of all the points of that rigid body generate a strip (in hatching on fig. 8) which is bordered by Δ_v and by the parallel to Δ_v at X'_v. The latter is the tangent to the curve H' at X'_v, which intersects Ox at the point A whose abscissa is $\frac{1}{\cosh \chi} < 1$. It is clear that the passage of the rigid body at time $t = o$ in the rest system occupies the segment $[OA]$: the apparent contraction of length of the moving rigid body is therefore equal to

$$\delta(v) = 1 - \frac{1}{\cosh \chi} = 1 - \sqrt{1 - v^2}.$$

2.6 Lorentz transformations in the Minkowskian plane and two-dimensional Lorentz frames

To summarize the previous constructions, we can say that the light world-lines C_R and C_L separate the Minkowskian (vector) plane with origin O into four angular regions: the future and past time-like regions V^+, V^- are characterized by the *positivity* of the quadratic form $Q(X) \doteq t^2 - x^2$; the spacelike regions by the *negativity* of $Q(X)$. Up to a sign, $Q(X)$ gives the square of the distance between

O and X, but this distance is either time-like (measured by a clock) or spatial (measured by a rod). Here is the full meaning of the *non-positive-definite character* of the Minkowskian quadratic form $Q(X)$. In contrast with the Euclidean case, the set defined by the equation $Q(X) = 0$ does not reduce to O but is the union of the light world-lines C_R and C_L: two events separated by the propagation of a light ray have a mutual Minkowskian distance equal to zero.

We are now going to transfer to the Minkowskian plane some basic notions of the Euclidean plane: there is a dictionary between the languages of these two worlds, but also big differences due to the privileged role of the pair of straight lines (C_R, C_L) in the Minkowskian case. (Note however that in the Euclidean case, a similar structure would also be recovered by a *complexification* of the coordinates: the pair of "isotropic lines" with equations $x = \pm iy$ then plays the same role as the pair (C_R, C_L)).

In the Euclidean vector plane, the elementary notion of angle is complementary to the notion of norm (or distance) in the following sense. The circles centered at the origin O are invariant under the rotations with center O and arbitrary angle θ. These rotations $R(\theta)$ form a commutative group: $R(\theta')R(\theta) = R(\theta + \theta')$. Each system of orthonormal axes (Δ, Δ') is transformed by any rotation $R(\theta)$ into another orthonormal system $(\Delta_{(\theta)}, \Delta'_{(\theta)})$. The corresponding two coordinatizations of the Euclidean plane, denoted respectively by $[OX] = (x, y)$ and $[OX] = (x_{(\theta)}, y_{(\theta)})_\theta$, are such that the Euclidean quadratic form $Q(X)$, identified with the *squared norm* of the vector $[OX]$, is invariant:

$$Q(X) \doteq [OX]^2 = x^2 + y^2 = x^2_{(\theta)} + y^2_{(\theta)}.$$

In the Minkowskian vector plane, it is the notion of "rapidity" or "hyperbolic angle" χ which plays the role of the angle θ. One can in fact also introduce a *commutative group* of transformations $L(\chi)$ called *"the Lorentz group in the plane"*; in the spirit of this paper, it is also suggestive to call it "the group of hyperbolic rotations". It acts in such a way that all the branches of hyperbola centered at the origin with asymptotes (C_R, C_L) are invariant under all the transformations $L(\chi)$. Moreover all the previous statements of the Euclidean case remain valid, if one replaces the pairs of orthonormal axes by *pairs of conjugate axes* (normalized by the curves H and H' as it has been explained above) and if $Q(X)$ now denotes the non-positive-definite Minkowskian quadratic form, or *"squared (pseudo)norm"* of the vector $[OX]$.

The Lorentz group in the plane

One can give an elementary presentation of the action of the transformations $L(\chi)$. These transformations of the plane are linear; so it is sufficient to know their action on two independent vectors OM, ON and convenient to choose the latter *lightlike*, namely along the lines C_R and C_L. We put:

$$L(\chi)[OM] = e^{\chi}[OM], \quad \text{for } M \text{ in } C_R,$$

$$L(\chi)[ON] = e^{-\chi}[ON], \quad \text{for } N \text{ in } C_L.$$

The lines C_R and C_L (and thereby the set with equation $Q(X) = 0$) are separately conserved by these transformations: in fact, they provide two one-dimensional representations of the multiplicative group ($e^{\chi}e^{\chi'} = e^{\chi+\chi'}$). Now every vector $[OX]$ can be decomposed in the form $[OX] = [OM] + [ON]$, with respect to the pair (C_R, C_L), so that we can define by linearity:

$$[OX_{(\chi)}] \doteq L(\chi)[OX] = e^{\chi}[OM] + e^{-\chi}[ON].$$

That means that the coordinates $u(\chi) = e^{\chi} > 0, \quad v(\chi) = e^{-\chi} > 0$ of the point $X_{(\chi)}$ with respect to the ("light"-)basis $([OM], [ON])$ satisfy the equation

$$u(\chi) \times v(\chi) = 1,$$

which represents a branch of hyperbola with asymptotes $(C_R,\ C_L)$. It then follows:

Basic geometrical property of the Lorentz transformations

All the level curves of $Q(X)$, either in the time-like or in the spacelike regions and including also the light-like world-lines ($Q(X) = 0$), are left invariant by the action of all the transformations $L(\chi)$.

One also checks the commutativity property of this group, namely the validity of the relation $L(\chi')L(\chi) = L(\chi+\chi')$ for all χ, χ', which is built-in in the previous definition.

Transforms of conjugate axes

Let us now consider the pair of unit vectors $[OX_0] = (0, 1), [OX_0'] = (1, 0)$ of the coordinate axes at rest. We will show that each transformation $L(\chi)$ transports this pair into the corresponding pair of unit vectors $[OX_v], [OX_v']$ of conjugate coordinate axes (Δ_v, Δ_v') such that $v = \tanh\chi$. To see this, we introduce the two lightlike vectors $[OM_0] = (\frac{1}{2}, \frac{1}{2})$ and $[ON_0] = (-\frac{1}{2}, \frac{1}{2})$ such that $[OX_0] = [OM_0] + [ON_0]$ and $[OX_0'] = [OM_0] - [ON_0]$. In view of the previous definition of the action of $L(\chi)$, we thus have

$$L(\chi)[OX_0] = e^{\chi}[OM_0] + e^{-\chi}[ON_0] = (\sinh\chi, \cosh\chi) = [OX_v],$$

$$L(\chi)[OX_0'] = e^{\chi}[OM_0] - e^{-\chi}[ON_0] = (\cosh\chi, \sinh\chi) = [OX_v'].$$

One can also compute similarly the action of another transformation $L(\chi')$ on the pair $([OX_v], [OX_v'])$; it gives another conjugate pair $([OX_w], [OX_w'])$ where $w = \tanh(\chi + \chi')$. In fact one has

$$L(\chi')[OX_v] = (\sinh(\chi + \chi'), \cosh(\chi + \chi')) = L(\chi + \chi')[OX_0] = [OX_w],$$

$$L(\chi')[OX_v'] = (\cosh(\chi + \chi'), \sinh(\chi + \chi')) = L(\chi + \chi')[OX_0'] = [OX_w'].$$

Additivity of rapidities

The previous computation shows that the action of the commutative group of "hyperbolic rotations" $L(\chi)$ on pairs of conjugate axes (Δ_v, Δ_v') (normalized by H and H') is similar to the action of the group of rotations $R(\theta)$ on pairs of orthonormal axes. A physical interpretation of the latter concerns the composition law of velocities: *the Galilean "law of additivity of velocities" is replaced by the Minkowskian "law of additivity of rapidities"*. If a relativistic particle **A** has the rapidity χ with respect to the earth and emits in the forward direction a particle **B** with rapidity χ' in its center of mass system, then **B** has the rapidity $\chi+\chi'$ with respect to the earth. The corresponding composition law for velocities is therefore:

$$w = \tanh(\chi+\chi') = \frac{\tanh\chi + \tanh\chi'}{1+\tanh\chi\ \tanh\chi'} = \frac{v+v'}{c(1+\frac{vv'}{c^2})}.$$

Lorentz frames and Lorentz invariance of $Q(X)$

Every vector $[OX] = t[OX_0] + x[OX_0']$ of the Minkowskian plane can be rewritten as

$$[OX] = t_v[OX_v] + x_v[OX_v']$$

for any choice of conjugate axes (Δ_v, Δ_v') with unit vectors $([OX_v], [OX_v'])$. We shall also write in short: $[OX] = (x,t) = (x_v, t_v)_v$. Choosing such a coordinatization is also called *"choosing a Lorentz frame with velocity v (or rapidity χ)"* in the Minkowskian plane.

The last point to be checked for completing the parallel between the Lorentz group in the Minkowskian plane and the rotation group in the Euclidean plane is the *"invariance property of the Minkowskian quadratic form by changes of Lorentz frame"*, namely the fact that for any Lorentzian coordinatization $X = (x,t) = (x_v, t_v)_v$, one has the invariance relation

$$Q(X) = t^2 - x^2 = t_v^2 - x_v^2.$$

To show this, we associate with $v = \tanh\chi$ the Lorentz transformation $L(-\chi)$ (namely the inverse of $L(\chi)$) which pulls the pair $[OX_v], [OX_v']$ back to the pair at rest $[OX_0], [OX_0']$. With every vector $[OX] = (x,t) = t_v[OX_v] + x_v[OX_v']$ we can then associate its transform $[OX(-v)] \doteq L(-\chi)[OX] = t_v[OX_0] + x_v[OX_0'] = (x_v, t_v)$. Then according to the basic geometrical property of Lorentz transformations, the points $X(-v)$ and X belong to the same level-curve of $Q(x)$, which proves the invariance relation written above.

Change of Lorentz frame in the light-cone coordinatization

For the rest-frame as well as for the Lorentz frame with rapidity χ, it is convenient to introduce the corresponding *light-cone coordinates* of the point $X = (x,t) = (x_v, t_v)_v$, namely

$$(U \doteq t+x,\ V \doteq t-x), \quad (U_v \doteq t_v + x_v,\ V_v \doteq t_v - x_v).$$

Let us treat the case when X belongs to the region V^+ (i.e. $U > 0, V > 0$); the three other regions would be treated similarly. The invariance property of Q now takes the very simple form

$$Q(X) = UV = U_v V_v \doteq a^2,$$

which allows one to parametrize that region in terms of "hyperbolic polar coordinates" (a, Ψ) (resp. (a, Ψ_v)) with $a > 0$, namely:

$$U = ae^{\psi},\ V = ae^{-\psi};\ \ U_v = ae^{\psi_v},\ V_v = ae^{-\psi_v}.$$

But we know that $\psi = \psi_v + \chi$ (this is the action of the "hyperbolic rotation with rapidity" χ that has been presented above). One thus obtains the very simple relation

$$\frac{V_v}{U_v} = \frac{V}{U} \times e^{2\chi}$$

which defines completely the change of Lorentz frame in this light-cone coordinatization.

2.7 The four-dimensional Minkowski's spacetime; tetrads, Lorentz group and Poincaré group

Up to now we have concentrated on relativistic motions along a single direction of space (Ox), which allowed us to construct a two-dimensional section (Ox, Ot) of Minkowski's spacetime and to introduce the corresponding group of Lorentz transformations in this Minkowskian plane.

We shall now show how the geometrical exploitation of the five postulates (stated in Sec.2-1) can be extended so as to construct the full four-dimensional Minkowski's spacetime. This can be performed in four steps:

i) Use of the rotational symmetry for the observer $\mathcal{O}_0$

According to our first postulate, the observers at rest can represent each event X as follows:

$$X \doteq (x_1, x_2, x_3, ct) \doteq (\mathbf{x}, ct) \doteq (|\mathbf{x}|\ \mathbf{j},\ ct),$$

where $\mathbf{j}$ denotes a spatial unit vector ($|\mathbf{j}| = 1$) which may serve to indicate a direction of motion. In fact, if we consider uniform motions passing at O with velocity $\mathbf{v} \doteq v\mathbf{j}$ oriented in a given spatial direction $\mathbf{j}$, we can reproduce all the previous considerations (from Sec.2-2 to Sec.2-6) for representing these motions in a Minkowskian plane generated by the axis with unit vector $\mathbf{j}$ and Ot. By analogy with geographical representations of space, such planes can be called *meridian planes of spacetime with respect to the observers at rest.*

So one can say that by rotational symmetry (all the directions $\mathbf{j}$ being equivalent), the union of anniversary curves in all meridian planes generate an "anniversary hypersurface", still denoted by H. This is the set of events $X_{\mathbf{v}}$ reached by all observers $\mathcal{O}_{\mathbf{v}}$ starting together from O towards all possible directions $\mathbf{j}$ of space, after one year has elapsed *at their own clock*. H is *a sheet of a two-sheeted hyperboloid* whose equation is

$$(ct)^2 - (x_1^2 + x_2^2 + x_3^2) \doteq (ct)^2 - \mathbf{x}^2 = c^2;\ \ t > 0$$

Here we have restored the unit-independent notation including c. We shall generally keep it also in the next sections in order to always exhibit explicitly the physical dimensionality of the quantities involved.
This anniversary hypersurface H can be seen as providing by itself a geometrical characterization of all the uniform motions. In fact, one can say that any pointlike object in uniform motion is characterized by the Minkowskian vector $[OX_{\mathbf{v}}] \doteq cu$ whose tip ($X_{\mathbf{v}}$) belongs to H. Putting $u \doteq (u_1, u_2, u_3, u_0) \doteq (\mathbf{u}, u_0)$, one then has:

$$u^2 \ \doteq\ u_0^2 - u_1^2 - u_2^2 - u_3^2 \ \doteq\ u_0^2 - \mathbf{u}^2 = 1, \quad \text{with } u_0 > 0.$$

In the latter u^2 denotes what we call *the squared Minkowskian pseudonorm of* u, and u is also called a *timelike unit vector* (Note that the anniversary hypersurface H is now normalized at c).

Equivalently u can be characterized by the pair $(\chi, \mathbf{j})$, where χ is the rapidity (such that $v = c \tanh \chi$) and $\mathbf{j}$ specifies the direction of the motion, according to the following formulae

$$u_0 = \cosh \chi = \frac{1}{[1 - \frac{v^2}{c^2}]^{\frac{1}{2}}},$$

$$\mathbf{u} = \sinh \chi \, \mathbf{j} = \frac{\frac{v}{c}}{[1 - \frac{v^2}{c^2}]^{\frac{1}{2}}} \, \mathbf{j}.$$

This leads one to call the Minkowskian vector $cu = (c\mathbf{u}, cu_0)$ *the "relativistic velocity vector"* since its space-component admits a *small-velocity expansion*

$$c\mathbf{u} = \mathbf{v}(1 + \frac{v^2}{2c^2}) + \cdots,$$

which reproduces the velocity vector $\mathbf{v}$ in the first-order Galilean (or "non-relativistic") approximation. The unit vector u can then be called the *"dimensionless" relativistic velocity vector* of the uniform motion.

The same considerations of rotational symmetry lead one to introduce the *one-sheeted hyperboloid* with equation

$$(ct)^2 - (x_1^2 + x_2^2 + x_3^2) \doteq (ct)^2 - \mathbf{x}^2 = -c^2,$$

which is obtained as the union of all branches of hyperbola H' in the meridian planes generated by a space axis with unit vector $\mathbf{j}$ and Ot. This hypersurface, still denoted by H' is the set of all points $X'_{\mathbf{v}}$ such that the pair of axes $(\Delta_{\mathbf{v}}, \Delta'_{\mathbf{v}})$ are conjugate with respect to the light world-lines inside the corresponding meridian plane. The (hyper)surfaces H and H' are represented on fig. 9.

The Minkowskian quadratic form $Q(X)$

In view of the fundamental role played by H and H', we are led to introduce the following quadratic form on the four-dimensional Minkowski's spacetime:

$$Q(X) \doteq (ct)^2 - x_1^2 - x_2^2 - x_3^2,$$

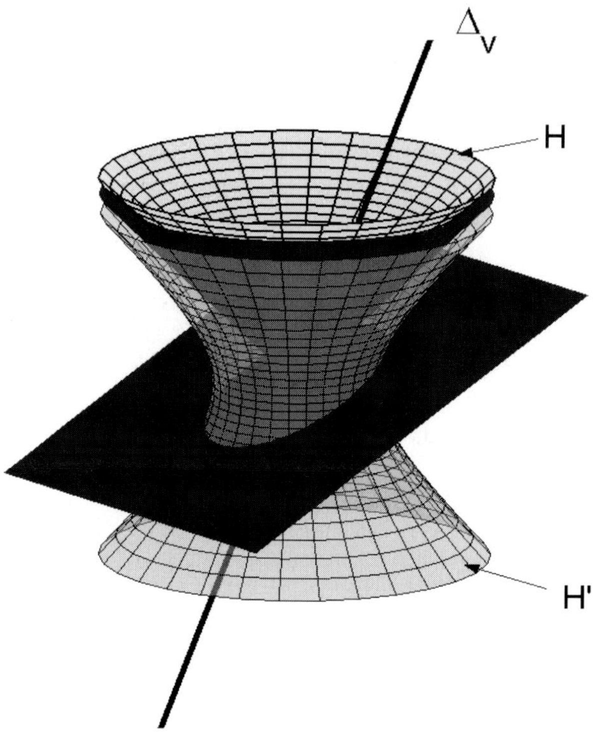

Figure 9: A representation of the four-dimensional Minkowski's spacetime: Level surfaces H, H' of $Q(X)$ and a conjugate pair $(\Delta_{\mathbf{v}}, \Pi_{\mathbf{v}})$ are indicated. The upper part of the light-cone is visible as an elliptic arc between H and H'.

whose level (hyper)surfaces are described as follows:

a) all the sheets of hyperboloids centered at O which are homothetic to H and lie either in V^+ or in V^-. They correspond to $Q(X) > 0$.

b) all the one-sheeted hyperboloids centered at O which are homothetic to H'. They correspond to $Q(X) < 0$.

c) the light-cone C whose equation is $Q(X) = 0$.

We shall denote by $\hat{H}$ anyone of these level hypersurfaces of $Q(X)$.

ii) Conjugacy properties: the space hyperplanes $\Pi_{\mathbf{v}}$

The analysis of all simultaneous events with respect to any given observer $\mathcal{O}_{\mathbf{v}}$ can be performed along the same line as in Sec.2-2, even if the geometry is a bit more complicated than in the Minkowskian plane. In fact, the principle is always the same, being based on the second postulate which settles the light-cone C as the primary absolute element of spacetime. Being given an observer $\mathcal{O}_{\mathbf{v}}$ with world-line $\Delta_{\mathbf{v}}$ (inside the light-cone C) and a space-direction Δ' (i.e. by definition outside the light-cone C), these two straight lines determine a plane P which intersects C along a pair of light-lines. Now we can say that *Δ' is a direction of simultaneity for $\mathcal{O}_{\mathbf{v}}$ if, in the plane P, $\Delta_{\mathbf{v}}$ and Δ' are conjugate with respect to the light-lines of P* : that means that by performing the parallelogram construction of Sec.2-2 (fig. 3), in the plane P, with $\Delta_{\mathbf{v}}$ as the given diagonal, one obtains Δ' as the direction of the second diagonal. In view of the universality of the light-velocity (fourth postulate) completed again by "isochronousness" (third postulate), this geometrical construction remains the *universal criterion of simultaneity with respect to $\mathcal{O}_{\mathbf{v}}$*. We shall now show the following:

Linearity property: The set of all directions of simultaneity Δ' for $\mathcal{O}_{\mathbf{v}}$ is a three-dimensional linear subspace. This hyperplane $\Pi_{\mathbf{v}}$ is physically interpreted as providing the space-slices at constant time $t_{\mathbf{v}}$ for $\mathcal{O}_{\mathbf{v}}$.

Let us show that if Δ'_1 and Δ'_2 are directions of simultaneity for $\mathcal{O}_{\mathbf{v}}$, then any direction Δ' in the plane determined by these two directions is also a direction of simultaneity for $\mathcal{O}_{\mathbf{v}}$. Given the planes P_1 and P_2 determined respectively by $(\Delta_{\mathbf{v}}, \Delta'_1)$ and $(\Delta_{\mathbf{v}}, \Delta'_2)$ and given any point X of $\Delta_{\mathbf{v}}$ in V^+ , one can construct the corresponding parallelograms (OA_1XB_1) and (OA_2XB_2) whose all sides are light-like segments (as in fig. 3 of Sec.2-2) and whose diagonals A_1B_1 and A_2B_2 are respectively parallel to Δ'_1 and Δ'_2 and intersect at the middle of OX. Since the four-points A_1, A_2, B_1, B_2 all belong to the future light-cone C^+, as well as to the past light-cone $C^-(X)$ with apex X, they belong to their intersection which is an ellipse E: A_1B_1 and A_2B_2 are diameters of this ellipse. If we now consider any direction Δ' in the plane determined by Δ'_1 and Δ'_2, which is parallel to the plane of E, we see that the diameter of E parallel to Δ' intersects E in two points A and B such that $(OAXB)$ is a lightlike-sided parallelogram: therefore Δ' is a direction of simultaneity for $\mathcal{O}_{\mathbf{v}}$. This proves that the set of directions of simultaneity for $\mathcal{O}_{\mathbf{v}}$ is a linear subspace of the spacetime. The fact that this subspace $\Pi_{\mathbf{v}}$ is three-dimensional is easy to see: Assuming that it were two-dimensional, it would determine with $\Delta_{\mathbf{v}}$ a three-dimensional subspace S of spacetime outside which no spacelike direction Δ' could be a direction of simultaneity for $\mathcal{O}_{\mathbf{v}}$. But let us then pick up any spacelike direction Δ' outside S. It determines with $\Delta_{\mathbf{v}}$ a plane P' which intersects C along two light-lines and therefore allows one to construct a direction of simultaneity Δ" for $\mathcal{O}_{\mathbf{v}}$ inside P'. Since P' can intersect S only along $\Delta_{\mathbf{v}}$ (if not, it would be contained in S and Δ' would be contained in S), the assumption cannot be true.

To summarize, we have associated with each world-line $\Delta_{(u)} \doteq \Delta_{\mathbf{v}}$ with timelike unit vector u, a corresponding spacelike hyperplane $\Pi_{(u)} \doteq \Pi_{\mathbf{v}}$ which can be called the *conjugate hyperplane to $\Delta_{(u)}$*. The intersection of $\Pi_{(u)}$ with the one-sheeted hyperboloid H' is an *ellipsoid $\mathcal{E}_{(u)} \doteq \mathcal{E}_{\mathbf{v}}$* which is *the set of all events X in $\Pi_{(u)}$ such that $Q(X) = -c^2$*. This hyperplane and the corresponding ellipsoid are tentatively illustrated on fig. 9 (as a plane and an ellipse represented in perspective).

At that point of our study, it remains to show the following essential property.

All the events in $\mathcal{E}_{(u)}$ are interpreted by the observer $\mathcal{O}_{(u)}$ with world-line $\Delta_{(u)}$, as all the simultaneous events at zero time which take place at (lightyear) unit distance from the origin in all possible directions of space. In other words, *the ellipsoid $\mathcal{E}_{(u)}$* (as well as all the homothetic ellipsoids having their centers on the axis $\Delta_{(u)}$) are *perceived as spheres centered at the origin* by the corresponding observer $\mathcal{O}_{(u)}$. This will be displayed in the next step.

iii) Four-dimensional Lorentz transformations, tetrads and the invariant forms of $Q(X)$

We consider the conjugate pair $(\Delta_{(u)}, \Pi_{(u)})$ associated with a certain observer $\mathcal{O}_{(u)}$; $[OX_{(u)}]$ is the unit vector of the time-axis $\Delta_{(u)}$ of $\mathcal{O}_{(u)}$. We are looking for coordinatizations of the Minkowskian spacetime *adapted to that observer.* Such coordinatizations can be defined by choosing triplets of unit spatial vectors $[OX'_{(u),1}]$, $[OX'_{(u),2}]$, $[OX'_{(u),3}]$ in the hyperplane $\Pi_{(u)}$, and by decomposing any vector $[OX]$ of spacetime under the following form

$$[OX] = (ct_{(u)})[OX_{(u)}] + x_{(u),1}[OX'_{(u),1}] + x_{(u),2}[OX'_{(u),2}] + x_{(u),3}[OX'_{(u),3}].$$

However the remaining problem consists in determining all possible triplets $[OX'_{(u),1}]$, $[OX'_{(u),2}]$, $[OX'_{(u),3}]$ such that the Minkowskian quadratic form $Q(X)$ still has the same invariant form with respect to these new coordinates, namely:

$$Q(X) \doteq (ct)^2 - (x_1^2 + x_2^2 + x_3^2) = (ct_{(u)})^2 - (x_{(u),1}^2 + x_{(u),2}^2 + x_{(u),3}^2).$$

In fact, if the latter is valid, it follows that the events X in the ellipsoid $\mathcal{E}_{(u)}$ satisfy the equations

$$t_{(u)} = 0, \quad x_{(u),1}^2 + x_{(u),2}^2 + x_{(u),3}^2 = c^2$$

and are therefore perceived by the observer $\mathcal{O}_{(u)}$ as covering the whole (lightyear-)unit sphere centered at the origin. In particular, the corresponding triplet $[OX'_{(u),1}]$, $[OX'_{(u),2}]$, $[OX'_{(u),3}]$ will be interpreted as *a spatial orthonormal system for $\mathcal{O}_{(u)}$*. The tips of these three vectors themselves belong to the ellipsoid $\mathcal{E}_{(u)}$ and their mutual orthogonality (which corresponds to a certain "conjugacy property with respect to the ellipsoid $\mathcal{E}_{(u)}$") will be fully clarified in the last step iv).

For the moment, *we take the previous invariance property of Q as a basic criterion to be satisfied by an orthonormal triplet in $\Pi_{(u)}$*, and we say that *the linear transformation $L_{(u)}$ which transforms the unit vectors of the rest-frame $[OX'_{0,1}]$, $[OX'_{0,2}]$, $[OX'_{0,3}]$, $[OX_0]$, into the "tetrad" ($[OX'_{(u),1}]$, $[OX'_{(u),2}]$, $[OX'_{(u),3}]$, $[OX_{(u)}]$), is a Lorentz transformation of Minkowski's spacetime.* One also says that *this tetrad is affiliated to the conjugate pair* $(\Delta_{(u)}, \Pi_{(u)})$ and is *admissible for the observer $\mathcal{O}_{(u)}$*. Each tetrad and the corresponding coordinatization define a *Lorentz frame.*

The construction of general Lorentz transformations relies on two basic special classes:

a) The group $\mathcal{L}_{ort}$ of orthogonal transformations at rest

We consider the group of transformations which transform the initial rest-frame into another rest-frame whose spatial axes form a new orthonormal (positively oriented) system of the space (Ox_1, Ox_2, Ox_3), while the time-axis Ot is preserved. Since these transformations preserve the value of the spatial Euclidean (squared) norm

$$x_1^2 + x_2^2 + x_3^2 = {x'_1}^2 + {x'_2}^2 + {x'_3}^2,$$

it is clear that they transform every point X of spacetime into a point X' such that $Q(X) = Q(X')$.

b) The group $\mathcal{L}_{hyp}$ of "pure Lorentz transformations"

Let us fix $\mathbf{j}$ along Ox_1 and the vector $\Delta_{(u)}$ with unit vector $u \doteq u_{(1)}$ in the Minkowskian plane (Ox_1, Ot). Then it is easily checked that the conjugate hyperplane $\Pi_{(u_{(1)})}$ is generated by the conjugate axis $\Delta'_{(u_{(1)})}$ in the plane (Ox_1, Ot) (see fig. 8) together with the spatial plane (Ox_2, Ox_3). We then consider the linear transformation which keeps all the vectors in the plane (Ox_2, Ox_3) fixed, and acts as a two-dimensional hyperbolic rotation with rapidity χ in the plane (Ox_1, Ot). This transformation is called a *pure Lorentz transformation* of Minkowski's spacetime. The corresponding change of coordinates is of the form

$$[OX] \doteq (x_1, x_2, x_3, ct) \rightarrow [OX'] \doteq (x'_1, x'_2 = x_2, x'_3 = x_3, ct'),$$

where the passage from (x_1, ct) to (x'_1, ct') has been given in Sec.2-6. It then follows from the "basic geometrical property of the Lorentz transformations in the plane" (see Sec.2-6) that one has:

$$Q(X) \doteq c^2t^2 - x_1^2 - x_2^2 - x_3^2 = c^2t'^2 - {x'_1}^2 - {x'_2}^2 - {x'_3}^2 = Q(X').$$

It also results from the study of Sec.2-6 that these transformations form a commutative group.

The most general Lorentz transformations

In order to construct the most general Lorentz transformation, we shall compose special transformations of the previous groups $\mathcal{L}_{ort}$ and $\mathcal{L}_{hyp}$. We also keep in mind that when such special Lorentz transformations act on any point X of spacetime, the transform remains on the corresponding level hypersurface $\hat{H}_X$ of $Q(X)$ passing at X: either on a spherical horizontal section of $\hat{H}_X$ in the former case, or in a hyperbolic section of $\hat{H}_X$ parallel to the plane (Ox_1, Ot) in the latter case.

Now we proceed as follows. Being given any conjugate pair $(\Delta_{(u)}, \Pi_{(u)})$, one can find a transformation L_1 in $\mathcal{L}_{ort}$ which transforms that pair into a pair $(\Delta_{(u_{(1)})}, \Pi_{(u_{(1)})})$, with $u_{(1)}$ in the plane (Ox_1, Ot). (It must transform the unit vector $\mathbf{j}$ of the horizontal component of u into the unit vector of Ox_1). Then there exists a unique transformation L_2 in $\mathcal{L}_{hyp}$ which transforms the pair $(\Delta_{(u_{(1)})}, \Pi_{(u_{(1)})})$ into the pair at rest $(Ot, (Ox_1, Ox_2, Ox_3))$.

Let us now consider an arbitrary transformation L_0 in $\mathcal{L}_{ort}$ and define the composition product

$$L_{(u)} \doteq L_1^{-1} \circ L_2^{-1} \circ L_0.$$

We call $([OX'_{(u),1}], [OX'_{(u),2}], [OX'_{(u),3}], [OX_{(u)}])$ the image by $L_{(u)}$ of the orthonormal system (or "reference tetrad") $([OX'_{0,1}], [OX'_{0,2}], [OX'_{0,3}], [OX_0])$, the last vector $[OX_{(u)}]$ being (by construction) the time unit vector for the given observer $\mathcal{O}_{(u)}$. We then claim that this image is a general admissible tetrad affiliated with the given pair $(\Delta_{(u)}, \Pi_{(u)})$. This can be seen by an argument similar to the one given at the end of Sec.2-6 for the two-dimensional case. With every vector

$$[OX] = (ct_{(u)})[OX_{(u)}] + x_{(u),1}[OX'_{(u),1}] + x_{(u),2}[OX'_{(u),2}] + x_{(u),3}[OX'_{(u),3}],$$

one associates its "pull-back transform"

$$[OX_{pb}] \doteq L_{(u)}^{-1}[OX] = (ct_{(u)})[OX_0] + x_{(u),1}[OX'_{0,1}] + x_{(u),2}[OX'_{0,2}] + x_{(u),3}[OX'_{0,3}].$$

Then since $L_{(u)}^{-1} = L_0^{-1} \circ L_2 \circ L_1$, one can make use of the fact that the successive images X_1, X_2 and finally X_{pb} of X by the sequence of transformations L_1, L_2 and L_0^{-1} remain on the same level hypersurface of $Q(X)$. This entails that

$$Q(X) = Q(X_{pb}) = (ct_{(u)})^2 - x_{(u),1}^2 - x_{(u),2}^2 - x_{(u),3}^2.$$

Conversely, one sees by the same argument that any tetrad admissible for $\mathcal{O}_{(u)}$ is transformed by $L_2 \circ L_1$ into a tetrad admissible for $\mathcal{O}_0$, which is thereby the image by some transformation L_0 in $\mathcal{L}_{ort}$ of the reference tetrad defined by the coordinate axes.

iv) Pseudoorthogonality and the group property of Lorentz transformations
Being given any pair of events X, X' in spacetime, let us define the following symmetric expression

$$[OX].[OX'] \doteq \frac{1}{2}[Q(X+X') - Q(X) - Q(X')] = (ct)(ct') - x_1x'_1 - x_2x'_2 - x_3x'_3,$$

in which the event $X+X'$ denotes the tip of the vector $[OX]+[OX']$. This algebraic expression is similar to the one which defines the scalar product of two vectors $\mathbf{x}$, $\mathbf{y}$ in terms of the squared norms of $\mathbf{x}$, $\mathbf{y}$ and $\mathbf{x}+\mathbf{y}$ in Euclidean space. By analogy, we shall say that the vectors $[OX]$ and $[OX']$ are *pseudoorthogonal* if

$$[OX].[OX'] \doteq (ct)(ct') - x_1x'_1 - x_2x'_2 - x_3x'_3 = 0.$$

It is easy to check that the vectors of the reference tetrad are mutually pseudo orthogonal.

We know that the images of any event X by the transformations L in $\mathcal{L}_{ort}$ or in $\mathcal{L}_{hyp}$ remain on the level hypersurfaces of $Q(X)$. Then it follows from the previous definition that the images of all pseudoorthogonal pairs by all these transformations are pseudoorthogonal pairs. This is therefore also true for all the Lorentz transformation $L_{(u)}$ constructed in the previous paragraph. So by applying this result to the reference tetrad, we conclude that in every tetrad affiliated with any possible conjugate pair $(\Delta_{(u)}, \Pi_{(u)})$, all the vectors of the tetrad are mutually pseudoorthogonal: so for the spacelike triplet in the tetrad, pseudoorthogonality coincides with the Euclidean notion of orthogonality inside $\Pi_{(u)}$, while the pseudoorthogonality of this triplet with respect to $[OX_{(u)}]$ is identical with the property of conjugacy introduced earlier. Taking into account the fact that all the vectors $[OX]$ of a tetrad are unit timelike or spacelike vectors (i.e. such that either $Q(X) = c^2$ or $Q(X) = -c^2$), we can say that all tetrads are *systems of pseudoorthonormal vectors with respect to Q*.

In view of this characteristic property of tetrads, we can thereby conclude that *the action of any Lorentz transformation $L_{(u)}$ on any tetrad gives another tetrad.*

It follows that the composition product of two Lorentz transformation $\mathcal{L}_{(u_1)} \circ \mathcal{L}_{(u_2)}$ is another Lorentz transformation (since it transforms the reference tetrad into a tetrad). The definition of inverse transformations being obvious, we conclude that all the transformations $L_{(u)}$ form a group, called the *Lorentz group of the four-dimensional Minkowski's spacetime.*

By adjunction of the translations of space and time, one obtains the more general *"inhomogeneous Lorentz transformations"* which act on any vector $[OX]$ as follows:

$$[OX] \ \rightarrow (L_{(u)}, a)[OX] = L_{(u)}([OX]) + a;$$

in the latter, a denotes a given four-vector which specifies a translation T_a of spacetime. The set of all the inhomogeneous Lorentz transformations form a group which is called the *Poincaré group.*

Remark on the rest-frame and on the distorted appearance of the general Lorentz frames

We note that among all the conjugate pairs $(\Delta_{\mathbf{v}}, \Pi_{\mathbf{v}})$, one and only one is orthogonal in the usual sense. The familiar choice of this orthogonal pair (e.g. vertical-horizontal) for representing the rest-frame is a manifestation of our biased geometrical perception which privileges orthogonality and sedentarity. But as in the Galilean case, the observer at rest enjoys no special physical properties with respect to any other observer in uniform motion (that's again the *"principle of relativity"*). So the verticality of the time-axis and the horizontality of space could have been chosen for representing the Lorentz frame of any given uniform motion: there is nothing deep in that choice. One can also say that the Minkowskian representation of the spacetime of special relativity is defined for $\mathcal{O}_0$ (as well as for any observer $\mathcal{O}_{\mathbf{v}}$) *up to the arbitrariness in the choice of the Lorentz frame* or in short *up to a Lorentz transformation*: it is the equivalence class of all these representations. But any chosen representation provides an absolute and faithful description of the events of the universe. Another aspect of all that which deserves to be pointed out again concerns the unavoidable "distorted visual perception" introduced by the conjugacy property. We mean the fact that we have an *ellipsoidal representation* of the surfaces which are actually perceived as *spheres* by observers in uniform motion. Probably the best way for becoming familiar with that strange aspect of the Minkowskian representation consists again in using the metaphor of geographical maps. One can always represent a land on a map equipped with oblique coordinates and different scales of length on the two coordinate axes. That's awkward for our perception, but it remains an absolute and faithful description of the land. In the Minkowskian representation of spacetime, this is the price to pay for having a global geometrical description of all the "spatial slices", corresponding to all possible observers in uniform motion !!

3 Accelerated motions and curved world-lines

The only motions that have been considered for stating the postulates of special relativity and for constructing Minkowski's spacetime are uniform motions. Their world-lines are oriented straight lines whose direction belongs to the cone V^+ and one also call them *inertial motions* by referring to the fact that no force is acting on a pointlike object whose motion is of that type. Under the name of *accelerated* (or *noninertial) motions* we shall denote the most general type of motion; such a motion is geometrically represented by a curved world-line in Minkowski's spacetime. A curved world-line is smooth if it is an oriented smooth curve admitting at each point a tangent whose direction belongs to V^+. A general world-line can be considered as an oriented union of smooth curved world-line segments. From the physical viewpoint, objects endowed with motions of such a general type are submitted to the action of a time-dependent force and to additional shocks which produce possible discontinuities in the direction of the tangent to the corresponding world-line. Here we shall keep outside the treatment of dynamical problems of special relativity (except for the special case of *uniformly accelerated motions* considered in Sec.3-2 and Sec.3-3). In fact, we shall only concentrate on the kinematical aspects of these motions, which can be presented in terms of the Minkowskian geometry of curved world-lines by pursuing our analogy with Euclid's geometry.

3.1 Curvilinear distances and the slowing down of clocks

Recall on Euclidean space

Let γ be any curved path with end-points A and B in Euclidean space $\mathbf{R}^3$; we suppose it to be smooth or composed of a finite succession of smooth paths. Mathematically, the *length* $d_\gamma(A,B)$ *of the path* γ is defined by the theory of curvilinear integrals as

$$d_\gamma(A,B) = \int_\gamma ds,$$

where ds denotes the Euclidean length element

$$ds = [dx_1^2 + dx_2^2 + dx_3^2]^{\frac{1}{2}}.$$

This theory involves the following ideas:
i) *conceptually*, $d_\gamma(A,B)$ appears as the limit for N tending to infinity of the length d_N of an approximate polygonal path composed of j_N successive small linear paths of equal lengths $\frac{1}{N}$, whose end-points A_j all belong to γ, with $A_1 = A$ and $d(A_{j_N}, B) \leq \frac{1}{N}$. The points A_j can be constructed recursively by the following rule: A_j is at the intersection of γ with the sphere of radius $\frac{1}{N}$ centered at A_{j-1} (and such that $A_j \neq A_{j-2}$).
ii) *physically*, the length of the path γ can be measured by using a flexible graduated ribbon.
iii) *numerically*, the previous curvilinear integral can be computed by introducing any parametrization of the form $\mathbf{x} \doteq (x_1, x_2, x_3) = \mathbf{x}(t)$ of γ, where t is a parameter varying between t_A and t_B, such that $\mathbf{x}(t_A) = A$ and $\mathbf{x}(t_B) = B$. One then has:

$$d_\gamma(A,B) = \int_{t_A}^{t_B} \frac{ds}{dt} dt.$$

The Minkowskian length or "proper time" of a curved world-line
The previous Euclidean considerations admit a close parallel for curved world-lines in Minkowski's space.

Let γ be any general curved world-line with initial and final events A and B in Minkowski's spacetime $\mathbf{R}^4$: the event B lies in the future of A (namely in the future cone $V^+(A)$). Mathematically, the *Minkowskian length $d_\gamma(A,B)$ of the world-line γ* is again defined by the theory of curvilinear integrals as

$$d_\gamma(A,B) = \int_\gamma ds,$$

but ds now denotes the Minkowskian length element or *"proper-time element"*

$$ds = [(c\ dt)^2 - dx_1^2 - dx_2^2 - dx_3^2]^{\frac{1}{2}}.$$

This theory involves the same ideas as in the Euclidean case, but their physical interpretation in terms of time-measurements must now be kept in mind:
i) *conceptually*, $d_\gamma(A,B)$ again appears as the limit for N tending to infinity of the Minkowskian length d_N of an approximate polygonal path. This path is composed of successive small linear paths of equal Minkowskian lengths or *time-like distances* $\frac{1}{N}$, whose end-points A_j all belong to γ, with $A_1 = A$ and $d(A_{j_N}, B) \leq \frac{1}{N}$. The points A_j can now be constructed recursively by the following rule: A_j is at the intersection of γ with the sheet of hyperboloid $H^+_{A_{j-1}}(\frac{1}{N})$ centered at A_{j-1} and whose all points lie in the future of A_{j-1} and at the time-like distance $\frac{1}{N}$ from A_{j-1}: this sheet of hyperboloid is homothetic to the anniversary surface of A_{j-1}, and obtained from the latter by applying to it the scaling ratio $\frac{1}{N}$.
ii) *physically*, the *(time-like) length* of the path γ can be measured by using a *clock* which has to be as much insensitive to accelerations as possible. The fact that atomic clocks satisfy such requirements with a high degree of robustness against

strong accelerations has been established experimentally in various works around 1960 (see in particular the article by Sherwin [5]).
iii) *numerically*, the previous curvilinear integral can again be computed by introducing any relevant parametrization of the path γ, but a specially significant parametrization results in a very nice formula due to Einstein.

Einstein's formula for the slowing down of clocks
One assumes that the events A and B occur at the same point $\mathbf{x}_A = \mathbf{x}_B$ in the rest system, so that physically the path γ may represent any motion starting from $\mathbf{x}_A$ at time t_A and coming back to the same point at time t_B.

Let us now choose precisely the time-coordinate t in the rest system as a relevant parameter for the description of γ; the latter is thus given by a parametrization of the following form:

$$(\mathbf{x}, ct) \doteq (x_1, x_2, x_3, ct) = (\mathbf{x}(t), ct), \quad \text{with} \quad t_A \leq t \leq t_B.$$

One then has:

$$\frac{ds}{dt} = c\left[1 - \left(\frac{dx_1}{cdt}\right)^2 - \left(\frac{dx_2}{cdt}\right)^2 - \left(\frac{dx_3}{cdt}\right)^2\right]^{\frac{1}{2}} = c\left[1 - \left(\frac{\mathbf{v(t)}}{c}\right)^2\right]^{\frac{1}{2}},$$

where $\frac{d\mathbf{x}(t)}{dt} \doteq \mathbf{v}(t)$ represents the *instantaneous velocity* of the motion in the rest-frame at the rest-time t. By plugging the latter expression of $\frac{ds}{dt}$ in the curvilinear integral for $d_\gamma(A, B)$, one thus obtains:

$$d_\gamma(A, B) = c\int_{t_A}^{t_B}\left[1 - \left(\frac{\mathbf{v(t)}}{c}\right)^2\right]^{\frac{1}{2}} dt \ \leq \ c(t_B - t_A).$$

This formula thus exhibits the general phenomenon of *"slowing down of the clock attached to the world-line γ"* with respect to the clock at rest. It provides a quantitative physical formulation of the following geometrical statement (namely the most general form of the Minkowskian triangular inequality):

"IN MINKOWSKI'S SPACETIME, ANY TIME-LIKE STRAIGHT-LINE SEGMENT IS *LONGER* THAN ANY CURVED SEGMENT WITH THE SAME END-POINTS."

Remark. The previous computation provides an expression for the *slowing down*

$$\sigma_\gamma \doteq (t_B - t_A) - \frac{1}{c}d_\gamma(A, B)$$

which exhibits a *very simple first-order approximation at low velocities* ($\frac{v}{c}$ small). One gets:

$$\sigma_\gamma = \int_{t_A}^{t_B} \frac{1}{2}\frac{\mathbf{v(t)}^2}{c^2} dt = (t_B - t_A)\frac{v_M^2}{2c^2},$$

where v_M^2 denotes the *mean squared velocity* of the motion with world-line γ between the initial and final times. This formula is remarkably interesting for performing experimental checks of the slowing-down phenomenon, since v_M may for example be related to the temperature of atoms in thermal motion (see [5] and references therein).

3.2 Minkowski's description of accelerations

The instantaneous relativistic velocity vector for a general motion

We have seen in Sec.2-7 that any pointlike object in uniform motion is intrinsically characterized by its *normalized relativistic velocity vector* u, which is a unit vector in the Minkowskian sense: $u^2 \doteq u_0^2 - \mathbf{u}^2 = 1$. We can then pursue the parallel between smooth Euclidean curved lines and Minkowskian world-lines by considering in both cases the notion of *unit tangent vector* $u(X_0)$ at any point X_0 of the line. If the line is parametrized by the length parameter s via a vector equation of the form $X = X(s)$, one then defines $u(X_0)$ at $X_0 = X(s_0)$ by the equation:

$$u(X_0) = \frac{d}{ds}X(s)_{|s=s_0}.$$

In both cases the squared norm or pseudonorm of $u(X_0)$ is equal to 1, since one has in view of the definition of ds^2:

a) in three-dimensional Euclidean space (as an example)

$$u(X_0)^2 = \left(\frac{dx_1}{ds}\right)^2 + \left(\frac{dx_2}{ds}\right)^2 + \left(\frac{dx_3}{ds}\right)^2 = 1.$$

b) similarly in Minkowskian spacetime:

$$u(X_0)^2 = \left(c\frac{dt}{ds}\right)^2 - \left(\frac{dx_1}{ds}\right)^2 - \left(\frac{dx_2}{ds}\right)^2 - \left(\frac{dx_3}{ds}\right)^2 = 1 \quad \text{with} \quad \frac{dt}{ds} > 0.$$

In the latter case, $cu(X_0)$ will be called the *instantaneous relativistic (or Minkowskian) velocity vector* of the motion $(X = X(s))$ at the event X_0. $u(X_0)$ can be called the *dimensionless* instantaneous velocity vector.

The acceleration vector

According to Minkowski, one defines the acceleration vector $\gamma(X_0)$ at X_0 as

$$\gamma(X_0) \doteq c^2 \frac{du(X(s))}{ds}|s = s_0.$$

In the latter, the normalization factor c^2 ensures the right dimensionality LT^{-2} of acceleration. Then by taking the derivative with respect to s of the equation $u(X(s))^2 = 1$, we obtain the pseudoorthogonality relation

$$\gamma(X).u(X) \doteq \gamma_0 u_0 - \gamma_1 u_1 - \gamma_2 u_2 - \gamma_3 u_3 = 0$$

which is valid for all points $X = X(s)$ of the world-line. In other words: *The Minkowskian acceleration* $\gamma(X)$ *is always a spacelike vector which is conjugate to* $u(X)$. The physical interpretation of the latter is that at any event X_0 of the world-line, the vector $u(X_0)$ indicates the corresponding time-axis $\Delta_{(u(X_0))}$ of the traveler, while the acceleration vector $\gamma(X_0)$ is contained in the conjugate hyperplane $\Pi_{(u(X_0))}$, interpreted by the traveler as the Euclidean space at time zero. Then the Euclidean norm of this vector defines the intensity of the acceleration which is felt by the traveler at the event X_0. In view of the sign convention for defining the squared Minkowskian pseudonorm of $\gamma(X_0)$, which is negative, it is given by

$$|\gamma(X_0)| = (-\gamma(X_0)^2)^{\frac{1}{2}}.$$

Uniformly accelerated motions

We shall now present the Minkowskian treatment of *one-dimensional uniformly accelerated motions.* Under this name, we now mean the motions represented by a world-line in a Minkowskian two-dimensional plane (Ox, Ot), whose acceleration's intensity $|\gamma(X)|$ is a constant γ. That means that the tip of the spacelike vector $\gamma(X)$ varies on a branch of hyperbola centered at O and homothetic either to the curve H' or to its opposite in that plane (see Sec.2-5).

We will check that *all such branches of hyperbola* together with those obtained from the latter by spacetime translations *are themselves the world-lines of uniformly accelerated motions.* (For simplicity, we shall skip the proof of the fact that they represent *all* the one-dimensional uniformly accelerated motions). We introduce such hyperbolic world-lines by the following parametrization in which the parameter τ will be seen to be the proper time of the motion (the notation τ being thus substituted to the length notation $s = c\tau$ of the previous paragraph).

$$X = X(\tau) \doteq \quad (x(\tau), ct(\tau)):$$

$$x(\tau) = a\cosh\frac{c\tau}{a} + x_0, \quad ct(\tau) = a\sinh\frac{c\tau}{a} + ct_0.$$

We just have to compute successively:

$$u(X(\tau)) = \frac{d}{d(c\tau)}X(\tau) = (u_1(\tau), u_0(\tau)):$$

$$u_1(\tau) = \sinh\frac{c\tau}{a}, \quad u_0(\tau) = \cosh\frac{c\tau}{a},$$

which shows that $u(X(\tau))^2 = 1$.

$$\gamma(X(\tau)) = c^2\frac{d}{d(c\tau)}u(X(\tau)) = (\gamma_1(\tau), \gamma_0(\tau):$$

$$\gamma_1(\tau) = \frac{c^2}{a}\cosh\frac{c\tau}{a}, \quad \gamma_0(\tau) = \frac{c^2}{a}\sinh\frac{c\tau}{a},$$

from which it follows that $\gamma(X(\tau))^2 = -\frac{c^4}{a^2}$ is constant and yields the value $\gamma = \frac{c^2}{a}$ for the acceleration. So one can say that the acceleration is proportional to the "time-curvature" $\frac{1}{a}$ of the world-line.

Remarks.

a) *Non-relativistic (or Galilean) approximation.* It is clear that the hyperbolic world-line with equation $x^2 - (ct)^2 = a^2$ or $x = [(ct)^2 + a^2]^{\frac{1}{2}}$ admits as a second-order approximation near the event $x = a, t = 0$ the familiar parabola with equation

$$x = a + \frac{c^2}{2a}t^2 = a + \frac{1}{2}\gamma t^2,$$

b) In Euclidean geometry, the "osculating circle" at a point X of a Euclidean curve is obtained as the limit of the circle containing three neighbouring points of the curve, when these three points tend together to X. Minkowski introduced similarly (in [3]) a notion which can be called the *"osculating uniformly accelerated motion"* of a general motion at the event X: its world-line is the limit of a hyperbolic world-line containing three neighbouring events of the general motion, in the limit when these three events tend to X.

3.3 A comfortable trip for the "Langevin traveler"

The standard presentation of the "twin paradox" (or "Langevin traveler"), which amounts to a direct trip with return between a point of the earth and some far-distant space station S, with large uniform velocity v in both directions, is remarkable by its beautiful pedagogical simplicity. In fact, we have seen in Sec.2-4 that it exactly illustrates what we called in geometrical terms the Minkowskian triangular inequality. However, since it appeared in the literature, various objections have been raised whose point was generally to conclude that this was a school example, which was probably physically incorrect or at best unrealistic. This type of opinion has also been often endorsed by vulgarizers of special relativity, as a reassuring thought with respect to what looks like a scandal for the common sense.

The main objection was about the instantaneous passage from velocity v to velocity $-v$ when reaching the term of the travel. Such passage had to be produced by a shock, or even if smoothened by some decelerating device, it seemed to involve so large accelerations that certainly the biological organisms and maybe the clocks themselves could not stand such constraints. Now in view of Minkowski's study of uniformly accelerated motions (presented above in Sec.3-2), one can actually show the possibility of organizing a more comfortable trip for the Langevin traveler, in which the latter would be submitted to a *constant acceleration (or deceleration)* We even impose (for making the acceleration biologically normal) that its value be precisely equal to the value of the gravity acceleration g on the earth. Of course, we admit that the whole travel will take place in the vacuum, far from any gravitational source, in such a way that the flat Minkowskian spacetime remains a reasonably good approximation to the real spacetime.

After having specified an appropriate class of world-lines for that space-traveler, the problem, which is purely geometrical, consists in comparing the length of proper time τ (namely the timelike Minkowskian length) of the traveler's world-line with the corresponding time t that will have elapsed on the earth between the traveler's departure and return. A table of the corresponding values of τ and t will be given below. Its result is striking: while the maximal value of τ fits with a reasonably long life-time for a human being (let us say eighty-six years), the corresponding value of t reaches about five billions of years, namely the age of the earth !!

Of course, a second problem (which has a touch of dream as in anticipation novels...) concerns the production of the constant acceleration for the spaceship on which the traveler is going to live. If the acceleration is produced by either expelling or disintegrating a mass of matter aboard the spaceship, as in conventional rockets, one can make a simple computation of the minimal mass consumption based on the relativistic law of energy-momentum conservation (see Sec.5 below). Assuming that all the disintegrated mass is transformed into photons (which is the most favorable process) it is possible to compute the ratio between the remaining mass $M(\tau)$ at proper time τ and the initial mass M_0 of the spaceship. The set of values which are listed in the table indicate that that for τ larger than twenty years, the procedure becomes radically unrealistic. In fact, the mass to be loaded aboard the spaceship then becomes a non-negligible fraction of the mass of the earth (which also means that gravitational effects have to be taken into account; the use of flat Minkowski's spacetime is no longer justified). But the limitations of this procedure do not exclude the consideration of other types of possible propulsions, which could make use for instance of energies available in the cosmic medium.

Choice of the motion

The trajectory is along a straight line joining the earth, denoted by S_0, and the space station S considered as at rest with respect to the earth. The travel which is proposed is composed of

i) a uniformly accelerated motion with acceleration g from S_0 to the middle M of S_0S;

ii) a uniformly accelerated motion with acceleration $-g$ from M to S (namely a phase of deceleration);

iii) the acceleration $-g$ is maintained as in ii) and produces the first half of the returning trip (i.e. from S to M);

iv) the acceleration is shifted from $-g$ to g for producing a uniformly decelerated motion from M to S_0.

It is clear that the discontinuity of the acceleration (from g to $-g$) produced at M is bearable by the physical and biological systems in the spaceship: it is just felt as a sudden inversion of the direction of the normal gravity g on the earth.

The spacetime representation of this motion is a world-line composed of three successive arcs of hyperbolae with centers a, d and b (see fig. 10), namely:

i) an arc AC joining the departure event A on S_0 to the end of the acceleration

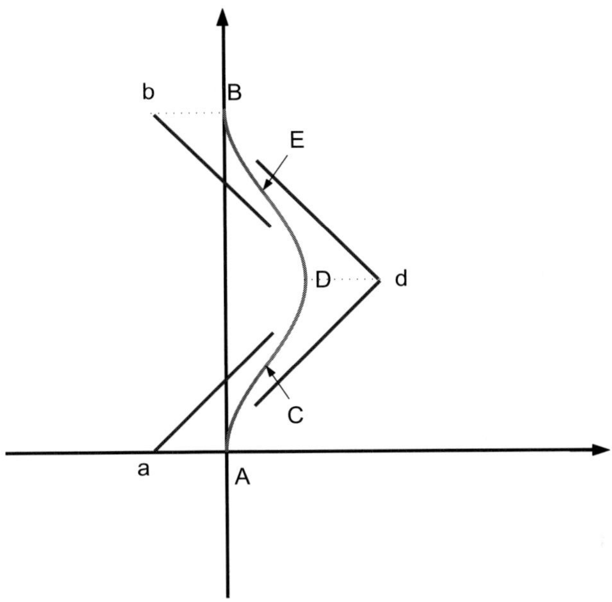

Figure 10: A comfortable world-line for the Langevin traveler

phase C at the point M; this arc is parametrized by the proper time τ of the spaceship according to the following equations:

$$x = \frac{c^2}{g}(\cosh \frac{g}{c}\tau \ - 1), \quad t = \frac{c}{g}\sinh \frac{g}{c}\tau.$$

ii) an arc CDE where D denotes the passage on S (no stop being expected there) and the end-point E denotes the passage at M on the way back.

iii) an arc EB representing the last deceleration phase whose end-event B represents the arrival on S_0.

As it is visualized on fig. 10, the arcs CD, DE, and EB of the traveler's world-line are obtained from the arc AC by obvious symmetries and it is clear that the total traveler-time length τ_B of the travel as well as the corresponding earth-time length t_B are respectively equal to four times the traveler-time length τ_C and the corresponding earth-time length t_C that have elapsed between A and C. In view of the equations of AC this yields the following relation between t_B and τ_B:

$$t_B = 4\frac{c}{g}\sinh \frac{g}{c}\frac{\tau_B}{4}.$$

It is pleasant to notice that with our choice of units (i.e. years and lightyears) not only $c = 1$ but also the earth's value of g is very close to 1. We thus obtain the

The "Langevin traveller" in uniformly accelerated motion

traveller's proper time τ (in years)	earth's proper time t (in years)	$\frac{M(\tau)}{M_0}$
1	1 and 4 days	0.37
2	2 and 1 month	0.13
4	4.7	0.02
8	14.5	4×10^{-4}
12	40.1	8×10^{-6}
16	104	1.6×10^{-7}
20	297	3.2×10^{-9}
28	$2,200$	1.3×10^{-12}
32	$5,960$	
40	$44,000$	
48	$326,000$	
60	6.54×10^6	
72	131×10^6	
84	2.64×10^9	
86	5 billions	

very simple formula

$$t_B = 4 \sinh \frac{\tau_B}{4}$$

whose numerical application can be found in the table.

We notice that for small values of the travel's length of time τ_B, namely between zero and four years, the corresponding values of the earth-time length t_B is not very different; this is because τ_B is the first-order approximation of $4 \sinh \frac{\tau_B}{4}$ at small τ_B. But for larger travel's lengths of time, the exponential character of the sinh function becomes preponderous, which yields such overwhelming discrepancies as two-thousand years of earth's time for twenty-eight years of travel's time and... geologicallike ages for seventy years of travel's time !

Mass decrease required for the spaceship's propulsion

The equation for the rate of mass decrease will be fully justified in Sec.5 on the basis of the relativistic law of conservation of the total energy-momentum of the system. This equation is

$$\frac{d}{d\tau} M(\tau) = -\frac{g}{c} M(\tau) = -M(\tau),$$

which therefore yields the formula

$$M(\tau_B) = M_0 e^{-\tau_B}$$

illustrated numerically in the table.

4 On the visual appearance of rapidly moving objects: Lorentz contraction revisited

Although being valid as a two-dimensional geometrical property of Minkowski's spacetime in a plane (Ox, Ot) , the property of "contraction of lengths" described in Sec.2-5 differs from what would actually be seen by an observer (or a camera) at the passage of a rapidly moving object. As a matter of fact, according to the original Terrell's work [6] (see also [7]) the analysis of the actual physical phenomenon can be summarized as follows.

i) Even if the moving object S is one-dimensional, namely is an infinitely thin rod oriented along the motion trajectory Ox (as considered in Sec.2-5), one must consider the observer at rest $\mathcal{O}$ as situated *at a certain distance* d of Ox. Therefore the actual visual appearance of the rod for such an observer at a certain time $t = t_0$ is obtained by determining the set of light world-lines which have been emitted from all the points of the rod in the past of t_0 and which converge at the corresponding "reception event" $O \doteq O(d, t_0)$ of the observer $\mathcal{O}$. This determines the "photograph" of the rod at time t_0. When the value of t_0 varies, the geometrical construction of the relevant light world-lines results in modifications of the direction of observation and of the apparent length of the rod; these modifications of the visual appearance of the object for the observer $\mathcal{O}$ at rest will thus accompany the motion of the object. In other words, the aspect of the rod on the photograph will vary with time *by combining the relativistic property of contraction of lengths together with perspective effects*; the latter are comparable to those which occur in ordinary space when changing the direction of observation in order to catch successive situations of the moving object (in a purely Galilean treatment with infinite light-velocity).

ii) The previous type of analysis being taken into account, a realistic study still has to be done for the case of a three-dimensional object. For instance, it is interesting to consider a cube-shaped or spherical object S whose center moves along the axis Ox and whose size may be considered as small with respect to the distance d from the observer to Ox. It turns out that the visual appearance of such thick objects *never exhibits* the phenomenon of contraction of lengths in the direction Ox as it was pictured in Gamov's famous book ("The adventures of Mr Tompkins in the land of special relativity"). As a matter of fact, the observed appearance of an object at successive times exhibits a perspective effect whose corresponding ("virtual") direction of observation is *shifted* with respect to the real direction of observation, as though the perspective were accompanied by an "anomalous rotation effect". This apparent change of direction of observation is a typical geometrical effect of Minkowski's spacetime: it is characterized by an angle called *"the relativistic aberration"*. It is interesting to note that for the special case of a spherical object, the disk-shaped appearance remains for all the directions of observation which accompany the object's motion.

The relativistic aberration

Let S and O represent two given events of spacetime corresponding respectively to the emission of a light beam by a pointwise object and to the reception of this light beam: O belongs to the future light-cone $C^+(S)$ of S. The object is in uniform motion with respect to the rest-frame of an observer $\mathcal{O}$ who will observe the reception event at O. This uniform motion is characterized by its world-line $\Delta_{(u)}$ which we choose to belong to the plane (Sx, St) (the point S is contained in $\Delta_{(u)}$; it now plays the role of the origin of Minkowski's spacetime, called O in Sec.2). χ and $v = \tanh \chi$ will denote the rapidity and velocity of the motion; d denotes the distance from the observer $\mathcal{O}$ to the motion's line Sx of the object. At O, the light beam coming from the object is received by the observer $\mathcal{O}$ from a direction which includes the angle θ with the axis Sx in the coordinate-plane (Sx, Sy) and t_0 denotes the corresponding reception time. From these data, we can express the coordinates of the reception event $O \doteq O(d, t_0)$ in the rest frame as follows

$$(x = ct_0 \ \cos\theta, \ y = d = ct_0 \ \sin\theta, \ z = 0, \ t = t_0).$$

(Note that throughout the whole argument the scenario remains in the three-dimensional space-time (Sx, Sy, St)).

With [6] we now introduce another observer $\mathcal{O}'$ who is *at rest in the frame of the moving object* and whose world-line (parallel to $\Delta_{(u)}$) contains the point O: that means that this moving observer $\mathcal{O}'$ "shares with $\mathcal{O}$" the reception event O of the light beam emitted by the object at S, although the latter is now seen as "at rest" by $\mathcal{O}'$. At this event O, $\mathcal{O}'$ receives the light beam from a direction which includes the angle θ' with the corresponding space-axis Sx' of the object's Lorentz frame: this axis Sx' is conjugate of $\Delta_{(u)}$ in the plane (Sx, St). The space hyperplane $\Pi_{(u)}$ of $\mathcal{O}'$ is in fact generated by the three axes Sx', Sy, Sz, the coordinates y and z being unchanged with respect to those of the rest-frame of $\mathcal{O}$. We can now express the coordinates of the reception event O in the Lorentz frame of $\mathcal{O}'$ as follows

$$(x' = ct_0' \ \cos\theta', \ y = d = ct_0' \ \sin\theta', \ z = 0, \ t' = t_0').$$

It is the difference $\alpha \doteq \theta' - \theta$ which is called the *relativistic aberration*; then the basic computation which remains to be done is to compute θ' and thereby α as a function of θ and of the rapidity χ (or velocity v) of the object.

Comparing the two representations of O leads one to introduce at first the ratio

$$M \doteq \frac{t_0}{t_0'} = \frac{\sin\theta'}{\sin\theta}.$$

We shall now make use of the formula for the change of Lorentz frames in the light-cone coordinates (see the end of Sec.2-6). Let us introduce the light-cone coordinates of the event O (or more properly to its projection onto the plane (Sx, St)), namely

$$U = ct + x,\ V = ct - x,\ \ U' = ct' + x',\ V' = ct' - x';$$

$$\text{we thus get}: \ \frac{V}{U} = \tan^2\frac{\theta}{2},\ \ \frac{V'}{U'} = \tan^2\frac{\theta'}{2}.$$

Then the last formula of Sec.2-6 readily yields:

$$\tan\frac{\theta'}{2} = \tan\frac{\theta}{2} \times e^{\chi}.$$

The latter relation defines a function $\theta' = \theta'(\theta, \chi)$ which enjoys the following properties:

a) for fixed χ, θ and θ' tend together either to zero or to infinity;

b) for $\theta = \frac{\pi}{2}$ (resp. $\theta' = \frac{\pi}{2}$), one obtains $\sin\theta' = \frac{1}{\cosh\chi}$ (resp. $\sin\theta = \frac{1}{\cosh\chi}$), where $\frac{1}{\cosh\chi} = (1 - \frac{v^2}{c^2})^{\frac{1}{2}}$ is the *Lorentz contraction factor* (see Sec.2-5).

The visual appearance of extended objects

Let us now suppose that the moving object is *extended* instead of being pointwise, but that *its extension is small* with respect to the distance d at which the observer $\mathcal{O}$ is standing, and to begin with, that it is *"flat for the observer $\mathcal{O}$"* (and therefore also for $\mathcal{O}'$): that means that the set of its world-lines form a small cylinder parallel to $\Delta_{(u)}$ in the subspace (Sx, Sy, St); there is no extension in the third direction Sz.

We now consider the small angles $d\theta$ and $d\theta'$ subtended by the object, as they are seen from O respectively by the observers $\mathcal{O}$ and $\mathcal{O}'$, namely in the planes respectively parallel to (Sx, Sy) and (Sx', Sy). It is clear that the relation between these two angles is obtained by differentiating (at fixed χ) the previous relation between θ, θ' and χ. The result is:

$$\frac{d\theta'}{d\theta} = \frac{\sin\theta'}{\sin\theta} = M$$

This ratio M of the subtended angles, or of the apparent dimensions of the object when passing from the observer $\mathcal{O}$ to the observer $\mathcal{O}'$, can thus be called the *magnification*. What is remarkable in that relation between $d\theta$ and $d\theta'$ is that (eventhough θ' is a function of θ and χ), it does not depend explicitly of the rapidity χ.

As a matter of fact, one can even give a still nicer interpretation of it by introducing the distances r and r' at which the (small) object is seen respectively by $\mathcal{O}$ and $\mathcal{O}'$. Concerning r' it is of course a fixed distance, since the object is at rest for $\mathcal{O}'$ and one has (in the plane parallel to (Sx', Sy) by O)

$$d = r'\sin\theta'.$$

Concerning r, it is the distance from $\mathcal{O}$ (in the plane (Sx, Sy)) to the position occupied by the object at the emission event S and one thus also has

$$d = r \sin \theta.$$

It then immediately follows from these relations that one has:

$$r d\theta = r' d\theta',$$

which means that *the dimensions of the object transversally to the directions of observation of* $\mathcal{O}$ *and* $\mathcal{O}'$ *are equal.* One can now see very simply that *a similar result is valid for general small objects having also an extension in the direction* Sz. In fact, the component along Sz of the object is the same in the rest frame as in the Lorentz frame where the object is at rest; it therefore has equal transversal extensions $dz = dz'$ along Oz for both observers $\mathcal{O}$ and $\mathcal{O}'$, which entails:

$$r d\theta dz = r' d\theta' dz'$$

This means that the surface transversal dimensions of the object with respect to the directions of observation of $\mathcal{O}$ and $\mathcal{O}'$ are equal: the *perspectival shapes and dimensions* of the object are the same when the object is moving as when it is fixed, *provided one replaces the actual direction of observation of the moving object, namely the angle* θ, *by the "virtual" direction* $\theta' = \theta'(\theta, \chi)$ *corresponding to its observation as a fixed object.*
However, in view of the different distances r and r' from $\mathcal{O}$ and $\mathcal{O}'$ to the object, this identity of the perspectival shapes and dimensions is *modified from the angular viewpoint by the magnification factor*

$$M = \frac{d\theta'}{d\theta} = \frac{r}{r'},$$

whose expression as a function of θ and χ is:

$$M(\theta, \chi) = \frac{\sin \theta'(\theta, \chi)}{\sin \theta}.$$

The result can then be expressed alternatively as follows. From the angular viewpoint, the visual appearance of the object for $\mathcal{O}$ can be deduced from the corresponding one for $\mathcal{O}'$ by a *"conformal" transformation on the unit sphere* , namely a transformation which dilates *small* subtended solid angles by the amplification factor $M(\theta, \chi)$. (In this description, the direction of motion defines a pole on the sphere, while θ represents the corresponding azimutal angle of the direction of observation of $\mathcal{O}$).
Practical Geometrical Construction:

In order to represent how the object with rapidity χ is seen by the observer $\mathcal{O}$ from the direction with angle θ, one determines the direction with angle $\theta' = \theta'(\theta, \chi)$ from which it is seen by $\mathcal{O}'$ as a fixed object. One then applies to the

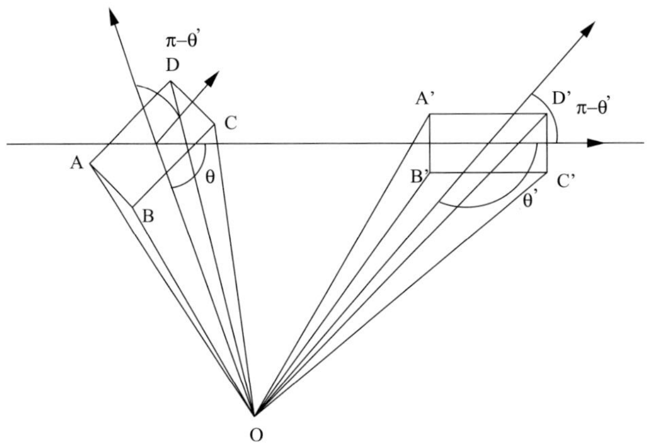

Figure 11: Passage of a “relativistic bus”: The relativistic aberration and the apparent rotation

fixed object (with its true dimensions) a rotation with angle $\alpha = \theta' - \theta$ (i.e. the relativistic aberration) before settling it at the point where $\mathcal{O}$ expects to see it from the direction θ. This is the *correct perspective* under which $\mathcal{O}$ will see the object from that direction. This procedure has been illustrated on fig. 11 by taking the example of a “relativistic bus”. It is now clear that if the object is spherical, its disk-shaped appearance and dimension are preserved for all possible directions of observation.

What about the “hidden Lorentz contraction” ?

Having obtained the previous general result, let us come back to our very first case of an infinitely thin rod with length l, oriented along Sx and moving along Sx with rapidity χ. Assume that one fixes $\theta = \frac{\pi}{2}$, which means that the observer $\mathcal{O}$ at rest looks at the rod from the direction Sy where he or she is sitting. By referring to the geometrical argument of Sec.2-5, one easily checks that in that case the observer *does* observe a Lorentz contracted rod with apparent length $\frac{l}{\cosh\chi}$. Now let us look at it from the viewpoint of the general result. This rod is seen by $\mathcal{O}'$ as a fixed rod from a direction defined by the angle θ' such that $\sin\theta' = \frac{1}{\cosh\chi}$ (see above the property b) of the function $\theta'(\theta,\chi)$). Then by applying the previous Practical Geometrical Construction, one sees that the observer $\mathcal{O}$ must see the rod as if it were rotated by the angle $\alpha = \theta' - \frac{\pi}{2}$, so that its perspectival length is

$$l \times \sin\theta' = \frac{l}{\cosh\chi},$$

the corresponding angle subtended by the object being equal to $\frac{l}{d}\frac{1}{\cosh\chi}$.

The rotation has exactly reproduced the Lorentz contraction !!

Observing the object without perspective effect

In the Galilean treatment (with infinite velocity of light), the object is observed by $\mathcal{O}$ without perspective effect when the direction of observation is perpendicular to the line of motion, namely when $\theta = \frac{\pi}{2}$. In the case of Minkowski's spacetime, the corresponding phenomenon is obtained when $\theta' = \frac{\pi}{2}$, namely when the observer $\mathcal{O}'$ sees the object without perspective effect. Then the identical effect is obtained by $\mathcal{O}$ provided his or her direction of observation includes an angle θ_0 with the motion's axis. According to property b) of the function $\theta'(\theta, \chi)$, this angle θ_0 is such that

$$\sin\theta_0 = \frac{1}{\cosh\chi}.$$

For the case of the infinitely thin rod, we see that it appears to the observer $\mathcal{O}$ *with its exact length* l when looked at in that direction, but from the angular viewpoint the subtended angle remains (because of the "magnification factor") $\frac{l}{d}\frac{1}{\cosh\chi}\ldots$ i.e. the same as for the Lorentz contracted appearance at $\theta = \frac{\pi}{2}$!

In conclusion, the effects of *perspective modified by the relativistic aberration, which acts as a rotation,* are clearly defined for describing the visual appearance of moving objects of general shape. The concept of "Lorentz contraction", although perfectly clear in two-dimensional spacetime, then becomes hidden as far as the observation of three-dimensional objects is concerned; it may be restored in the special case of thin objects, but the term is of subtle use and semantically confusing...

5 The Minkowskian energy-momentum space: $E = mc^2$ and particle physics

In the Newtonian dynamics, based on the Galilean conception of spacetime, one introduces for each massive pointlike object with mass m and constant velocity $\mathbf{v}$ its momentum $\mathbf{p} = m\mathbf{v}$. For any isolated dynamical system composed of such objects, their velocities and momenta depend on time, but the total momentum, namely the vector sum $\mathbf{P}$ of all the corresponding momenta, must be conserved at all times. The other quantities which have to be conserved at all times are a) the total energy E of the system, and b) the masses of the various objects, since the latter are supposed to conserve their individualities for all times.

In the framework of special relativity, each massive pointlike object with mass m in uniform motion is now characterized by its relativistic (or Minkowskian) velocity vector cu. According to Einstein's beautiful idea, one can now associate with it a *relativistic four-momentum vector* $p = mcu$, which can be represented in the coordinates of the rest-frame as follows:

$$p = (\mathbf{p}, p_0); \quad \mathbf{p} = (mc \sinh \chi)\, \mathbf{j} = m\mathbf{v}\left[1 - \frac{v^2}{c^2}\right]^{-\frac{1}{2}},$$

$$p_0 = mc \cosh \chi = mc\left[1 - \frac{v^2}{c^2}\right]^{-\frac{1}{2}}.$$

The (tip of the) vector p thus belongs to the upper sheet of hyperboloid H_m^+ with equation

$$p_0^2 - p_1^2 - p_2^2 - p_3^2 = m^2c^2, \quad p_0 > 0.$$

The space-component $\mathbf{p}$ of p admits a small-velocity expansion of the following form

$$\mathbf{p} = m\mathbf{v}\left(1 + \frac{v^2}{2c^2}\right) + \cdots,$$

which therefore reproduces the Newtonian momentum $m\mathbf{v}$ at the first-order approximation. As for the time-component p_0, its small-velocity expansion gives

$$p_0 = mc\left[1 - \frac{v^2}{c^2}\right]^{-\frac{1}{2}} = mc\left(1 + \frac{v^2}{2c^2}\right) + \cdots.$$

Multiplying both sides of the latter by c in order to get the dimensionality of an *energy*, i.e. $\mathrm{ML^2T^{-2}}$, one then obtains:

$$p_0 c = mc^2 + \frac{1}{2}mv^2 + \cdots.$$

While the second term of this expansion is clearly identified as the kinetic energy of the massive object in the Newtonian formalism, the first term $E_0 = mc^2$ *is the "internal energy at rest" of the massive object, identified (up to the dimensionality factor* c^2*) with its mass* m. In fact, when the velocity v vanishes, the four-vector pc is along the time-axis and reduces to its time-component $E_0 = mc^2$. One can then also say that for an arbitrary uniform motion with velocity $\mathbf{v}$, the time-component p_0c of the four-vector pc is the *complete relativistic energy of the moving object*, whose value is

$$E \doteq p_0 c = mc^2\left[1 - \frac{v^2}{c^2}\right]^{-\frac{1}{2}} = |\mathbf{p}|\frac{\mathbf{c^2}}{\mathbf{v}}.$$

This is why the four-momentum vector pc or p is also called the *energy-momentum vector* of the object (the identification being often made, in view of the convenient choice of units such that $c = 1$).

Remark. It is very important to note that in units where $c = 1$, the squared mass $m^2 = (mc)^2$ of the object is equal to the squared pseudonorm of the four-momentum vector p. *In special relativity theory, the concept of "mass" is therefore* *(like the proper time of a motion) a relativistic invariant: its value is independent of the Lorentz frame which has been chosen for describing the object.*

Massive and massless free particles

In microphysics, the theoretical treatment of particles requires a quantum-mechanical framework. However, this framework makes use basically of the Minkowskian space of four-momenta of point-like massive objects that we have just described. As a matter of fact, the quantum elementary particles *with mass* m are described as "wave-packets" which are probabilistic superpositions of *"classical" four-momentum configurations* $p = (\mathbf{p}, p_0)$ satisfying the so-called *"mass shell" condition*:

$$p \text{ belongs to } H_m^+, \text{ i.e. } p_0^2 - \mathbf{p}^2 = (mc)^2 \text{ with } p_0 > 0.$$

Photons are similarly treated as *massless particles* ($m = 0$). The latter are thereby characterized by a four-momentum vector p which belongs to the light-cone C^+:

$$p_0 = |\mathbf{p}|.$$

The concepts of massive pointlike object and of relativistic four-momentum thus keep some meaning for describing the *free particles* of microphysics, namely non-interacting particles. However, it becomes meaningless for describing particles in mutual interaction, in contrast with the case of Newtonian objects, whose momenta and energies keep their meaning as functions of the time during the interaction.

The simplest thing that can be done *a priori* for describing the mutual interactions in particle physics is to describe the relations between the four-momentum configurations of free particles *before* the interaction and those which occur *after* the interaction; in fact, for the interactions of nuclear type, the latter always take place in a very short time. Then there is a basic relativistic law, which generalizes the Newtonian laws of conservation of the total momentum and of the total energy of the system. This law is

The law of conservation of the total energy-momentum vector for systems of free particles

This law states that if a set of several (let us say n) free particles with initial four-momentum vectors $p^{(1)}, p^{(2)}, \ldots, p^{(n)}$ meet together in some region of Minkowski's spacetime where they interact, then another set of free particles will emerge in the future of that region and their number n' is not necessary equal to n. However, the four-momentum vectors $p'^{(1)}, p'^{(2)}, \ldots, p'^{(n')}$ of these final particles are such that the following vector equality holds in the Minkowskian four-momentum space:

$$p^{(1)} + p^{(2)} + \cdots + p^{(n)} = p'^{(1)} + p'^{(2)} + \cdots + p'^{(n')}.$$

Of course, this implies that in contrast with the case of Newtonian pointlike objects, the particles of microphysics do not conserve their individualities throughout the interaction. However the vector conservation law which they obey puts some strong constraints which are *consequences of the Minkowskian triangular inequality.*

Let us consider for example the case of a system of two initial particles with four-momenta $p^{(1)}$, $p^{(2)}$ (which is physically the generic case for the collisions produced in the accelerators). Let us call m_1 and m_2 the masses of these particles; one thus has:

$$p^{(1)^2} = m_1^2,\ p^{(2)^2} = m_2^2.$$

Then the total four-momentum is

$$P = p^{(1)} + p^{(2)},$$

whose squared pseudonorm $P^2 \doteq M^2$ is interpreted as the *squared total mass of the "composite system" of these two particles.* M is of course a relativistic invariant, independent of the Lorentz frame. In practice one often chooses a frame in which P is along the time-axis, which one calls the *center-of-mass frame.* Now , we see that because of the Minkowskian triangular inequality applied to the following triplet of vectors

$$[OQ_1] = p^{(1)},\ [Q_1Q_2] = p^{(2)},\ [OQ_2] = P,$$

one has necessarily

$$M \geq m_1 + m_2,$$

the equality being valid if and only if $p^{(1)}$ and $p^{(2)}$ are collinear; this means that the two particles are both at rest in the center-of-mass frame. If they are not, the difference $Mc^2 - m_1c^2 - m_2c^2$ represents *the (relativistic) kinetic energy of the system.*

Unstable and stable particles

Being given such a composite system, one may ask whether there may exist a corresponding "elementary system" (or "elementary particle") with the same mass M. One then sees that the existence of the latter would *a priori* be restricted by the kinematical possibility of its decomposition into two particles of masses m_1 and m_2 (according to the equation $P = p^{(1)} + p^{(2)}$). If this decomposition is not forbidden by some imperative rule (like for instance the conservation of the electric charge), one will say that such a particle of mass M exists *as an "unstable particle"* and can disintegrate in the pair of *stable* particles of masses m_1 and m_2. The fact that the latter are called "stable" corresponds to assuming the impossibility for each of them of similar decompositions into pairs of particles of smaller masses obeying the corresponding Minkowskian inequalities.

Elastic and inelastic collisions

Let us consider for example the case of equal masses $m_1 = m_2 \doteq m$. Then one has $M \geq 2m$. Now, let us ask ourselves what can be the constraints on the number of final stable particles emerging from the interaction between two initial stable particles of mass m. By iterating the previous geometrical argument with Minkowskian triangles, one gets the following result.

For $M < 3m$,, two and only two stable final particles can be produced; one will then speak of an *"elastic" collision of two particles.* For $3m \leq M < 4m$, either two or three can be produced; both processes are geometrically possible. More generally, if $(n-1)m \leq M < nm$, all processes including the production of any number of final particles *smaller than or equal to* $n-1$ are possible. For the production of three or more particles, one also speaks of *"inelastic" collision of two particles.*

One can of course generalize the previous geometrical argument to the case of particles of different masses: note that the values of the masses of the existing particles of microphysics is a discrete set whose determination requires the treatment of quantum relativistic dynamical theories such as *Quantum Field Theories* (a very hard program which is by far outside the scope of this paper).

Inclusion of the photons

It is important to note that massless particles such as photons can be included in the previous geometrical arguments. In particular one can check (by drawing the corresponding triangles) that

i) From the collision of two photons, one can obtain a total momentum whose mass M can be arbitrarily large, so that any number of final massive particles can a priori be produced throughout the interaction of these two photons: *"pure light can create matter"*

ii) Together with the elastic collision of two massive (electrically charged) particles, one can always expect *a priori* the additional production of any number of photons, (called *"soft photons"*) even if the total mass M ($> 2m$) of the system is not very much larger than $2m$.

An exercise on four-momentum conservation: "the propulsion of the Langevin-traveler's spaceship" (see Sec.3-3)

Let us assume that at time τ (in its proper time), the spaceship's mass is $M(\tau)$ and that, in its restframe, it is submitted to a strength-vector $\mathbf{F}(\tau)$ which produces a uniform acceleration equal to g in the fixed direction $\mathbf{j}$. Since the spaceship's velocity is equal to zero in that frame, Newton's fundamental principle of dynamics applies and gives:

$$\mathbf{F}(\tau) = M(\tau)g\,\mathbf{j} = \frac{d\mathbf{P}}{d\tau}.$$

The strength-vector $\mathbf{F}(\tau)$ is produced by the expulsion of a part of the mass by unit of time, namely $\frac{dM(\tau)}{d\tau}$, whose associated variation of momentum must be equal in intensity and opposite to $\mathbf{F} = \frac{d\mathbf{P}}{d\tau}$ (in view of the relativistic law of conservation of momentum).

From a relativistic viewpoint, the rate of loss of mass must in fact be identified (up to the factor c^2) with a rate of loss of energy

$$\frac{dM(\tau)}{d\tau} = \frac{1}{c^2}\frac{dE}{d\tau},$$

which may be produced under either form of an emission of matter (with relativistic velocity $v < c$) or of an emission of light (i.e. photons with velocity c).

In the case of matter, the energy and momentum losses are related to the velocity v of the emitted matter by the relativistic formula (given at the beginning of this subsection):

$$\left|\frac{d\mathbf{P}}{d\tau}\right| = \frac{v}{c^2}\left|\frac{dE}{d\tau}\right|,$$

which therefore yields the differential equation

$$\frac{dM}{d\tau} = -\frac{g}{v}M(\tau).$$

In the case of photons, one has a similar relation (with $v = c$):

$$\left|\frac{d\mathbf{P}}{d\tau}\right| = \frac{1}{c}\left|\frac{dE}{d\tau}\right|,$$

which yields

$$\frac{dM}{d\tau} = -\frac{g}{c}M(\tau).$$

One concludes that the loss of mass is minimized when $v = c$, namely if one can dispose of an engine which transforms matter into pure radiation.

6 Toward simple geometries of curved spacetimes

In spite of its non positive-definite distance, Minkowski's spacetime still shares with Euclidean space the property of being "flat", namely an affine space. But in the same way as the Euclidean plane must be replaced by a sphere (as a first approximation) for the observer who wishes to represent the surface of the earth, the four-dimensional Minkowskian spacetime must be replaced by a *curved* spacetime for the observer of the universe who wishes to describe the inclusion of matter submitted to gravitational attraction and the evolutional properties of the universe at astronomical scales of lengths and times. What we are mentioning here concerns the second big revolution of theoretical physics in the twentieth century: according to the principles of general relativity introduced by Einstein in 1916 (and also independently by Hilbert in a more mathematical formulation), local curvature of spacetime around an event X is caused by the presence of a density of matter at that point. But there is also another type of global curvature which is linked to expansion or contraction properties of spatial sections of spacetime; this type of curvature is characterized by what is called a "cosmological constant".

The general mathematical theory of curved spacetimes is outside the scope of the present pedagogical essay and we shall only indicate here some hint about the primary concepts involved in that theory. In mathematics, the notion of *manifold* introduces an additional abstraction to geometry. In the same way as the two-dimensional surface of the earth is perceived by us as "embedded in the ambient three-dimensional spacetime", a model of curved spacetime can reasonably be conceived as a "surface of dimension four embedded in a flat space of larger dimension" (for example five). As a matter of fact, this type of geometrical representation in terms of an "ambient space of higher dimension" is not necessary for defining the relevant mathematical notion of "manifold", which has been inspired by the geographical notion of *"atlas"*. In a world atlas, one is given a set of *planar* representations of various regions of the surface of the earth, in such a way that: a) each region is represented by precise geometrical rules encoded in a lattice of level curves representing parallels and meridians which constitute a map of that region; b) whenever two regions overlap, there are consistent geometrical rules which exhibit the correspondence between the two corresponding maps in their representations of the overlapping region; c) the union of all maps cover the whole surface of the earth. Such a type of collection of local data, which provides a faithful representation of a curved surface without requiring an embedding in a higher-dimensional ambient space, is used in the general mathematical definition of "abstract manifolds". The concept of atlas is thus often used for representing various models of curved spacetimes, thereby defined as *"abstract Minkowskian (or Lorentzian) manifolds"*. In such an atlas, each map is then specified by what one calls *a system of local coordinates of space and time*. The *Minkowskian local structure* is specified in each given map, by prescribing in terms of the corresponding local coordinates what are the cones of light world-lines passing at each given event X: these light world-lines will in general appear as curved lines, constituting a *"light-conoid" C_X with apex X, composed of the union of a future conoid C_X^+ and of a past conoid C_X^-*.

From the physical viewpoint, one can say that the conceptual advantage of this "atlas-representation" of a curved space or spacetime is to make the economy of an "ambient space", which has a priori no physical interpretation. As a matter of fact, the problem of the physical interpretation of additional dimensions introduced for mathematical reasons currently appears in various up-to-date investigations of theoretical physics.

However for certain models, a representation making use of an embedding of spacetime in a five-dimensional flat ambient space can be very illustrative and useful. Here of course, the word "ambient space" is of pure mathematical nature. These models correspond to "quadratic spacetimes" represented by appropriate *quadrics* (i.e. second-degree hypersurfaces) which enjoy the following simple geometrical property with respect to the ambient space. At each event X of the spacetime, the light-conoid C_X is *the cone of all linear generatrices of the quadrics* passing at X.

These models of quadratic spacetimes have in common to be "pure-cosmological-constant models", which means that no density of matter is incorporated there. They enter in two classes with rather different mathematical properties and physical interest, which are called "de Sitter" and "anti-de Sitter spacetimes": they are presented in Ugo Moschella's paper.

References

[1] A. Einstein, *On the electrodynamics of moving bodies*, Annalen der Physik **17**, 891 (1905).

[2] A. Einstein, *Does the inertia of a body depend on its energy content?*, Annalen der Physik **18**, 639-641 (1905).

[3] H. Minkowski, *Space and Time*, Cologne Conference, September 1908.

[4] H. Poincaré, *Sur la dynamique de l'électron*, Note a l'Académie des Sciences, Paris, et Rendiconti del Circolo Matematico di Palermo (1905).

[5] C.W. Sherwin, *Some Experimental Tests of the "Clock Paradox"*, Phys. Rev. **120**, 17–21 (1960).

[6] J. Terrell, *Invisibility of the Lorentz Contraction*, Phys. Rev. **116**, 1041–1045 (1959).

[7] V.F. Weisskopf, *The visual appearance of rapidly moving objects*, Physics Today, September 1960, p.24-27.

[8] D. Weiskopf, *An Immersive Virtual Environment for Special Relativity*, WSCG Conference Proceedings, V. Skala (ed.) University of West Bohemia, Pilsen, February 2000, 337–344.

[9] E.C. Zeeman, *Causality Implies the Lorentz Group*, Journal Math. Phys. **5**, 490–493 (1964).

The de Sitter and anti-de Sitter Sightseeing Tour

Ugo Moschella

Introduction

While celebrating the 100th anniversary of the discovery of special relativity [1], it may not be inappropriate to open a window on the de Sitter universes, as their importance in contemporary physics is gradually increasing. Just to mention two examples, the astronomical evidence for an accelerated expansion of the universe gives a central place to the de Sitter geometry in cosmology [2] while the so-called AdS/CFT correspondence [3] supports a major role for the anti-de Sitter geometry in theoretical physics.

From the geometrical viewpoint, among the cousins of Minkowski spacetime (the class of Lorentzian manifolds) de Sitter and anti-de Sitter spacetimes are its closest relatives. Indeed, like the Minkowski spacetime, they are maximally symmetric, i.e. they admit kinematical symmetry groups having ten generators[1]. Maximal symmetry also implies that the curvature is constant (zero in the Minkowski case).

Owing to their symmetry, it is possible to give a description of the de Sitter universes without using the machinery of general relativity at all. However, it is worth saying right away that, even if they share important features with Minkowski spacetime, their physical interpretation is quite different and the technical problems to be solved in order to merge de Sitter spacetimes with quantum physics are considerably harder.

The aim of this note is to give a simple and short geometrical introduction to the de Sitter and anti-de Sitter universes and to briefly comment on their physical meaning.

1 An analogy: non-Euclidean spaces of constant curvature

One easy way to replace the usual flat geometry of the Euclidean physical space $\mathbb{R}^3$ with some curved geometry consists in moving to a fictitious four-dimensional flat world and considering there the geometry of convenient three-dimensional hypersurfaces. The simplest curved model of space is the surface of a hypersphere embedded in a four-dimensional Euclidean flat space $\mathbb{R}^4$:

$$\mathbb{S}^3 = \{x \in \mathbb{R}^4, \;\; x_1^2 + x_2^2 + x_3^2 + x_4^2 = a^2\}. \tag{1}$$

$\mathbb{S}^3$ is homogeneous, isotropic and has positive curvature with value $6/a^2$. The six-dimensional invariance group of $\mathbb{S}^3$ is simply the rotation group SO(4) of the

[1] A four dimensional Riemannian manifold has an isometry group with at most ten generators. In the Minkowski case the isometry group is the Poincaré group and the ten independent transformations have a familiar physical interpretation: one time translation, three spatial translations, three rotations and three boosts.

four-dimensional ambient space; it can be interpreted as the group of motions of the spherical space in the same way as the Euclidean group E(3) (translations and rotations) is the group of motions of $\mathbb{R}^3$. The main difference is that there are no commutative "translations" on $\mathbb{S}^3$.

All the non-Euclidean geometrical properties of the hypersphere come by restriction to it of the Euclidean geometry of the fictitious ambient space. In particular all geodesics, that are the analog in the curved geometry of what are straight lines in the flat case, can be obtained by intersecting the hypersphere with two-planes passing through the geometrical center of the sphere (see Figure 1). One recognizes immediately that in this geometry "straight lines" are maximal circles. The second possibility is a bit more complicated and produces a space with neg-

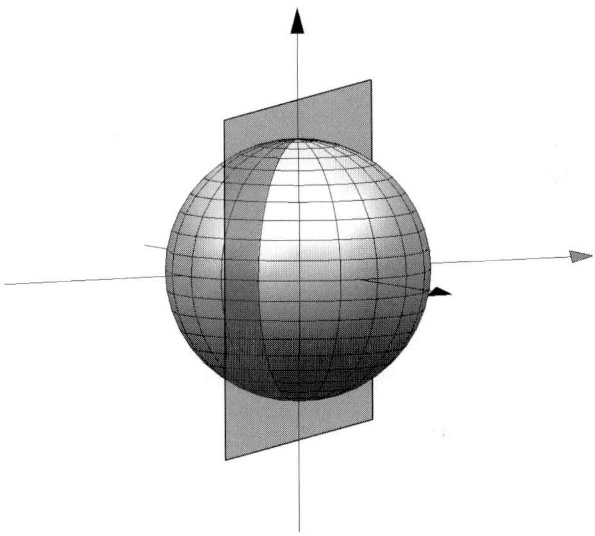

Figure 1: *A spherical model of space (positive curvature). Geodesics are maximal circles and are obtained by intersecting the sphere with two-planes passing through the center of the sphere in the ambient space.*

ative curvature. One moves again to a fictitious four-dimensional world, but now this is a four-dimensional Minkowski spacetime $\mathbb{M}^4$ (loosely speaking, a timelike direction has been added to the Euclidean $\mathbb{R}^3$, while in the previous case a spatial direction was added). Here, a model of space with negative constant curvature is the upper sheet of the two-sheeted hyperboloid $\mathbb{H}^3$:

$$\mathbb{H}^3 = \{x \in \mathbb{M}^4, \quad x_0^2 - x_1^2 - x_2^2 - x_3^2 = a^2\}. \tag{2}$$

As shown in Figure 2 the lightcone emerging from any point of $\mathbb{H}^3$ does not meet the surface anywhere else. This means that, in the ambient spacetime, the surface is *spacelike* and, as such, it is a good model for a space. As before the geometry

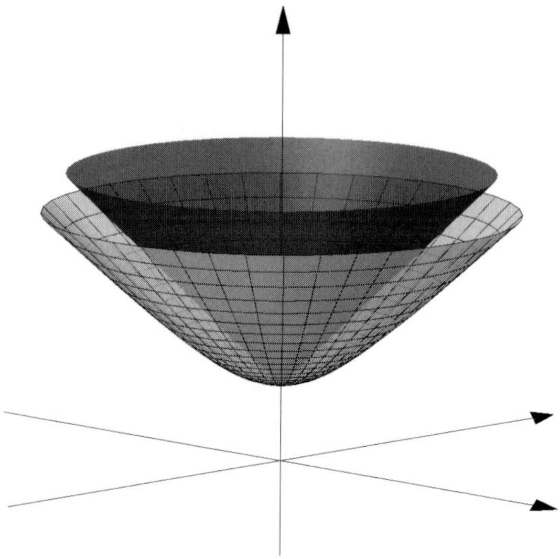

Figure 2: *A hyperbolic model of space (negative curvature).* $\mathbb{H}^3$ *(the outer surface) is spacelike in the ambient Minkowski spacetime. Geodesics are branches of hyperbolae.*

of $\mathbb{H}^3$ is constructed by restriction of the Lorentzian geometry of the ambient Minkowski spacetime $\mathbb{M}^4$. In particular, the six-dimensional isometry group of $\mathbb{H}^3$ is the Lorentz group $\mathrm{SO}(1,3)$ of the ambient spacetime. Geodesics are branches of hyperbolae, obtained as before by intersecting $\mathbb{H}^3$ with two-planes containing the center.

2 The de Sitter universe

Let us now introduce a five-dimensional Minkowski spacetime $\mathbb{M}^5$ by adding a spacelike direction to $\mathbb{M}^4$ (just as we did in the spherical case). In $\mathbb{M}^5$ we consider the hypersurface with equation

$$dS_4 = \{x \in \mathbb{M}^5,\ x_0^2 - x_1^2 - x_2^2 - x_3^2 - x_4^2 = -R^2\}. \tag{3}$$

This is the de Sitter spacetime [5] (see Figures 3 and 4). It has constant negative curvature $-12/R^2$ (the sign depends on conventions) and reproduces (after a renormalization) Minkowski spacetime in the limit of zero curvature (i.e. when the radius R tends to infinity).

The causal structure of dS_4 is induced by restriction of the Lorentzian geometry of the ambient Minkowski spacetime $\mathbb{M}^5$ exactly as the geometry of the sphere was determined by the Euclidean geometry of the ambient $\mathbb{R}^4$. In particular, the de Sitter line element is obtained concretely by restricting the five-dimensional

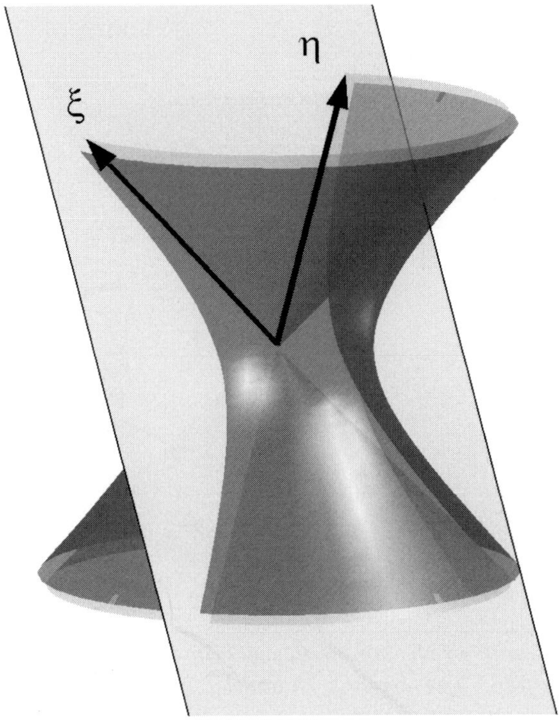

Figure 3: *The light surface represents the de Sitter universe. The cone is the light-cone of the five-dimensional ambient spacetime, asymptotic to the de Sitter hyperboloid. Timelike geodesics are hyperbolae and are obtained by intersecting the hyperboloid with two-planes passing through the origin of the the ambient spacetime. Any two-plane associated with a timelike geodesic can be identified by specifying two null vectors* ξ *and* η *that can be used also to parametrize the geodesic itself. In flat spacetime geodesics are labeled by their four-momentum. By analogy, the light-cone* C *can be interpreted as de Sitter momentum space. In particular, de Sitter plane waves are constructed using vectors belonging to the lightcone* C *[7].*

invariant interval to the manifold dS_4:

$$ds^2 = \left[(dx_0)^2 - (dx_1)^2 - (dx_2)^2 - (dx_3)^2 - (dx_4)^2\right]\big|_{dS_4} \tag{4}$$

This line element is the most symmetrical solution of the field equations written down by Einstein in 1917, where he introduced the famous cosmological constant Λ [4]. The radius R corresponding to a given value of Λ is

$$R = \sqrt{\frac{3}{\Lambda}}.$$

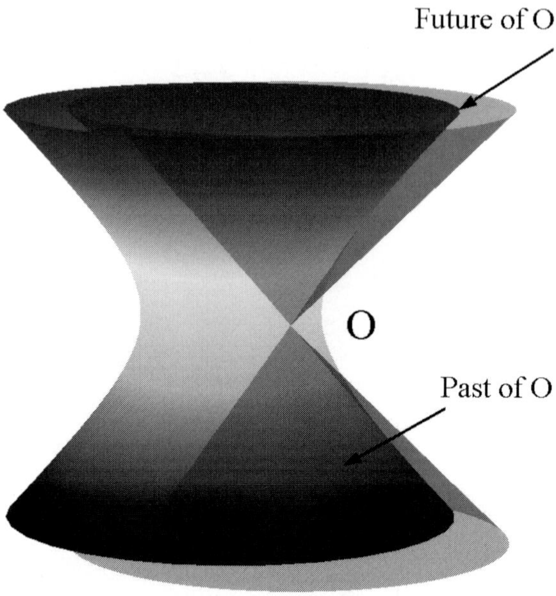

Figure 4: *The lightcone of the ambient spacetime induces the causal ordering of the de Sitter manifold. The regions shadowed by the five-dimensional lightcone emerging from the event O are the past and the future of O. In this figure the choice of grays shows a contraction era (blueshift) followed by an expansion era (redshift)*

A pivotal role is played by the five-dimensional lightcone of the ambient spacetime:

$$C = \{\xi \in \mathbb{M}^5,\ \xi_0^2 - \xi_1^2 - \xi_2^2 - \xi_3^2 - \xi_4^2 = 0\}. \tag{5}$$

The cone C induces the causal ordering of the events on the de Sitter manifold; it also plays the role of de Sitter momentum space (see Figure 4). The de Sitter spacetime has a boundary at timelike infinity (while timelike infinity of the Minkowski manifold is a point). The cone C also provides a description of this boundary, which may be used instead of a Penrose diagram.

The de Sitter kinematical group coincides with the Lorentz group of the ambient spacetime $\mathrm{SO}(1,4)$. As for the sphere, there are no commutative translations on the de Sitter manifold. This fact is source of considerable technical difficulties in the study of de Sitter Quantum Field Theory.

The relationship between the de Sitter universe and the geometry of the sphere is deeper than a mere analogy. Indeed, for imaginary times

$$x_0 \to ix_0$$

the (Euclidean) de Sitter manifold (see Figure 5) is a sphere and the Euclidean

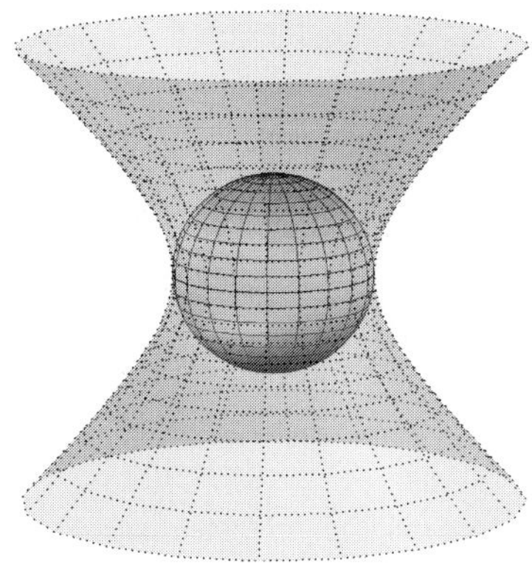

Figure 5: *A pictorial representation of the Euclidean de Sitter manifold.*

de Sitter group is the rotation group SO(5). A study of the complex de Sitter manifold with applications to Quantum Field Theory has been described in [7].

The de Sitter geometry finds its most important physical applications in cosmology. In cosmology one usually "breaks" the general relativistic covariance and singles out a special coordinate system: there is a natural choice of "cosmic time" that makes the universe appear spatially homogeneous and isotropic at large scales. This property is mathematically encoded in the Friedmann-Robertson-Walker line element:

$$ds^2 = dt^2 - a(t)^2 dl^2. \tag{6}$$

The spatial distance dl^2 describes the geometry of a homogeneous and isotropic space manifold: either $\mathbb{S}^3$, $\mathbb{R}^3$ or $\mathbb{H}^3$.

In this respect the de Sitter geometry is rather special: due to the maximal symmetry and the topology of the de Sitter manifold, all three possible FRW cosmologies can be realized on de Sitter by suitable choices of the cosmic time coordinate (see Figure 6).

The simplest choice of time is the coordinate x_0 (see Figure 7).

$$\begin{cases} x_0 &= R\sinh\left(\frac{t}{R}\right) \\ x_i &= R\cosh\left(\frac{t}{R}\right)\,\omega_i \qquad i = 1,2,3,4 \end{cases} \tag{7}$$

with $\omega_1^2+\omega_2^2+\omega_3^2+\omega_4^2 = 1$, so that Equation (3) is easily satisfied. The hypersurfaces of constant time thus are spheres $\mathbb{S}^3$ and the coordinate system covers the whole universe. With this choice the de Sitter line element describes a closed FRW model:

$$ds^2 = (dx_0^2 - dx_1^2 - \ldots dx_4^2)\big|_{dS_4} = dt^2 - R^2\cosh^2\left(\frac{t}{R}\right)d\omega^2. \tag{8}$$

Another possible choice of time is $x_0 + x_4$ (see Figure 8). The time parameter is introduced by the relation $x_0 + x_4 = Re^{\frac{t}{R}}$; with this coordinate only one half of

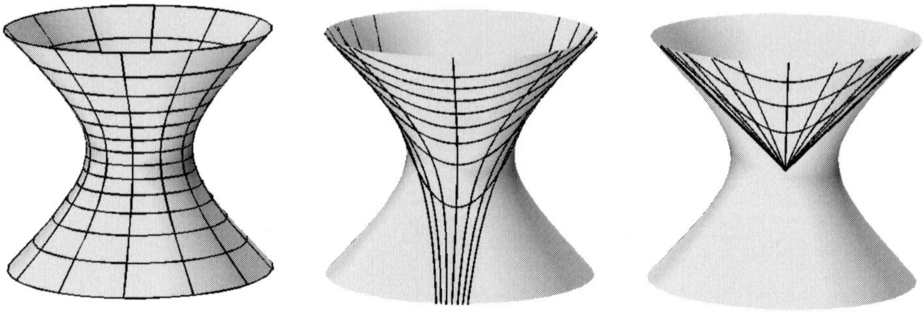

Figure 6: *Various choices of cosmological coordinates. Black curves represent hypersurfaces of constant cosmic time. Blue curves are timelike geodesics. The red manifold represents a closed FRW model with a contraction epoch followed by an expansion epoch. The light blue manifold is an exponentially expanding flat model. The yellow represents a hyperbolic open model.*

the manifold is covered. Hypersurfaces of constant cosmic time are copies of $\mathbb{R}^3$. In these coordinates the de Sitter line element appears as a flat FRW model with exponentially growing scale factor:

$$ds^2 = dt^2 - \exp\left(\frac{2t}{R}\right) d\mathbf{x}^2. \tag{9}$$

This form of the de Sitter line element was introduced by Lemaître in 1925 [6]. It is interesting to note that the first coordinate system used by de Sitter himself was a static coordinate system with closed spatial sections. De Sitter was following Einstein's cosmological idea of a static closed universe, the idea that led to the introduction of the cosmological term in Einstein's equations. A static coordinate system (i.e. a coordinate system where nothing depends explicitly on time) is not the most natural to describe an expanding universe, but it has other interesting properties, mainly in relation to black hole physics (horizons, temperature and entropy).

Static closed coordinates are represented in Figure 9. The Lemaître form of the de Sitter line element is the most useful in cosmological applications. Recent observations point towards the existence of a nonzero cosmological constant and a flat space. For an empty universe (i.e. a universe filled with a pure cosmological constant) this would correspond precisely to the above description of the de Sitter universe.

3 Anti-de Sitter

Let us now introduce a flat five-dimensional space $\mathbb{E}^{(2,3)}$ by adding a timelike direction to $\mathbb{M}^4$ (as we did in the hyperbolic case). $\mathbb{E}^{(2,3)}$ has two timelike directions

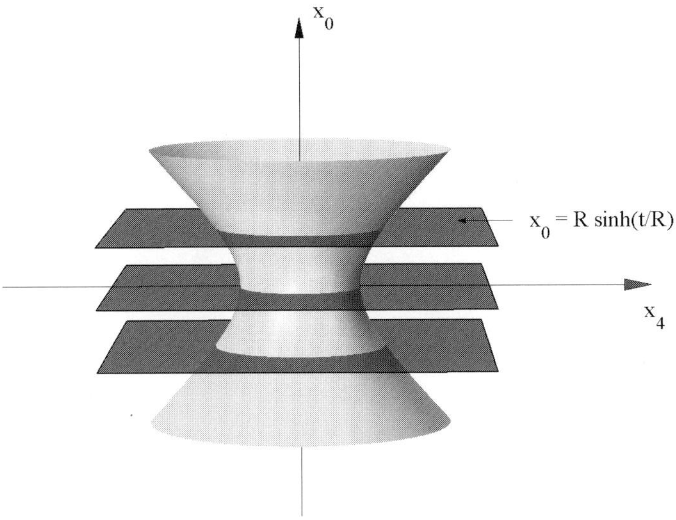

Figure 7: *Construction of the coordinate system representing the de Sitter geometry as closed FRW model. Hyperurfaces of equal cosmic time are intersection of the de Sitter manifold with hyperplanes* x_0 *=const.*

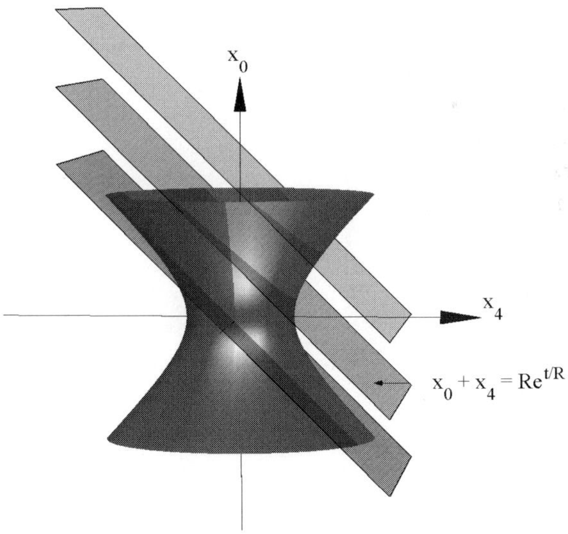

Figure 8: *Construction of the coordinate system representing the de Sitter geometry as a flat FRW model. Hyperurfaces of equal cosmic time are intersection of the de Sitter manifold with hyperplanes* $x_0 + x_4 =$ *const. Only one half of the manifold is covered since it has to be* $x_0 + x_4 > 0$.

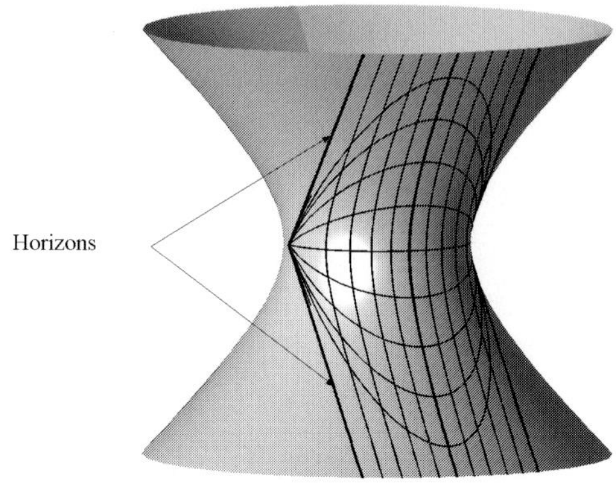

Figure 9: *A chart representing static closed coordinates. This is the coordinate system originally used by W. de Sitter in 1917. Vertical timelike curves are obtained by intersecting the hyperboloid with parallel two-planes. Only the central hyperbola is a geodesic because it is the only one lying on a plane that contains the origin of the ambient spacetime. The other timelike curves are accelerated trajectories. There is a redshift for light sources moving along these world-lines; this effect was called the de Sitter effect and was thought to have some bearing on the redshift results obtained by Slipher.*

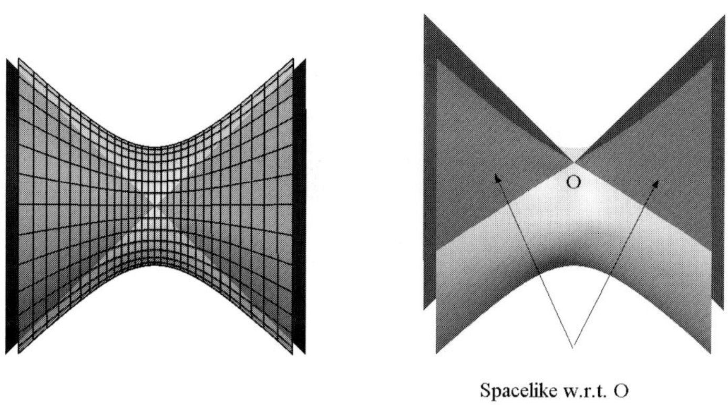

Figure 10: *A visualization of the anti-de Sitter universe. The asymptotic cone plays a crucial role exactly as in the de Sitter case. The regions of* AdS_4 *that are in the shadow of the five-dimensional cone emerging from an event O are the regions that are not causally connected to the event O. The asymptotic cone in the ambient space can be regarded as a representation of the boundary at spacelike infinity of the AdS manifold and carries a natural action of the conformal group that is the group-theoretical foundation for the AdS-CFT correspondence.*

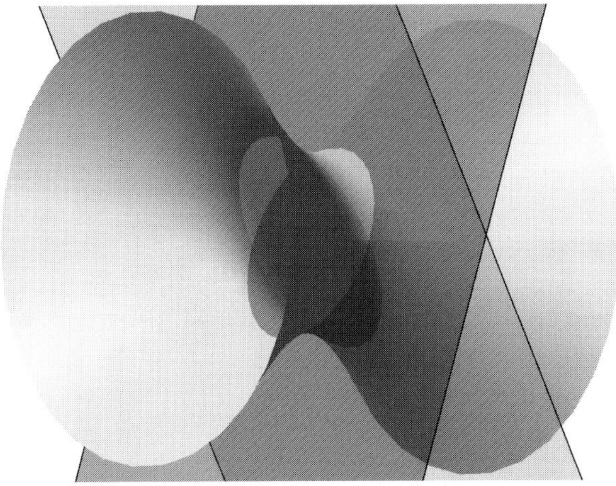

Figure 11: *Anti-de Sitter timelike geodesics are ellipses and are obtained by intersecting the hyperboloid with two-planes passing through the center of the the ambient space. The geodesics passing through a certain event all meet at the antipodal point. The focusing of geodesics remains true also in the covering space.*

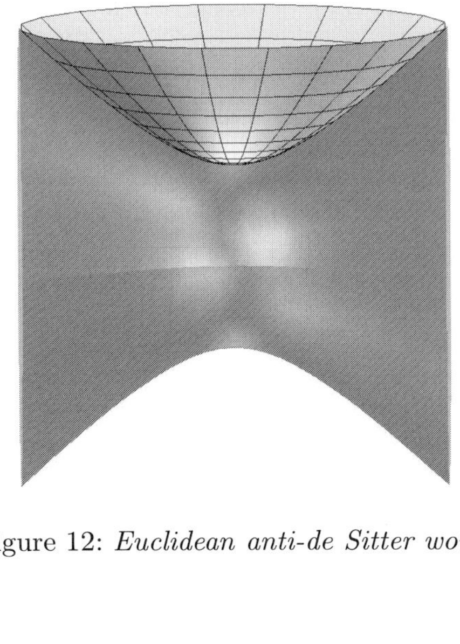

Figure 12: *Euclidean anti-de Sitter world.*

and three spacelike directions and therefore it is not a spacetime in the ordinary sense (a Lorentzian manifold with one temporal and three spatial dimensions). However, the hypersurface with equation

$$AdS_4 = \{x \in \mathbb{E}^{(2,3)},\ x_0^2 - x_1^2 - x_2^2 - x_3^2 + x_4^2 = R^2\}, \tag{10}$$

is a spacetime: this is the anti-de Sitter universe (see Figure 10). It has constant positive curvature and reproduces (after a renormalization) the Minkowski spacetime in the limit when the curvature tends to zero.

The causal structure of AdS_4 is induced by restriction of the geometry of the ambient space $\mathbb{E}^{(2,3)}$ (the analogy is now with the geometry of $\mathbb{H}^3$ that is determined by the causal structure of the ambient spacetime $\mathbb{M}^4$). As before the null cone of the ambient space

$$C = \{\xi \in \mathbb{M}^5,\ \xi_0^2 - \xi_1^2 - \xi_2^2 - \xi_3^2 + \xi_4^2 = 0\} \tag{11}$$

induces the causal ordering on the anti-de Sitter manifold (see Figure 10).

Owing to the existence of closed timelike curves (see Figure 11) the causal ordering is only local. One may construct a globally causal manifold by considering the covering of the anti-de Sitter manifold (recall that the covering of a circle is a line). However even the covering of the anti-de Sitter remembers the "periodicity in time" of the original manifold: geodesics issued from an event meet again infinitely many times in the covering.

The anti-de Sitter line element is constructed by restricting the five-dimensional invariant "interval" of the ambient space to the manifold AdS_4:

$$ds^2 = \left[(dx_0)^2 - (dx_1)^2 - (dx_2)^2 - (dx_3)^2 + (dx_4)^2\right]\big|_{AdS_4} \tag{12}$$

This line element is the maximally symmetrical solution of the cosmological Einstein equations when the cosmological constant Λ is negative. The anti-de Sitter kinematical group coincides with the isometry group $\mathrm{SO}(2,3)$ of the ambient space.

The relationship between the anti-de Sitter universe and the geometry of $\mathbb{H}^3$ is deeper than a mere analogy. Indeed, for imaginary time

$$x_4 \to i x_4$$

the (Euclidean) anti-de Sitter manifold (see Figure 12) is a copy of $\mathbb{H}^4$ and the Euclidean de Sitter group is $\mathrm{SO}(1,4)$. A study of the complex anti-de Sitter manifold with applications to Quantum Field Theory has been described in [8].

AdS is not a globally hyperbolic spacetime. In non-globally hyperbolic manifolds knowledge of equations of motion and of initial data is not enough to determine the time evolution of physical quantities. In the anti-de Sitter case, the lack of global hyperbolicity is due to the existence of a boundary at spacelike infinity: information can flow in from infinity. This fact is source of difficulties in quantizing fields on the anti-de Sitter manifolds. However this is also an opportunity since

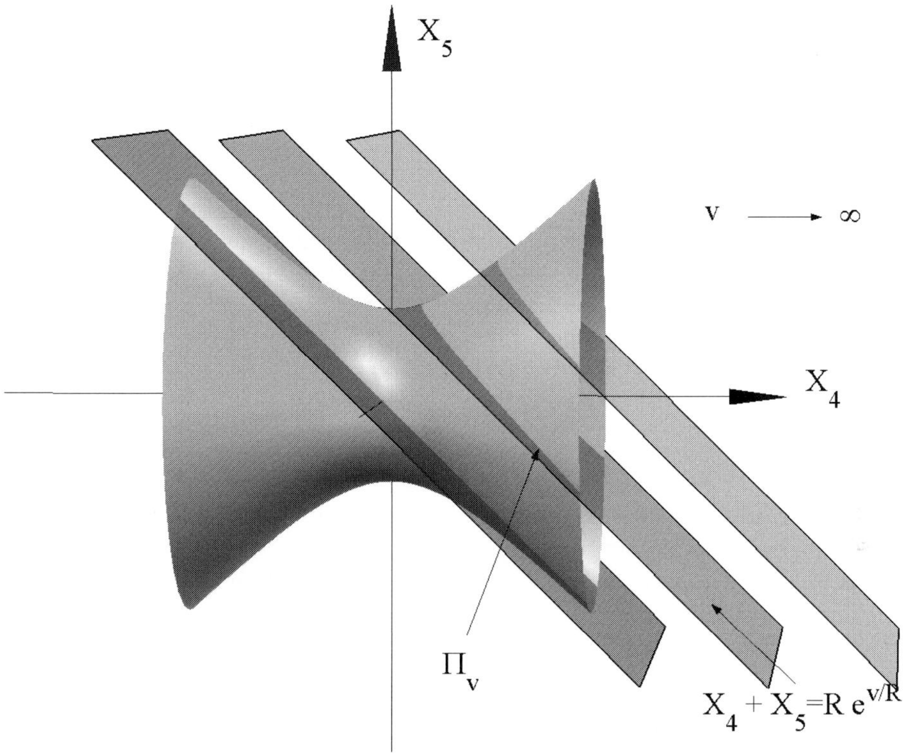

Figure 13: *Construction of the AdS-Poincaré coordinates. The limit $v \to \infty$ describes the boundary of the AdS manifold.*

this boundary at infinity offers the very possibility for formulating the famous AdS/CFT correspondence [3].

To present an intuitive idea of this topic let us introduce coordinates on a five-dimensional anti-de Sitter manifold AdS_5 (embedded in a six-dimensional space $\mathbb{E}^{(2,4)}$) obtained by intersecting AdS_5 with hyperplanes $\{X_4 + X_5 = Re^{v/R}\}$ (see Figure 13). Each slice Π_v of AdS_5 is a copy of Minkowski spacetime $\mathbb{M}^4$. Points in each slice Π_v can be thus parametrized by Minkowskian coordinates x_0, x_1, x_2, x_3 (scaled by $e^{v/R}$). This explains why the anti-de Sitter coordinates (v, x_0, x_1, x_2, x_3) are also called Poincarè coordinates.

The coordinate system covers only one-half of the anti-de Sitter manifold; the anti-de Sitter metric takes the following form:

$$\begin{aligned} ds^2 &= \left[(dX_0)^2 - (dX_1)^2 - (dX_2)^2 - (dX_3)^2 - (dX_4)^2 + (dX_5)^2\right]\big|_{AdS_5} \\ &= e^{\frac{2v}{R}}(dx_0^2 - dx_1^2 - dx_2^2 - dx_3^2) - \mathrm{d}v^2. \end{aligned} \tag{13}$$

The use of this parametrization is crucial in a recent approach to the mass hierarchy problem [9] and to multidimensional cosmology. In this context the slices Π_v are called branes. The Minkowskian geometry of the brane is induced by the ambient anti-de Sitter metric: for instance space-like separation in any slice Π_v can be understood equivalently in the Minkowskian sense of the slice itself or in the sense of the ambient anti-de Sitter universe.

When we consider the limit $v \to \infty$ we arrive at the anti-de Sitter boundary at spacelike infinity, which therefore may (essentially) be thought as a four-dimensional Minkowski spacetime. The AdS-CFT correspondence establishes an equivalence between a theory on the five-dimensional AdS_5 and a relativistic theory on the boundary $\mathbb{M}^4$ (this is an instance of another popular idea in contemporary theoretical physics: the *holographic principle*). The theory on the boundary is conjectured to have a larger symmetry group, namely the conformal group [3, 10, 11].

4 Epilogue

The de Sitter and anti-de Sitter tour now comes to its end. Before concluding let us summarize the highlights to be retained.

De Sitter's geometry is the vacuum solution of Einstein's equations with a cosmological term and plays in contemporary physical cosmology a very important double role. First, the unifying aspect of the different inflation models consists in the fact that the primordial universe has undergone a phase of exponential expansion, approximately described by de Sitter's geometry. A possible theoretical understanding of the structure of the universe which is observable today is based on de Sitter geometry at the inflation epoch.

The second motivation of interest of de Sitter geometry lies in the observational data of the recent years, starting from the observations of distant type Ia supernovae and up to the data on the temperature fluctuations of the cosmic background radiation.These observations have upturned consolidated ideas, indicating that the gravitational effect of the greatest part of the energy content of the universe is similar to Einstein's cosmological constant. This form of energy is called "dark energy".

Thus de Sitter geometry seems to assume the role of reference geometry of the universe. In other words, it seems that it is de Sitter, and not Minkowski, the geometry of spacetime deprived of its content of matter and radiation (if one describes dark energy with a cosmological constant).

Once one admits the possible existence of a cosmological constant, it is also interesting to explore the consequences of a model in which the latter is negative.

In this case spacetime geometry is termed Anti-de Sitter. This geometry has strange properties which are in confilict with common sense, as the existence of closed timelike curves and of a boundary at spacelike infinity. Nonetheless, anti-de Sitter plays a central role in contemporary high energy physics with the formulation of the conjecture on the correspondence AdS/CFT (Anti-de Sitter/Conformal Field Theory).

In conclusion, there are still lots of rooms left in de Sitter worlds!

References

[1] A. Einstein, Zur Elektrodynamik bewegter Körper, *Annalen der Physik*, **17**, 891 (1905).

[2] N. Straumann, *On the cosmological constant problems and the astronomical evidence for a homogeneous energy density with negative pressure*, arXiv:astro-ph/0203330. In: B. Duplantier and V. Rivasseau (Eds): "Vacuum energy, renormalization", 7–51. Basel: Birkhäuser (2003).

[3] J.M. Maldacena, Black Holes and Holography in String Theory. In: B. Duplantier and V. Rivasseau (Eds): *Séminaire Poincaré*, 61–67. (2004). The Large N Limit of Superconformal Field Theories And Supergravity, *Adv. Theor. Math. Phys.* **2**, 231 (1998), hep-th/9711200.

[4] A. Einstein, Kosmologische Betrachtungen zur allgemeinen Relativittstheorie *Sitzungsber. Preuß. Akad. Wiss. Berlin*, 142 (1917).

[5] W. de Sitter, On the Curvature of Space, *Proc. Kon. Ned. Akad. Wet.* **20**, 229 (1917).

[6] C. Møller, *The theory of Relativity*, Oxford: Oxford University Press (1951).

[7] J. Bros and U. Moschella, Two-point Functions and Quantum Fields in de Sitter Universe, *Rev. Math. Phys.* **8**, 327 (1996). J. Bros, H. Epstein and U. Moschella, Analyticity properties and thermal effects for general quantum field theory on de Sitter space-time, *Commun. Math. Phys.* **196**, 535 (1998).

[8] J. Bros, H. Epstein and U. Moschella, Towards a general theory of quantized fields on the anti-de Sitter space-time, *Commun. Math. Phys.* **231**, 481 (2002).

[9] L. Randall and R. Sundrum, A large mass hierarchy from a small extra dimension, *Phys. Rev. Lett.* **83**, 3370 (1999), [hep-ph/9905221].

[10] M. Bertola, J. Bros, U. Moschella and R. Schaeffer, A general construction of conformal field theories from scalar anti-de Sitter quantum field theories, *Nucl. Phys. B* **587**, 619 (2000).

[11] K.H. Rehren, Algebraic holography, *Annales Henri Poincaré* **1**, 607 (2000).

Jacques Bros
Service de Physique Théorique
C.E. Saclay
91191 Gif-sur-Yvette, France
email : Jacques.Bros@cea.fr

Ugo Moschella
Dipartimento di Fisica e Matematica
Università dell'Insubria, 22100 Como
INFN, Sezione di Milano, Italia
email : ugo.moschella@uninsubria.it

Einstein, 135 – 149

Poincaré Seminar 2005

Experiments with Single Photons

Philippe Grangier

1 Back to the beginning: Einstein's 1905 and 1909 articles

The birth of the light quanta – "licht quanten" in their original version – is rightfully associated with the article [1] published by Albert Einstein in 1905, "An heuristic point of view about the production and transformation of light". Interestingly, several points expressed in a very collegial style in this article were exposed again in a more direct, "einsteinian" style, in a conference [2] that Einstein gave in Salzburg on september 21, 1909. This conference, entitled "The evolution of our conceptions about the nature and the constitution of radiation" reveals, and to some extend completes, the way of thinking that lead Einstein to the 1905 papers on relativity and radiation.

For instance, in 1909 Einstein gives again the list of open problems in the radiation theory, which briefly alluded to in the 1905 article. These problems were:

1. why does the appearance of a photochemical reaction depends only on the colour of light, and not on its intensity ?

2. why is short wavelength radiation generally more active chemically than long wavelength radiation ?

3. why is the kinetic energy of cathode rays (electrons) produced by the photoelectric effect independent on the light intensity ?

4. how to explain the lack of "energy dispersion" observed with X rays: secondary X-rays, produced from electrons generated by primary X-rays, may recover almost all the initial energy, while this energy should be "spread out" in free space.

This last point appears so surprising to Einstein that he writes: "From this point of view, it seems that Newton's emission theory contains more truth than the wave theory, since it says that the energy given to a light particle when it is emitted is not spread out in infinite space, but remains available for an elementary absorption process." It is then clear that Einstein wants to show that all these effects become understandable, if one admits that "the energy of light is distributed in a discontinuous way in space, as localized quanta which can move without division, and which can be absorbed or emitted only as a whole".

Another point clearly apparent in 1909 is that Einstein, though he fully admitted that Planck's formula can only be true, was really shocked by any attempt to make Planck's hypothesis compatible with the classical theory of radiation. He writes for instance: "One might believe, by looking at this (Planck's) demonstration, that Planck's formula can be considered as a consequence of the present theory of radiation. However, this is not the case, for the following reason". Then he points out on a simple example that the energy quantum $h\nu$ may be much larger (6.7 10^7 times larger in his example) than the mean energy of one oscillator. It thus appears that the energy should only take the values zero, 6.7 10^7 times the mean energy, or a multiple of this quantity. This is clearly in plain – and even shocking – contradiction with Maxwell's electromagnetic theory. Einstein's conclusion is thus: "Would it be possible to consider that this formula is true, but to provide a demonstration that does not rely on an hypothesis which is so monstrous at first sight ?".

In order to solve the dilemma, Einstein uses again thermodynamics, one of his favorite tools, and he concludes that in the domain of validity of Wien's law (the "quantum" domain), a monochromatic radiation behaves as if it was composed of independent energy quanta with a size $h\nu$. Interestingly again, he goes even further in the 1909 conference (as well as in another article [3] published also in 1909), and identifies two basic contributions to the fluctuations of radiation: one is a "particle-like" contribution, that we would call now shot-noise, and the other one is a "wave-like" contribution, which is due to random interferences, and that we would call now speckle-like fluctuations, or the Hanbury-Brown and Twiss effect. It is also really remarkable that his paper of 1925 about a perfect gas obeying the Bose-Einstein statistics [4], he recovers the same two terms, with the same interpretation – except that it applies now to "particles" and not to "radiation". In that case, the "particle-like" term appears natural, while the occurrence of a "wave-like term" is used by Einstein as a basis to a introduce "a very remarkable publication" by Louis de Broglie, which shows "how to associate a (scalar) wave field to a material particle" !

To our modern eyes, it is thus clear that through his deep analysis of thermodynamical fluctuations, Einstein was able to capture the essential features of quantum objects, which, whatever they are "classically", can exhibit both "particle-like" and "wave-like" fluctuations. At the end of his 1905 article, Einstein moves finally to his initial motivation, which was to solve the mysteries on the photochemical and photoelectric effects by using the light quantum hypothesis. He can thus interpret Stokes' law, and he gives the famous formula for the kinetic energy of the electrons produced by the photoelectric effect, which will be verified in 1916 by Millikan.

Despite these very convincing arguments, the light quantum hypothesis was the less successful among the three 1905 papers, in the sense that it was quasi-unanimously rejected by the scientific community. Apparently, though Einstein has insisted very much that the contradiction with classical electromagnetism was already present in Planck's hypothesis, the blame was put on him for making it too "visible". Also, many physicists were advocating that the light might "trigger" the photoelectric effect, rather than directly induce it. Nevertheless, the minds slowly evolved, and the last enemies of the light quantum vanished after the experiments done by Compton at the beginning of the 20's, on the energy-momentum conservation in the collision between an electron and a X-ray photon. The Nobel prize was attributed to Einstein in 1921, "for his services to Theoretical Physics, and especially for his discovery of the law of the photoelectric effect". In 1926 Gilbert Lewis invented the name of "photons", by which the light quanta have been known ever since.

One century later, what can we learn from these old debates ? We may first remember Planck's famous quotation, "truth never triumphs, but its enemies eventually die". First, it is clear that Einstein's arguments on the fluctuations were extremely strong, and should have been enough to convince his colleagues. On the other hand, the situation about the photoelectric effect itself was actually not so clear. Actually, it has been shown later that photoemission, taken by itself, does not really "prove" the quantization of the light. This can be realized by calculating [5] the ionization probability of quantized atoms submitted to a classical (wave-like) field oscillating at frequency ν: one does find the energy threshold effect, and even Einstein's formula. But then $h\nu$ appears from Fermi's golden rule, due to Bohr's formula $\nu = (E_{initial} - E_{final})/h$, rather than from the field quantization. Though the consistency of such a "semi-classical" model can be questioned, a full proof of the quantization of the field from photocounting events had yet to come. Of course, isolating a single photon would have put this ambiguity to an end. But in spite of its early birth, a single photon had never been "seen" for the first eighty years of its existence, essentially because it had not been possible to control how individual photons are emitted by a light source.

2 Quantum optics and the photon

Things started to change between the late 1960's and early 1980's, with the emergence of quantum optics, a discipline dedicated to the study of the quantum properties of light and, of course, of photons. It was then realized that quantitative discrepancies between the fully quantized and semi-classical descriptions of light-matter interaction can hardly be found by looking at single photodetection events, but that they appear straightforwardly when looking at correlations between several – in practice, at least two – photodetection events.

Since the proof of the photon is the "seeing", the first question that could be asked was "if we somehow can isolate a single photon, how can we see that we actually have one and only one photon?" A clever trick is to send that unknown state of light onto a beamsplitter (i.e. a half-silvered mirror), so that half the intensity is reflected and half is transmitted. Since a single photon cannot be split into two halves, it will **either** be reflected **or** transmitted with 50/50 probabilities, but will never go both ways at once. So, if sensitive photodetectors are set in

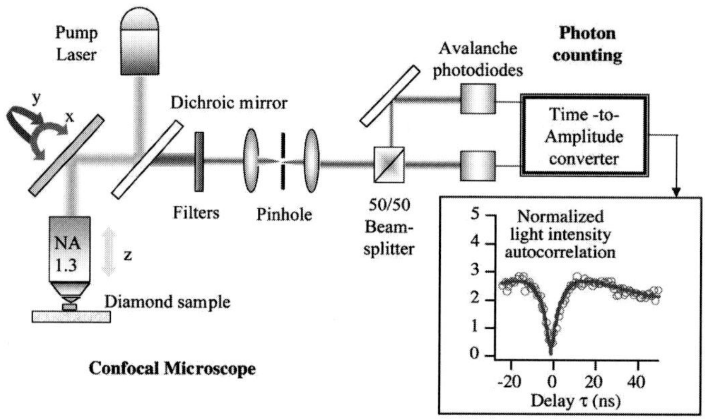

Figure 1: Modern version of an antibunching experiment: A single emitting dipole (here a colour center in a diamond nanocrystal) is irradiated by a continuous-wave green laser. The red fluorescence from the center is collected and split towards two photon-counting detectors (avalanche photodiodes). The number of coincidence counts vanishes at zero delay (i.e. for simultaneous detections), and increases at later times: this "antibunching effect" is the signature of the quantum character of the light emitted by a single dipole.

each of the two outputs of the beamsplitter, the probability of both detectors producing an electric pulse simultaneously will be at a minimum, in other words the two pulses will never be bunched. A first experiment [6] along these lines was realized by John Clauser in 1974, and then the "antibunching" effect [7] itself was observed in 1976 by Leonard Mandel and coworkers in Rochester (fig. 1). It clearly appeared as a phenomenon that is truly due to the quantum mechanical nature of light, since only quantum mechanics could provide a consistent explanation of the observed results.

Shortly after this experiment, scientists started playing with it to illustrate and verify all the strange things taught in elementary quantum mechanics courses, many of which had remained for all these decades as unchecked articles of faith. Beyond the antibunching effect, an important goal was to generate a "single photon state", that is the first excited state of the quantized radiation field, containing only one quantum of energy. Such states were produced simultaneously in 1986 in Rochester by Chung Ki Hong and Leonard Mandel [8], and in Orsay by Philippe Grangier, Gérard Roger and Alain Aspect [9], by using light sources which emit *pairs* of photons. The detection of the first photon in the pair "heralds" the second one, and at that instant the electromagnetic field is prepared in a "single photon state". For an ideal single photon state, the probability of joint detection on both sides of the beamsplitter is strictly zero – the photon does not split (see fig. 2). In

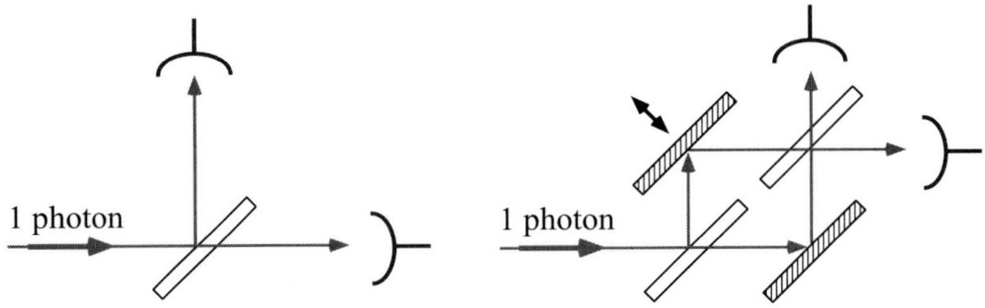

Figure 2: Wave-particle duality for a single photon: A one-photon state of the light is prepared and sent towards a beamsplitter. In the left part of the figure, the single photon exhibits a particle-like behaviour: it is detected by either one of the detectors, but there is never a "double click". One would conclude classically that the photon "chooses its way" on the beamsplitter. In the right part of the figure, the output beams are recombined to form a Mach-Zehnder interferometer. For a single-photon input, the photon output channel can now be controlled by moving any of the two mirrors (double arrows on the figure): for instance, one can adjust the mirror's position so that the photon always goes to the upper channel (with probability one). This is the single-photon equivalent of having a totally destructive interference in the lower channel ("real" fringes can also be reconstructed by sending many individual photons, for various mirrors positions). Classically, one would conclude that each photon has to go through both ways like a wave, but this conclusion is contradictory with the previous one. Only the quantum theory of light is able to give a consistent description of both experiments.

addition, the Orsay team set out to illustrate the wave-particle duality of quantum mechanics. They reasoned that the photon behaves like a particle because, by determining which detector got activated, we are actually answering a particle-like question, namely "which way did the photon go when it hit the beamsplitter ?" But by putting a second half-silvered mirror to make a Mach-Zehnder interferometer (fig. 2) they could see the interference of the two paths that the single photon could take, thus bringing into evidence its wave-like nature. In other words, by not trying to answer the question of which way the photon went, they allowed it to go **both ways** at once and produce an interference pattern, just like any wave would do.

3 Using single photons: Quantum Key Distribution

In the meantime, scientists started thinking of how to exploit the quantum properties of the photon to do useful things. Transmitting information by coding it on

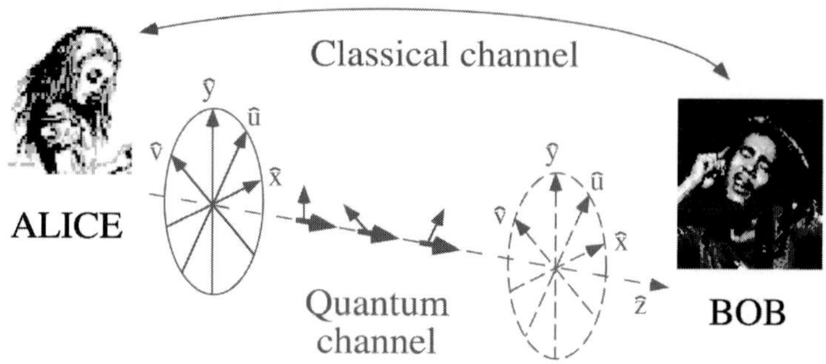

Figure 3: Quantum Key Distribution: Using the quantum channel, Alice sends to Bob a stream of photons that are individually polarized along any of the four directions $\hat{x}$, $\hat{y}$, $\hat{u}$,$\hat{v}$. By agreeing on the measurement basis after Bob has received the photons, and comparing a subset of the exchanged data through the public channel, Alice and Bob can extract a fully secure secret key.

a train of single photons is not such a good idea, since transmission losses would produce random deletions of photons, thus making any predetermined message unintelligible. However, a random number does not suffer from this disadvantage, since it remains random (but not the same) after a random decimation of its digits. And random numbers constitute a valuable resource, because they cannot be guessed and can therefore be used as cryptographic keys to encode messages for subsequent secure transmission. In 1984, Charles Bennett and Gilles Brassard proposed a protocol [10] (known as BB84), for sending a random number using a train of single photons. This turned out to be a very fruitful idea that gave birth to a new research field, often called "quantum cryptography", or more technically "quantum key distribution" (QKD). Over the years, a large number of groups explored both the theoretical and experimental sides of these ideas. The security proofs of QKD became more and more powerful and general, while hardware implementations of QKD systems made considerable progress.

The BB84 protocol for sending a random sequence of bits permits the authorized users (often named Alice and Bob) to detect any attack in which an eavesdropper (usually called Eve) tries to intercept the key, for instance by measuring each photon and then re-emitting it so as not to interrupt the transmission. The security of the transmission is unconditionally guaranteed by a strategy based on the quantum theory of measurement and the use of superposition states. For that purpose, the bits are coded by establishing a non-unique correspondence between a bit value and the polarization states of the photon. For example, the bit values 0 or 1 may be coded by emitting a photon polarized along $\hat{x}$ or along $\hat{y}$ respectively (fig. 3).

Alternatively, the "diagonal" basis may be used to encode 0 and 1 by polarizing the photon along $\hat{u}$ or $\hat{v}$ respectively. We may remark, however, that since the two bases are not orthogonal to each other, a definite bit value in one of them is expressed as a superposition state in the other, for example $\hat{u} = (\hat{x} + \hat{y})/\sqrt{2}$. During the transmission the two bases are interchanged randomly, so that a receiver who does not use the same basis as the emitter will receive a superposition state and thus get erroneous results half of the time. For example, if a 0 is coded by emitting a photon polarized along $\hat{x}$ but the measurement is carried out in the diagonal basis, the photon will be detected with equal probability to have a $\hat{u}$ or $\hat{v}$ polarization (thus interpreted as a 0 or a 1 with equal probability), producing an error half of the time. This is not a problem for Alice and Bob, because after the transmission is complete they can compare the basis sets used in emission and reception and discard the events in which the basis sets were different. When the eavesdropper, however, uses the wrong basis set in the course of the transmission (and this will occur statistically for half of the bits received) she has no way of comparing it with the basis used in emission, and thus the errors in her reception mean that she retransmits erroneous data 25% of the time. The legitimate users can then detect the presence of the eavesdropper simply by comparing a random sample of the bits received to obtain the error rate of the transmission.

In practice, there are always transmission errors, and merely interrupting the transmission as soon as the error rate increases (possibly due to Eve, but possibly not), would not be of great use to Alice and Bob. But a crucial point is that, as long as the error rate is not too large, the authorized parties are always able to extract from the exchanged quantum data a secret key that is *absolutely secure*. This is obtained by using provably secure classical algorithmic techniques, known as "privacy amplification", that rely on suitably designed hashing functions. As a result, the effect of an increase in the error rate will be to decrease the rate of transmission of the secret key, but not its security. Obviously, only a finite error rate is tolerable, and in practice the secret key rate drops to zero when the error rate goes above a value close to 15%.

Presently, several laboratories have demonstrated the quantum transmission of a cryptographic key in optical fibers, for distances up to 70 kilometers and transmission rates on the order of a few kbits/s [11]. Such systems are now commercially available, from companies such as "id Quantique" based in Geneva [12]. These devices may be relevant for specialized economic niches that require absolute security over concentrated areas, like business or management centers, and that are not too sensitive to cost and infrastructure complexity. There has also been proposals to implement global key distribution by using satellite-borne QKD.

Research on quantum key distribution has also stimulated interesting technological developments, in particular in the field of single photon detectors. Silicon avalanche photodiodes (APD) are sensitive enough to detect single photons in the visible and near-infrared range, and have found uses in many fields, for instance in single-molecule detection for biological applications. In the window of minimal attenuation in optical fibers (1550 nm) which is interesting for long distance telecom transmissions, QKD applications have pushed forward the development of InGaAs APDs, and although their performance does not match yet that of silicon APDs, complete photon-counting devices are now commercially available. QKD has also stimulated technological progress in other domains, such as non-linear optics (e.g. high-efficiency parametric fluorescence in periodically poled waveguides), and software (such as the full-size quantum cryptography software "QUCRYPT" designed by Louis Salvail, and now publicly available [13]).

4 Single photon sources

A research area on which QKD has had a particularly deep impact is the development of novel light sources. To date, most of the practical realizations of QKD have relied on strongly attenuated laser pulses, with an average number of photons per pulse much smaller than one. But in that case the Poisson photon statistics of laser light imposes two unwanted consequences: first, a fraction of the pulses contain two or more photons, and this is an open door to information leakage towards an eavesdropper; second, most of the attenuated laser pulses actually do not contain any photons at all, thus resulting in penalizingly low transmission rates. Clearly, an efficient source able to emit one, and only one, photon in each light pulse would considerably improve the performance of QKD systems, especially in high-loss situations, such as satellite communications. The need for such light sources, combined with the more fundamental interests of academic laboratories – improving our understanding and mastering of quantum optics – have given a strong impetus to research for sources capable of emitting single photons "on demand", and a great variety of approaches have been proposed and implemented in recent years [14].

At the heart of all single photon sources lies a single nanoscopic object, which is small enough so that a transition between its electronic states corresponds to light emission from a single Such is the case, for example, of an atom, a molecule or a semiconductor nano-aggregate. If such an emitting dipole is brought to an excited state, then from the mere conservation of energy it will emit one only photon. In general, spontaneous photon emission can occur in any direction, and thus usually only a very small fraction will go in a direction where it can be useful, making the emitter very inefficient. To increase efficiency, the nanoscopic emitter can be embedded in a high finesse optical cavity whose dimensions are of the order of the optical wavelength, that is a few hundred nanometers. Microscopic optical cavities are subject to "Cavity Quantum Electrodynamics" effects in which the structure of the electromagnetic field and the spontaneous emission are modified. In particular, in the so-called "Purcell effect", spontaneous emission into the cavity modes can be greatly enhanced, so that most emitted photons are funneled in one particular direction and thus generate a highly directional output beam. In addition, a "user-friendly" single photon source should preferably work at room temperature, it should have a high quantum efficiency, and it should be able to

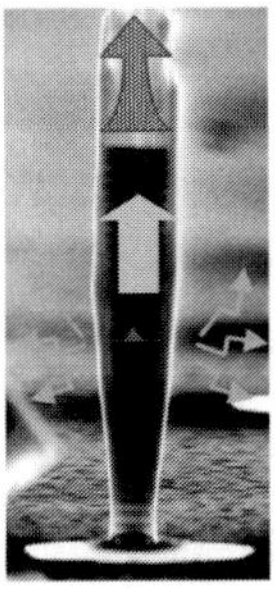

Figure 4: Electron micrograph of a GaAs micro-post cavity. The photons are emitted by an InAs quantum dot (depicted schematically by a triangle) embedded in the center of the microcavity, and resonant with the fundamental cavity mode. These photons will be channeled preferentially into that mode, and thus produce a highly directional beam (image: Izo Abram, LPN Marcoussis).

achieve a high pulse repetition rate without blinking or burning out.

Such single photon sources were achieved first by using single molecules, such as as terylene embedded in a crystal of para-terphenyl, which was used first at cryogenic temperatures, and then at room temperature. Other candidates, such as rhodamines or cyanines, have also been identified, but a significant drawback of molecules at room temperature is that they irreversibly turn off after some irradiation time. The exact mechanism responsible for this photobleaching is still under investigation, and improvements may occur in the future.

Another well-explored system, studied both in the United States and in Europe, is the single self-assembled semiconductor quantum dot, consisting of an InAs nano-aggregate embedded in GaAs. The single photon that is emitted when one electron hole pair is injected in the quantum dot can easily be identified thanks to its wavelength. In addition, in view of maximizing the collection efficiency of the single photon that is emitted, the InAs quantum dots can easily be incorporated in a microcavity (fig. 4) made of semiconductor through the standard processing technologies used for microelectronics. In such systems, cavity-enhanced spontaneous emission (Purcell effect) has been observed experimentally to be faster than in free space by a factor of up to 20, while factors of several hundred should be possible according to theory. Presently, quantum dots operation requires liquid helium temperatures, but this should improve in forthcoming years.

Another avenue is using individual nitrogen-vacancy (NV) color centers in diamond. The NV centers have many similarities with molecules but are extremely photostable, even at room temperature. Another advantage is that they appear both in bulk diamond or in diamond nanocrystals, and are therefore easy to manipulate (fig. 1). A stable source emitting single photon pulse trains based on an

NV center in a diamond nanocrystal excited by a small solid-state laser was recently implemented in Orsay [14]. The overall system is a reasonably compact, all-solid-state set-up operating at room temperature, that is probably the simplest single-photon source developed so far. Using this compact source delivering trains of single-photon pulses, Alexios Beveratos and his colleagues were able to demonstrate a complete quantum key distribution scheme [14, 15], where the rate of pulses containing two photons is strongly reduced with respect to an attenuated laser (by a factor 14 for the same rate of one-photon pulses). This makes interception by the so-called "two-photon attacks" virtually impossible. The cryptographic exchange is then more robust with respect to on-line losses, providing a clear advantage over an attenuated laser source for QKD applications. The performance of this set-up should improve further in a near future, providing a highly efficient, easy to use, and reliable single photon source that would constitute a basic piece of hardware for practical quantum key distribution (see fig. 5).

Another way to avoid two-photon attacks is to use the trick of "heralded" single photons, that was used in 1986 to produce single-photon states as said above. In the context of quantum cryptography, the experiment was realized e.g. in Geneva, by using pairs of twin photons, generated by a nonlinear optical process called "parametric downconversion", so that one member of the pair heralds its twin. The quantum mechanical "entanglement" that exists between the twins was also exploited with success. Though these schemes produce photons at irregular intervals, with effective counting rates that are subject to various technical limitations, they do provide also quite interesting QKD schemes [11].

Finally, schemes based on single trapped atoms or ions in high-finesse cavities are clearly more complex to implement, but might produce single photons with interesting spectral properties, as discussed in the section below. State-of-the-art results were obtained by dropping or trapping cold atoms through a high-finesse cavity: when going through the cavity each atom emits a burst of single-photon pulses. Each photon emission is triggered by a sequence of laser pulses, including excitation, emission in the cavity mode, and repumping to the initial level. Several recent results along these lines are described in ref. [14].

5 Coalescing photons

Looking further to the future, several recent proposals for all-optical quantum photonic networks have been advanced recently based on indistinguishable single photons acting as flying qubits, carrying information from node to node and interacting with each other. These ideas can even be extended towards the realization of a full-fledged quantum computer, using a scheme that was proposed recently by Emmanuel Knill, Raymond Laflamme and Gerald Milburn. For such schemes to work, photons must be indistinguishable, that is they must be in the same "single mode of the electromagnetic field". In should be noted that most of the single-photon sources described above produce photons that are incoherently spread over

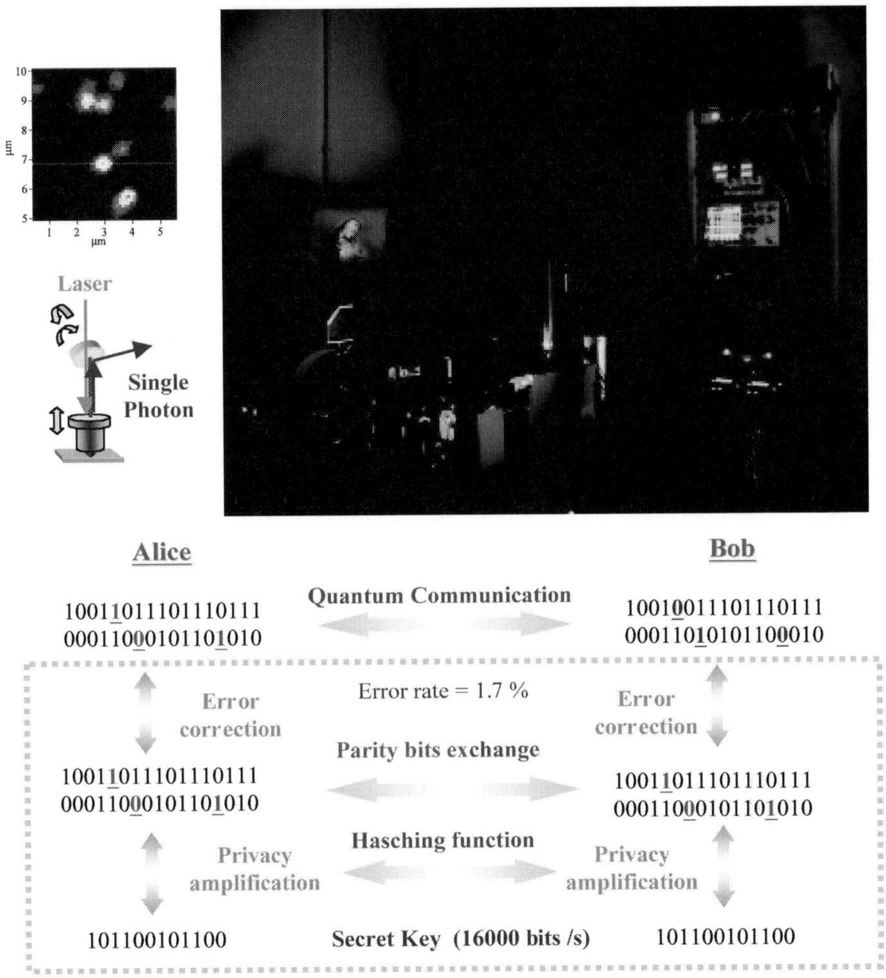

Figure 5: Quantum key exchange with a single photon source, obtained by exciting diamond nanocrystals by a pulsed laser. The upper left image shows light emission by the diamond nanocrystals (bright spots on the image). The upper right photograph shows the experimental set-up, where the photons are sent though a window to Bob's detection apparatus, located in another building. The lower part of the figure shows the various steps of the protocol which is used to extract the final secret key.

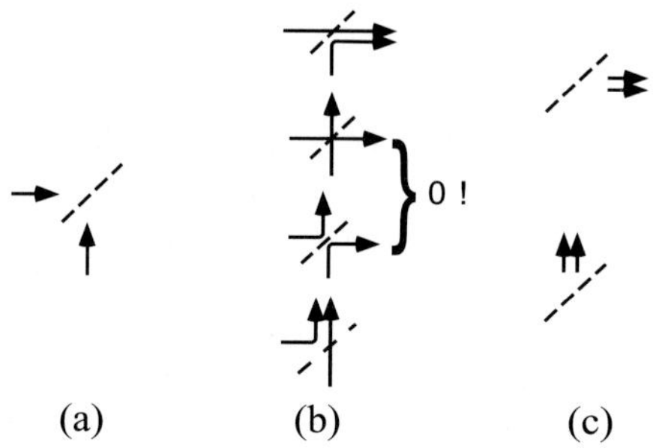

Figure 6: Coalescing photons on a beamsplitter: When two "single mode" (but otherwise independent) photons enter a 50-50 beamsplitter (a), they may be transmitted or reflected in various ways, as shown in (b). In particular, both photons may be transmitted, or both may be reflected, and it happens that the corresponding probability amplitudes cancel out. Then the two photons must go to the same output beam, as shown in (c): they "coalesce" on the beamsplitter.

many modes of the radiation field and, although they are usable in QKD, they do not have the appropriate properties for quantum computation.

In order to illustrate what is specific to indistinguishable photons, let us consider fig. 6(a): Two photons are sent onto a beamsplitter, in such a way that when one photon is transmitted it ends up in exactly the same mode as the other photon which is reflected. The four possible configurations for the two photons being transmitted or reflected are depicted in fig. 6(b). As it is usual in quantum mechanics, a probability amplitude in attached to each of these configurations, and it turns out that the amplitudes of the two diagrams in the middle of fig. 6(b) (both corresponding to one photon in each of the two output ports of the beamsplitter) have opposite signs. Clearly, if the two photons are indistinguishable (having exactly the same frequency, direction, and polarization) the two diagrams are identical and, since their amplitudes are of opposite sign, they cancel each other out! The immediate consequence of the two surviving diagrams is that the two photons must go to the same output beam: They "coalesce" as they meet on the beamsplitter to form a "two-photon state", that is the second excited energy state of the corresponding mode of the quantized electromagnetic field. This surprising quantum interference effect was first predicted and observed in 1987, by Leonard Mandel and coworkers. They used actually pairs of "twin photons", simultaneously produced in parametric down-conversion, so it was possible to argue that the two photons knew about each other before, since they were "twins" emitted in a single parametric fluorescence event. Would it be possible to get the same effect by using truly independently emitted (albeit indistinguishable) photons? The quantum answer to this question is yes, and it is not pure rhetoric, because interference between independently emitted photons is actually what is required for applications in quantum information processing, using the Knill-Laflamme-Milburn scheme.

The coalescence of two indistinguishable but independently generated photons, from a source consisting of a single quantum dot in a semiconductor microcavity, was experimentally demonstrated very recently in Stanford [16]. This experiment can be seen as a first step towards the realization of conditional quantum logic gates that would make photon-based quantum computing possible. But difficulties should not be underestimated: with present-day setups the error rates would be by orders of magnitude too large, compared with the range where quantum error-correcting codes can play an efficient role. Also, the number of interfering photons required to implement a useful computation is huge, and the integration of the devices would have to be pushed well beyond the present technological capabilities.

6 "En guise de conclusion": towards entangled photons on demand

Photon pairs emitted in parametric downconversion have often been mentioned above, because they have many applications in quantum optics: conditional preparation of single photon states, quantum key distribution, and last but not least, they can be prepared in an entangled state. When two photons are entangled, their states are always correlated no matter how we choose to measure them, as if the two photons constituted a single quantum object. For instance, a pair of polarization-entangled photons will exhibit correlations in every possible polarization basis, and performing polarization correlation measurements on the two photons once they are far apart leads to a violation of Bell's inequalities. This means that the correlations that appear between the results of the polarization measurements on the two remote photons are so strong, that no classical model based on "local realism" is able to explain them. In quantum information processing, such a quantum entanglement is a "resource", because it cannot be created by local actions on two remote photons, and it allows one to perform some specific tasks, such as quantum teleportation of the (unknown) polarization state of a third photon. Entangled photon pairs also provide a way towards the so-called "quantum repeaters", that would allow one to develop quantum key distribution schemes over arbitrarily long distances (it is noticeable that "classical repeaters", commonly used in optical telecommunication, do not preserve the quantum cryptographic security).

Presently, the main source of entangled photon pairs are parametric fluorescence events, but these events are essentially random, so that the pair production process obeys Poisson statistics. In the same way as deterministic single photon generation is useful, deterministic pair production would allow new quantum communication protocols to be developed. How can this be achieved? One may simply try to improve upon the old idea of the radiative cascade, that was used in the 70's and 80's for performing experimental tests of Bell's inequalities. But instead of using many-atom sources as it was done at that time, one should use a two-photon radiative cascade of a single emitter. Several groups have shown that a quantum dot does display such a cascade, corresponding to the radiative transitions between the electronic states of the quantum dot containing two, one, or zero electron-hole pairs. However, the first experiments did not produce the results hoped for: The photons exhibited correlations only for one polarization basis. In other words, they were correlated as if they were classical objects, and were not entangled quantum mechanically, because, apparently, decoherence processes in the quantum dot rapidly destroy the entanglement. Exploitation of the Purcell effect to reduce the radiative lifetime beyond the decoherence time should, in principle, permit the production of entangled photon pairs "on demand".

While the long-term goal of building a quantum computer is far-fetched, a medium-term goal for these experiments is to develop long-distance quantum communication networks, that would allow for the implementation of QKD systems over arbitrary large distances. One may think also about more elaborate protocols, able to share a quantum secret between many (rather than two) users. Such things are presently still far from being implemented, but this is one very fascinating aspects of quantum information: by exploiting the strangest properties of single photons and single atoms, it allows us to move continuously from science to science-fiction, and back.

References

[1] This article and the following ones are translated in french in "Albert Einstein: Quanta", by F. Balibar, O. Darrigol and B. Jech, Éditions du Seuil, Paris, 1989.

A. Einstein, Über einen die Erzeugung und Verwandlung des Lichtes betreffenden heuristischen Gesichtspunkt, *Annalen der Physik* **34**, 591 (1905).

[2] A. Einstein, Über die Entwicklung unserer Anschauungen über das Wesen und die Konstitution des Strahlung, *Physikalische Zeitschrift* **10**, 817 (1909)

[3] A. Einstein, Zum gegenwärtigen Stand des Strahlungsproblems, Physikalische Zeitschrift **10**, 185 (1909)

[4] A. Einstein, Quantentheorie des idealen gases, *Preussische Akademie des Wissenschaften, Phys.-math. Klasse, Sitzungberichte* 18 (1925).

[5] W.E. Lamb and M.O. Scully, *The Photoelectric Effect Without Photons*, in "Polarisation, Matière et Rayonnement", ed. A. Kastler, Presses Universitaires de France, 363–369 (1969).

[6] J.F. Clauser, Experimental Distinction Between the Quantum and Classical Field Theoretical Predictions for the Photoelectric Effect, *Phys. Rev. D* **9**, 85 (1974).

[7] H.J. Kimble, M. Dagenais and L. Mandel, Photon antibunching in resonance fluorescence, *Phys. Rev. Lett.* **39**, 691 (1977).

[8] C.K. Hong and L. Mandel, Experimental realization of a localized one-photon state, *Phys. Rev. Lett.* **56**, 58–60 (1986).

[9] P. Grangier, G. Roger and A. Aspect, Experimental evidence for a photon anticorrelation effect on a beam-splitter: a new light on single-photon interferences, *Europhysics Lett.* **1**, 173 (1986).

[10] C.H. Bennett and G. Brassard, Int. Conf. Computers, Systems and Signal Processing, Bangalore, India, 175 (1984).

[11] A review on quantum cryptography may be found in: N. Gisin, G. Ribordy, W. Tittel, and H. Zbinden, Quantum cryptography, *Rev. Mod. Phys.* **74**, 145 (2002).

[12] id Quantique SA, http://www.idquantique.com

[13] http://www.cki.au.dk/experiment/qrypto/doc/

[14] Many experimental results on single photons pulses can be found in three special issues:

"Quantum interference and cryptographic keys: novel physics and advancing technologies (QUICK)", in *The European Physical Journal D* **18** (2) (February 2002), edited by P. Grangier, J.G. Rarity and A. Karlsson;

"Proceedings of the Tenth International Conference on Modulated Semiconductor Structures (MSS 10)", in *Physica E* **13** (2-4) (March 2002), edited by G. Bauer;

"Focus on Single Photons on Demand", in New Journal of Physics **6** (2004), edited by P. Grangier, B. Sanders and J. Vukovic (free access at http://www.iop.org/EJ/abstract/1367-2630/6/1/E04).

[15] A. Beveratos *et al.*, Single photon quantum cryptography, *Phys. Rev. Lett.* **89**, 187901 (2002);

[16] C. Santori *et al.*, Indistinguishable photons from a single-photon device, *Nature* **419**, 594 (2002).

Philippe Grangier
Laboratoire Charles Fabry de l'Institut d'Optique
(UMR CNRS 8501)
Bât. 503, Centre Universitaire
91403 Orsay
email: `philippe.grangier@iota.u-psud.fr`

Einstein, 151 – 182

Poincaré Seminar 2005

Einstein 1905–1955: His Approach to Physics*

Thibault Damour

Abstract. We review Einstein's epistemological conceptions, and indicate their philosophical roots. The particular importance of the ideas of Hume, Kant, Mach, and Poincaré is highlighted. The specific characteristics of Einstein's approach to physics are underlined. Lastly, we consider the practical application of Einstein's methodological principles to the two theories of relativity, and to quantum theory. We emphasize a Kantian approach to quantum theory.

1 On Einstein's Epistemology

Some analysts of Einstein's thought, notably the historian Gerald Holton, as well as the physicist Max Born, suggested that Einstein had radically changed his epistemological approach from a hard-line Machian positivism in his youth (particularly in 1905) to a platonizing rationalism in his later years. By contrast, I think, in agreement with the fine analysis of Michel Paty [1], that Einstein always had in mind a multi-faceted and subtle view of the theory of knowledge, even if the discovery of the theory of General Relativity had the effect of partially reorienting his epistemology towards a more speculative rationalism. The best formulation I know of the subtlety and complexity of Einstein's ideas on epistemology is contained in a passage from "Reply to criticisms" which he wrote for the book *Albert Einstein: philosopher scientist* [2]. As this formulation is central to this article, let us cite it extensively :

"The reciprocal relationship of epistemology[1] and science is of a noteworthy kind. They are dependent upon each other. Epistemology without contact with science becomes an empty scheme. Science without epistemology is – insofar as it is thinkable at all – primitive and muddled. However, no sooner has the epistemologist[2], who is seeking a clear system, fought his way through to such a system, than he is inclined to interpret the thought-content of science in the sense of his system and to reject whatever does not fit into his system. The scientist[3], however, cannot afford to carry his striving for epistemological systematic[4] that far. He accepts gratefully the epistemological conceptual analysis ; but the external

*Based on translations from the French by Emily Parks (sections 1–6) and Eric Novak (sections 7–14).

[1] "Erkenntnistheorie".

[2] "Erkenntnistheoriker", literally the theoretician of knowledge.

[3] "... der Scientist".

[4] "... erkenntnistheorischer Systematik".

conditions, which are set for him by the facts of experience, do not permit him to let himself be too much restricted in the construction of his conceptual world by the adherence to an epistemological system. He therefore must appear to the systematic epistemologist as a type of unscrupulous opportunist : he appears as *realist* insofar as he seeks to describe a world independent of the acts of perception ; as *idealist* insofar as he looks upon the concepts and theories as the free inventions of the human spirit (not logically derivable from what is empirically given) ; as *positivist* insofar as he considers his concepts and theories justified *only* to the extent to which they furnish a logical representation of relations among sensory experiences. He may even appear as *Platonist* or *Pythagorean* insofar as he considers the viewpoint of logical simplicity as an indispensable and effective tool of his research."

To complete this citation, let us listen to what Einstein said in his Herbert Spencer Lecture, given at Oxford on June 10, 1933 [3] :

"If you want to learn something about the methods of theoretical physics from its practitioners, I suggest you to hold to the following principle : do not listen to what they say but look at what they do !"

The Moral :

- Whichever work of Einstein is considered, one should not try to interpret it only from a single epistemological approach. On the contrary, one should try to highlight his various components: empiricist, realist, idealist, speculative, ...; and
- to learn the richness of Einstein's approach to physics, it is best to consider specific examples, based on his works or on the texts where he explains himself in detail.

However, before considering explicit examples where Einstein puts his epistemology into action, it is important to have an idea of its sources.

2 Einstein and Philosophy

First, let us note that Einstein had always been very much interested in philosophy in general, and more particularly in the philosophy of knowledge (what he called *Erkenntnistheorie*, the theory of knowledge, that here we will call *epistemology*).

Sometime during 1902, in Bern, Maurice Solovine, who was then studying philosophy and physics, came to see Einstein after having seen a small advertisement in a Bern newspaper, where Einstein was offering his services as a private tutor in physics. On this occasion Einstein confided in him that, "when he was younger, he had a very lively taste for philosophy, but its vagueness and arbitrariness turned him off, and that now he was only concerning himself with physics." However, the physics lessons to Solovine transformed rapidly into discussions on the foundations of physics. These epistemological discussions were broadened to a small group of three friends : Einstein (1879–1955), Maurice Solovine (1875–1958) and Konrad Habicht (1876–1958). For fun, they gave their small discussion group the pompous name of the "Olympia Academy". They embarked on an ambitious program of reading and discussing works of philosophy, epistemology, criticism or history, notably: *The Analysis of Sensations* and *The Science of Mechanics: A Critical and Historical Exposition of its Principles* by Ernst Mach, *System of Logic* by John Stuart Mill, *A Treatise of Human Nature* by David Hume, *The Grammar of Science* by Karl Pearson, *Critique of Pure Experience* by Richard Avenarius, *Essay on the Philosophy of Science* by Ampère, *Science and Hypothesis* by Poincaré, Riemann's thesis on *The Foundations of Geometry*, the essay *On the nature of things-in-themselves* by Clifford, *What are Numbers and What are they for?* by Dedekind. To this they added a program of philosophical or literary works of general culture comprising: some dialogs of Plato, some works of Leibniz, *Antigone* from Sophocles, some tragedies of Racine and *Don Quixote* from Cervantes. Let us note also that Einstein had read, in his youth or when he was at the Zurich Polytechnic, other philosophers or scientific texts written about the foundations of science, namely Kant (read when Einstein was 16 and reread afterwards), Spinoza, Schopenhauer, Berkeley, Galileo, Boltzmann, Helmholtz and Hertz. Later, he continued to read philosophers or epistemologists (Russell, Bergson, Emile Meyerson) and had exchanges and discussions with Russell, with the neo-Kantian Ernst Cassirer, and with epistemologists from the Vienna Circle : Moritz Schlick, Rudolf Carnap, and Hans Reichenbach.

This long enumeration shows Einstein's profound interest for philosophy, and his peculiar attraction for a methodological reflection on the foundations of science. It is certain that such an epistemological reflection played a crucial role in his scientific work, by allowing him to bypass psychological blocks that limited the intellectual horizon of many other scientists at the beginning of the century. This is, however, a subtle and complex matter of which we will give just a short outline.

3 Hume, Kant, Mach and Poincaré

The thinkers who probably have most influenced the methodological reflection of Einstein and who, in one way or another, helped him in his scientific works are : David Hume, Immanuel Kant, Ernst Mach, and Henri Poincaré.

- David Hume (1711–1776) came after Newton, Locke and Berkeley. He asked the fundamental question : "How do we know?" He examined in a critical way, the origin and content of general notions, such as space, relation, substance, and causality. Being a skeptic, he put into doubt the usual idea that science draws general laws from experiment, by means of inductive logic. For example, he considered that causality is based not on logical necessity but on habit.

- Immanuel Kant (1724–1804) reflected on the ideas of Newton, Leibniz, and Hume among others. By going deeper into the nature of scientific knowledge, and the nature of the objects and structures of science (space, time, matter, force, causality), he introduced a deep conceptual revolution. Before him one thought that all knowledge, to be true, must adjust itself to the objects. He turned upside down this traditional view by introducing the idea that *objects must instead adjust themselves to human knowledge.* More precisely, he conceived the objectivity and certainty of knowledge as the result of the conditions that are imposed by the demands of the knowing subject. For example, for him space and time are not physical realities that exist before and besides matter, but rather *a priori* "forms" of human sensibility that serve as ideal foundations for conceiving and representing reality.

- Ernst Mach (1838–1916) was a physicist concerned with the historical criticism of the fundamental concepts of physics, and who got interested in the psychophysiology of sensations. He developed an epistemology which was empiricist, critical and positivist, which proposed a phenomenalistic reduction to sensations and the rejection of all "metaphysics". He brought up an abrasive criticism of the *a priori* of Newtonian mechanics (absolute time, absolute space, absolute motion) by insisting on the necessity to confront with experimental observations. He insisted on the reality of relative motions only.

- Henri Poincaré (1854–1912) thought in depth about the foundations of mathematics. He published the fruits of his reflections in articles of philosophical content and in his popular books, notably *Science and Hypothesis* (1902), which for several weeks left Einstein and his friends of the Olympia Academy breathless. The central element of Poincaré's scientific philosophy (called *conventionalism*) is the *free choice* that the scientist can make of his fundamental principles. Poincaré was struck, like many, by the discovery of the logical consistency of non-Euclidean geometries, and notably of geometries admitting (*à la* Klein) symmetry groups as large as the Euclidean-geometry one: like hyperbolic geometry (Lobatchevski) or elliptic (spherical) geometry (Riemann). He concluded that the choice of a particular geometry was an *arbitrary convention*, linked to the compensating choice of other conventions in the representation of physical phenomena.

As for Poincaré's influence over Einstein let us note that certain authors have suggested that Einstein: (1) must have read, before 1905, not only *Science and Hypothesis*, but also other publications of Poincaré (notably an article written in 1900 for Lorentz's Festschrift), (2) would have found in them useful ideas for his work on Special Relativity, and (3) would have then omitted to cite Poincaré. Considering that Einstein warmly mentions the influence, on his invention of General Relativity, of the reading of *Science and Hypothesis* by the "deep and subtle Poincaré"; considering the difficulty that Einstein had, when he worked in Bern, for consulting the scientific literature[5]; and considering the fact that Einstein's only citation of an article by Poincaré (Lorentz-Festschrift, 1900) dates from 1906 and only consists in citing the article's existence without using any result from it, it seems psychologically probable to me that Einstein in 1905 had only read *Science and Hypothesis.* In addition, it is likely that the reading of Poincaré's book was not as exhaustive as one would think. Indeed, Solovine wrote that when the Olympia Academy became impassioned by a book : "we read one page, half a page, sometimes only a sentence and the discussion, when the problem was important, was drawn out over several days" (cited by [1] p. 373). Clearly, in *Science and Hypothesis*, it was the discussion on the origin of geometric structures that had impassioned the members of the Olympia Academy, and it is plausible that they barely took note of the brief allusions of Poincaré to the problematics of relative motion or to the absence of "direct intuition of the simultaneity of two events" (see the citations in [4]).

4 Scientific Philosophy and Einstein's Conceptual Innovation

One can consider that Einstein's scientific philosophy was built in large part during his youth (say before 1905), as a personal synthesis of the philosophical and epistemological readings mentioned above. Among these readings, those of Hume, Kant, Mach and Poincaré played a particular role. In his *Remarks on the Theory of Knowledge by Bertrand Russell* [5], Einstein explains what Hume and Kant had brought to him :

"Hume saw that some concepts, that we deem as essential, such as, for instance, the link between cause and effect, could not be derived from the sensory data. [. . .] Man has an intense desire for assured knowledge. That is why Hume's clear message seemed crushing: the sensory raw material, the only source of our knowledge, through habit may lead us to belief and expectation but not to the knowledge and still less to the understanding of lawful relations. Then Kant took the stage with an idea which, though certainly untenable in the form in which he put it, signified a step towards the solution of Hume's dilemma: whatever in knowledge is of empirical origin is never certain (Hume). If, therefore, we have definitely assured knowledge, it must be grounded in reason itself. This is held to be the case, for example, in the principles of geometry and in the principle of causality. These and certain other types of knowledge are, so to speak, a part of

[5]Let us note in this respect that even though Poincaré had, contrary to Einstein, easy access to all the scientific literature it seems that he ignored the existence of Einstein's publications on Relativity until 1909, in spite of the fact that they were published in one of the most prestigious physics journals of the era. One should not over-estimate the knowledge of literature that scientists had at the time, especially in regard with articles published only in a Festschrift. There were no photocopiers at that time and so one could only consult articles in a library. We recall in passing that Einstein explained in a letter to Stark (September 1907) : " Unfortunately I am not able to consult all that has appeared on the subject, because the library is closed during my free time."

the implements of thinking and therefore do not previously have to be gained from sense data (i.e., they are a priori knowledge). Today everyone knows, of course, that the mentioned concepts contain nothing of the certainty, of the inherent necessity, which Kant had attributed to them. The following, however, appears to me to be correct in Kant's statement of the problem: in thinking we use, with a certain 'right,' concepts to which there is no access from the materials of sensory experience, if the situation is viewed from the logical point of view."

Elsewhere (in *Physics and Reality*, 1936, see [1]), Einstein recognizes that Kant's big achievement is to have affirmed that the intelligibility of the world was a necessary condition to its scientific representation : "It is one of the biggest accomplishments of Kant to have shown that it would make no sense to pose the existence of a real external world without this intelligibility."

Of course, one must remember that these texts were written by Einstein after the construction of the two theories of Relativity and after their first experimental verifications. These theories confirmed at once : (1) the necessity to pose *a priori* a logical framework defining the intelligibility of the world, and (2) the possibility to change this logical framework. However, in light of Einstein's numerous epistemological readings before 1905, it seems clear that his epistemology, as expressed in the texts above, is not an *a posteriori* rationalization, but played an important role in helping Einstein, the physicist, to set new logical frameworks defining a deepened intelligibility of the world.

More precisely, Einstein understood that :

- Hume's skepticism demystified the big conceptual absolutes and invited searching for the "habits" on which they were based on ;
- Kant's rationalism suggested to look for the origin of the fundamental scientific frameworks in the intelligibility power of the cognitive subject ;
- Mach's positivism invited the questioning of Newtonian absolutes and the need to express physics in terms of concepts linked to experimental observations ;
- Poincaré's conventionalism insisted on the free choice of the fundamental scientific concepts, and, at the same time, on the physiological-experimental origin of geometry.

In this way Einstein "went shopping for" his ideas among the great thinkers of science, and there found *liberating* elements for his research in physics. In doing so he avoided those elements that blocked his predecessors : such as universal skepticism, a mental block on the *a priori* character of space and time concepts, an exaggerated emphasis on the necessity to found each concept on observations, or an insistence on the *arbitrary* and *conventional* character of basic scientific principles.

More precisely, one can say that Einstein added to the useful messages of the epistemologies of Hume, Kant, Mach, and Poincaré summarized above, the following elements that are characteristic marks of his approach in physics :

(i) Einstein insists on the research of *general principles* of Nature, and on the fruitfulness of constraining the laws of Nature by imposing such principles as starting hypotheses.

(ii) Einstein explains that "the researcher must extract these general principles of nature by perceiving within a complex ensemble of experimental facts certain general characters that allow for a precise formulation."

(iii) He also explains that the choice of these general principles, and more generally of the fundamental scientific concepts, is a *free invention* of the mind that cannot be logically deduced from the bulk of experimental facts, but that is, however, strongly constrained by it.

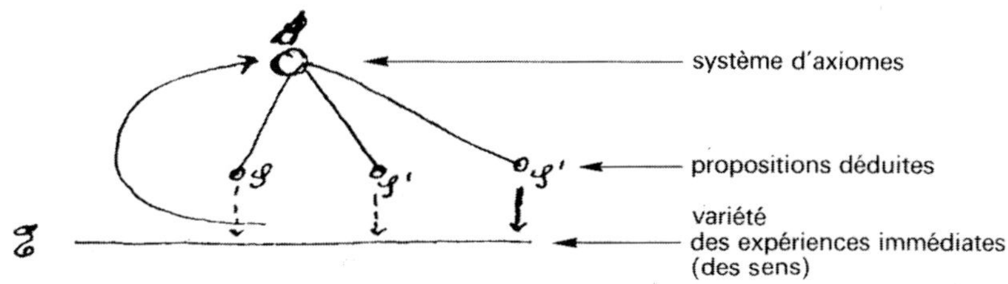

Let us also mention how Einstein explained his epistemological view to his old friend Maurice Solovine. It was summarized in a handmade drawing (see figure 4, taken from [1]), accompanied with the following explanations [6] :

"1. $\mathcal{E}$ (experienced facts) are given.

2. $\mathcal{A}$ are the axioms from which we draw our deductions. Psychologically, $\mathcal{A}$ is based on $\mathcal{E}$. However there is no logical path from $\mathcal{E}$ to $\mathcal{A}$ but only an intuitive (psychological) link of interdependence, always "revocable".

3. From $\mathcal{A}$ one logically deduces individual propositions $\mathcal{S}$, and these deductions may aspire to exactness.

4. $\mathcal{S}$ can be confronted with $\mathcal{E}$ (verification by experience). This procedure, at a closer look, also belongs to the extra-logic (intuitive) domain, because the links between the concepts in $\mathcal{S}$ and the experienced facts in $\mathcal{E}$ do not have a logical character.

But this relation from $\mathcal{S}$ to $\mathcal{E}$ is (pragmatically) much less uncertain than the relation of $\mathcal{A}$ to $\mathcal{E}$ (for example : the concept of a dog and the corresponding experienced facts). If one could not draw on such a near-assured correspondence, the logical machinery would be of no value for the "understanding of reality" (example : theology).

The quintessence is the eternally problematic interdependent relationship between thought and actual experience."

5 Einstein and the Theories of Relativity

Let us briefly comment on the "application" of the methodological principles (i), (ii), (iii) indicated above to the context of the discovery of the two theories of relativity by Einstein.

In the case of principle (i) and the theory of Special Relativity, let us recall that Einstein's attitude was very different from those of his "competitors" such as Lorentz or Poincaré. They thought of the existence of "corresponding states" (Lorentz) or of the "principle of relativity" (Poincaré) as *consequences* of an underlying *dynamics* to be understood in detail, rather than as a starting *axiom*, defining a new *kinematics*. This is clearly illustrated by the only discussion on Relativity that occurred between Einstein and Poincaré. This discussion took place in 1911 at the first Solvay Symposium in Brussels. It is reported by Maurice de Broglie [7] :

"I remember one day in Brussels, as Einstein was explaining his ideas [on the "new mechanics", meaning relativistic], Poincaré asked him, 'What mechanics do you adopt in your reasoning?' Einstein replied, 'No mechanics', which appeared to take his listener by surprise."

As a result, Lorentz (and probably also Poincaré) considered that Einstein was "cheating" by "turning the problem upside down", meaning by assuming (kinematically) what, for them, was to be proven (dynamically).

Einstein's formulation of the methodological principle (ii) above, sheds light on another aspect of his texts concerning the invention of Special Relativity. In fact, Einstein insisted on the abstraction of general principles from "complex ensembles of experimental facts". This explains the little importance, in Einstein's thinking, of the Michelson-Morley experiment, when considered in isolation. Contrary to the Relativity textbooks that emphasize this peculiar experiment because of its high precision and its sensitivity to v^2/c^2 order terms, it is clear that Einstein was especially sensitive to the existence of a whole "complex" of facts from experiment (going from the experiences of ordinary life, on Earth or on a train, to many electromagnetic or optical experiments) suggesting the impossibility to detect absolute motion. The Michelson-Morley experiment was probably for him only a particular example among a large set of experimental facts that made sense only considered all together.

Finally, about the methodological principle (iii), it is important to see that Einstein's *free invention* (from a logical point of view) contrast with Poincaré's *arbitrary convention.* Indeed, Poincaré concluded from the (logical) freedom of choice of fundamental scientific concepts that it was wiser not to modify the concepts inherited from the science of the past (namely Euclidean geometry, and Newtonian absolute time). By contrast, Einstein was sensitive to the suggestions (intuitively) drawn from the bulk of experimental facts and thought that, at each stage of development in physics, a particular logical framework had preeminence over the others. We clearly see the difference in methodological attitude between Einstein and Poincaré in 1912. In May of 1912, two months before his death, in a conference given in London on "Space and Time", Poincaré wrote about the conception (or "convention") of the space-time, *à la* Einstein-Minkowski, in the theory of Special Relativity[4] :

"Today some physicists want to adopt a new convention [...] those who are not of the same opinion can legitimately keep the old one so as not to trouble their old habits. I believe, between us, that they will do so for a long time."

At the same moment, Einstein, who in 1905 had convinced the elite of physicists of the necessity to pose the new kinematic framework of the special-relativistic space-time, was in the process[6] of blowing this framework up and of replacing it with a profoundly modified one: that of the deformed space-time of General Relativity.

6 Einstein and the Kantian Quantum

And what about the application of Einstein's methodological ideas to Quantum Physics? It is commonly believed that they found their limits right there, and that Einstein manifested a conceptual block, based on *a priori* ideas about "reality", that prevented him from accepting the "revolutionary" ideas of Heisenberg, Born, Jordan, Dirac, Schrödinger and Bohr. I think that this view is not correct, and does not do justice to the subtlety of Einstein's methodological approach, nor to his direct or indirect contributions to what constitutes the modern interpretation (*à la* Everett) of the quantum formalism. For a detailed discussion of the important methodological contributions of Einstein to the actual understanding of Quantum Theory I refer to a recent book [8]. The following and the end of this present text is made up of excerpts from this book. These excerpts concern several important moments in the understanding of the formalism of Quantum Theory. These moments are: (1) a conversation between Heisenberg and Einstein where one sees that Einstein's epistemological suggestions played a crucial role in Heisenberg's invention of his famous "uncertainty relations", (2) the important influence of Einstein's idea of a "ghost field" or "pilot field" on Born's interpretation of the

[6]At the risk of destabilizing his most convinced defendants, notably Planck who publicly criticized Einstein for daring to touch the principle of relativity (in the sense of special relativity).

"probability amplitude", (3) the collaboration with Boris Podolsky and Nathan Rosen, (4) the mail exchanges between Einstein and Schrödinger that led to the famous "Schrödinger's cat", (5) the influence that Einstein had on Everett, who was probably stimulated in thinking about the interpretation of quantum theory by hearing the last seminar of Einstein.

The main theme of these different moments is the idea that Einstein, far from being blocked by outdated *a priori*, had a deeper and more demanding vision of what must be a physical theory than many quantum physicists who were satisfied with a purely positivist vision of physics. This deeper vision can be called, when it is applied to quantum theory, "the Kantian Quantum" ("Le Kantique du Quantique"[7]), because it rests on an idea that goes back to the philosopher Immanuel Kant (but that Einstein had deduced from his work in General Relativity) : *It is the theory itself that defines "what is real".*

7 A Crucial Conversation

Berlin, Germany, early 1926

The young Werner Heisenberg was awed and impressed on entering the physics seminar room of the University of Berlin, on this day early in 1926[8]. He was only twenty-five years old, and had been invited to give a lecture on the "new quantum mechanics", which had just been born. While rather feverishly throwing a final glance at his notes, he saw, taking seats in the front row, the upper crust of the international physics community: Max Planck, Walther Nernst, Max von Laue, and others. The faces of these physicists, famous for their fundamental discoveries, held all of the seriousness and rigorous composure of German academic life. Then, just before the hour set for the beginning of the lecture, the physicist who impressed him most, he whose work he had admired since adolescence, when he had discovered the theory of general relativity in a book[9] entitled *Space, Time, Matter*, he whose letters were read aloud by his professor and thesis adviser in Munich, Arnold Sommerfeld, to illustrate his course: Albert Einstein entered the room and sat down in the front row, giving him a friendly smile, partly to excuse himself for nearly arriving late, and above all to put him at ease.

Thus given confidence, Heisenberg began to relate the physical concepts and mathematical formalism of the *new* quantum theory. Indeed, in the last few months there had developed, with unheard-of speed, a new mathematical formalism which was hoped to supplant the "old" *theory of quanta*. The old theory of quanta was

[7]I thank Malcolm MacCallum for useful suggestions about the various ways of translating into English this French pun on "the Song of Songs" ("Le cantique des cantiques").

[8]We are here inspired by the memoirs collected, much later, by Werner Heisenberg in his remarkable book *Physics and Beyond*, translated by A. J. Pomerans, Allen and Unwin, London, 1971.

[9]The book *Space, Time, Matter* by the mathematician Hermann Weyl, was one of the first books written on the theory of general relativity. The first edition dates from 1918.

that disparate collection of mutually contradictory ideas developed between 1900 and 1924, which attempted to describe the *quantum discontinuities* whose existence had slowly been revealed through the understanding of various physical phenomena. The discovery which had initiated the theory of quanta (the precise structure of black-body radiation) had been made here in Berlin itself, through the extremely precise measurements of Otto Lummer, Ernst Pringsheim, Heinrich Rubens, and Ferdinand Kurlbaum, and through Max Planck's theoretical "act of despair". But it was above all Einstein's collective work on quanta, between 1905 and December 1924, which had shown the need for a profound readjustment of physics. To which were added, starting in 1913, the innovative concepts of Niels Bohr who had shown how to apply the quantum ideas to atomic physics. The new quantum formalism which Heisenberg spoke of had come from some of Bohr's ideas on atomic structure, and some concepts introduced by Einstein in 1916 concerning the interaction between an atom and electromagnetic radiation. Einstein had introduced, among other things, some coefficients (denoted A), which measured the probability (per unit time) with which an atom, initially found in a certain (quantized) "state", could experience a "quantum transition" towards another quantized "state" with lower energy by emitting, at a random instant and in a random direction, a quantum of light[10]. Heisenberg had been initiated into the physics of these *quantum transitions* by his thesis advisor in Munich, Arnold Sommerfeld, and by Max Born, at Göttingen. After having completed his thesis at the age of twenty-two, he became Born's assistant at Göttingen in October of 1923. In 1923 and 1924, Heisenberg worked under Born's direction, and learned from him several crucial ideas and techniques, notably the idea to introduce new coefficients, denoted a, associated like Einstein's coefficients A to the quantum transition between two states of an atom. Roughly speaking, the new coefficients a, called "amplitudes of quantum transition[11]", were such that their squares were equal to Einstein's coefficients A.

The essential idea at the base of the new quantum theory had come to Heisenberg early in the month of June 1925, while he was recovering from a bad bout of hay fever by spending some time on the island of Heligoland, to the north of Germany. This idea consisted in replacing the usual notion of a continuous orbit describing the possible motion of an electron[12] around an atom by the collection of

---

[10] We recall that the possible energies, in quantum theory, for the "states" of an atom only take discontinuous values E_0, E_1, E_2, etc. The coefficient which Einstein associated to the quantum transition between the state of energy E_m and the state with (lower) energy E_n is denoted A_{nm}. Here m and n are indices which take the values 0, 1, 2, etc. If f_{nm} designates the frequency of light emitted during the transition between "the state m" and "the state n" (as we shall say for brevity), the energy of the quantum of light emitted is $E = hf_{nm} = E_m - E_n$, and its momentum takes the value $p = hf_{nm}/c$.

[11] The *amplitude* a_{nm} associated to the transition between the state m and the state n is a complex number ($a_{nm} = x_{nm} + iy_{nm}$ where $i = \sqrt{(-1)}$), of which the squared modulus ($|a_{nm}|^2 = |x_{nm}|^2 + |y_{nm}|^2$) is proportional to Einstein's coefficient A_{nm} associated to the same transition.

[12] Like Heisenberg in his first article, we here consider for simplicity an atom with only one electron.

amplitudes a, associated to the transitions between the atom's possible quantized states. Each transition amplitude is defined by supplying two numbers: the number fixing the initial energy state within the discontinuous list of possible quantum states of the atom, and the number fixing the final state. The total collection of amplitudes is thus analogous to a *checkerboard* or a *multiplication table*[13], of which each elementary square is fixed by supplying two numbers: one number fixing the "horizontal" projection of the square in question, the other fixing its "vertical" projection.

While Heisenberg was explaining the motivations which had led him to replace the description of the continuous orbit of an electron in an atom by such checkerboards of transition amplitudes, he looked with worry out of the corner of his eye to where Einstein was seated, to see how he was reacting to the introduction of such "witches' multiplication tables[14]". While not convincing him, Heisenberg succeeded in drawing Einstein's interest, particularly when, at the end of his lecture, he indicated that the new "rules of multiplication" of two amplitude tables, introduced by him and developed in recent work done in collaboration with Max Born and Pascual Jordan, permitted one to demonstrate, through detailed calculation, Einstein's result which said that the energy fluctuations of the radiation contained within a sub-volume were the sum of two separate terms: a term connected to the undulatory character of the radiation and a term connected to its corpuscular character. This result, concluded Heisenberg, showed that the new quantum formalism was capable of describing the undulatory and corpuscular aspects of a continuous field (such as the electromagnetic field) *at the same time.*

After the colloquium, Einstein came to congratulate Heisenberg on his remarkable results, and asked Heisenberg to accompany him home in order to discuss in more detail the new ideas at the base of the formalism which he had presented. Once arrived at his apartment, Einstein asked him to again explain the physical motivation leading to the replacement of the notion of a continuous orbit by that of an infinite table of transition amplitudes.

Let's listen to a central part of their dialog, such as it was later reconstructed by Heisenberg himself[15]:

HEISENBERG – ... Since it is reasonable to allow into a theory only directly observable quantities, I thought it more natural to restrict myself to these frequencies and amplitudes[16], bringing them in, as it were, as representatives of electronic orbits.

EINSTEIN – But all the same, you do not seriously believe that a physical theory should only include observable quantities?

---

[13]Born quickly realized that the "table" a_{nm} of (complex) amplitudes considered by Heisenberg could be identified with what the mathematicians called a "matrix", since the calculational rules introduced by Heisenberg, for physical reasons, were found to be the same as the rules for matrix calculations. We note however that the table of transition amplitudes a_{nm} is infinite, in general.

[14]To repeat an expression used by Einstein on December 25th 1925, in a letter to Besso.

[15]See chapter V of the memoirs of Heisenberg cited above.

[16]These are the table of values $f_{nm} = (E_m - E_n)/h$ and that of the values a_{nm} mentioned in the notes above.

HEISENBERG – I thought that it was you yourself who had made this idea the foundation of your theory of relativity. You stressed that one could not speak of an absolute time, since one cannot observe this absolute time. You said that only the readings of clocks, made in a system of reference either in motion or at rest, were able to determine the measurement of time.

EINSTEIN – Perhaps I used this sort of philosophy, but it is nonsense nevertheless. Maybe, to express myself more prudently, I will say that from a heuristic point of view, it could be useful to remember that which one really observes. But, at the level of principles, it is completely erroneous to want to found a theory uniquely on observable quantities. For, in reality, things happen in exactly the opposite way. *It is only the theory which decides what can be observed.*

We have emphasized the final sentence since it resonated for a long time in the young Heisenberg's mind, and played a crucial (and generally unknown) role in the later development of the quantum theory. Let us only here say that this "message" (it is the theory which decides what is observable) had been inculcated into Einstein by the years spent in the erratic construction of general relativity. For years, the connection (so clear in *special* relativity) between the coordinates of space and time and the measures of distance and duration had remained obscure in *general* relativity. Einstein had only worked his way free of confusion at the end of 1915 when he understood, after having constructed the theory, that it was the very mathematical formalism of general relativity which permitted one to define *a posteriori* that which was observable when space-time was deformed by matter.

8 "Waves Over Here, Quanta Over There!"

In the beginning of the year 1926, close to the time when Heisenberg had given his lecture at the Berlin colloquium, another mathematical formalism had been proposed, by the Austrian theoretical physicist Erwin Schrödinger, to supplant the "old" Planck-Einstein-Bohr theory of quanta. This formalism, called "wave mechanics", had, according to Schrödinger himself, taken root in the ideas of Louis de Broglie, and in the "brief but infinitely clairvoyant" remarks made by Einstein (within letters, and in the article from the end of 1924 discussed in the preceding chapter). Schrödinger's *wave mechanics* seemed completely different from the Born-Heisenberg-Jordan *matrix mechanics.* In the one, the state of the system considered (let's say electrons orbiting around the nucleus of an atom) was described by a *wave amplitude* $\mathcal{A}$, which was a *continuous* function[17] of time and of the coordinates of the electrons, while the other only considered the *discontinuous transitions* between the various possible stationary states of the atom, and described them by an *infinite checkerboard of transition amplitudes* a_{nm}. These two descriptions seemed to be antipodal to each other. The first gave a perfectly *continuous* image (in time, and in the space of configurations of the system), while

[17]More precisely, $\mathcal{A}$ is a *complex* function ($\mathcal{A} = \mathcal{A}_1 + i\mathcal{A}_2$). This wave amplitude is often denoted, following Schrödinger, by the Greek letter psi, ψ.

the second was only interested in the *discontinuous* transitions experienced by the system. Nevertheless, Schrödinger quickly enough showed that there was a *mathematical equivalence* between the two formalisms. More precisely, he showed that knowledge of the "wave equation" describing the propagation of the continuous amplitude $\mathcal{A}$ permitted the simultaneous calculation of the possible stationary states of the system, their quantized energies, and the infinite checkerboard of transition amplitudes between these stationary states. Roughly speaking, the possible stationary states are analogous to the series of pure vibrational states of an elastic object, like those of a piano string which can vibrate in its fundamental mode, or in the mode corresponding to the first harmonic (one octave higher than the fundamental mode) or even in the second harmonic (a fifth above the first harmonic), etc.

In fact, it seemed for a long time that Schrödinger's *wave* description was more complete than the Born-Heisenberg-Jordan *discontinuous* description. Above all, Schrödinger's description seemed to suggest that one could perhaps even get "rid of" the idea of quantum discontinuity (despite all that it had allowed to be understood, including Einstein's theory of atomic transitions), and describe reality uniquely in terms of a continuous wave phenomenon.

Einstein had initially welcomed, with satisfaction and some relief, Schrödinger's formalism, which seemed to him closer to his deeply rooted intuition about reality than the "witches' multiplication tables" used by Heisenberg and companions. But he was rather rapidly disenchanted. First, because the wave amplitude $\mathcal{A}$ was not propagating in the usual three-dimensional space but in a space of 6 dimensions for a system of two particles, 9 dimensions for a system of 3 particles, 12 dimensions for four, etc. And second, because wave mechanics had great difficulty in accounting for all of the experimental facts which had led Einstein and others, for around twenty years, to introduce the quantum discontinuities. In the month of August 1926, Einstein summarized his sentiments in a letter to Paul Ehrenfest:

"Waves over here, quanta over there! The reality of each has the solidity of rock. But the devil makes them rhyme together (and the rhyme is well and truly real)."

Einstein's dissatisfaction, on being confronted with the paradox that nature exhibits wave-like aspects and particle-like aspects at the same time, lasted until the end of his life. As we shall see, that which convinced most other scientists did not carry away Einstein's approval.

9 Einstein's "Ghost Field", Born's "Probability Amplitude", and Heisenberg's "Uncertainty Relations"

We shall not try to discuss, in even a slightly exhaustive way, the development of the physical interpretation of the mathematical formalism of quantum theory. We will only show the essential, though sometimes hidden, role played by some of

Einstein's ideas.

The first crucial advance dates from the summer of 1926, and is due to Max Born. As he explicitly wrote[18]: "I start from a remark by Einstein on the relation between [a] wave field and light-quanta. He [Einstein] said approximately that waves are there only to point out the path to the corpuscular light-quanta, and spoke in this sense of a 'ghost field' which determines the probability for a light-quantum ... to take a definite path ..." These remarks by Einstein on a "ghost field", or a "pilot field", were communicated verbally by him to several scientists (Max Born, Eugene Wigner, and others) in the 1920s, but he never published them. However that may be, it seems that they motivated Max Born to propose the interpretation of the wave amplitude $\mathcal{A}(t, \mathbf{q})$ of a certain physical system as an "amplitude of probability" to *find*, at the instant t, the system in the configuration described by the variables $\mathbf{q}$. [As mentioned previously, when we consider a single particle, $\mathbf{q}$ denotes its three coordinates in space; but, when we consider a system of *two* particles, $\mathbf{q}$ denotes the *six* coordinates necessary to fix the spatial position of two particles; etc.] Born further explained (in a footnote added during proofreading) that the *probability* of finding a system in a configuration $\mathbf{q}$ was proportional to the *square*[19] of the amplitude $\mathcal{A}(\mathbf{q})$. Max Born then summarized the essence of the interpretation of quantum theory which he was proposing:

"The motion of particles follows probability laws but the probability itself propagates according to a causal law."

The second part of Born's quote alludes to the fact that "Schrödinger's wave equation", written by the latter in early 1926, is a *deterministic* equation of propagation, which determines in a unique way the temporal evolution of the amplitude $\mathcal{A}$, once one knows its value at an arbitrary initial instant.

Born's "probabilistic interpretation" was an important conceptual advance, but it raised more questions than it answered. In fact, it was a mere hypothesis, while it should have been *derived* from the mathematical formalism of the quantum theory. This is what Heisenberg believed during the end of 1926 and the beginning of 1927. Werner Heisenberg was then working in Niels Bohr's group in Copenhagen. He held intense discussions with Bohr, which often lasted well past midnight, on the physical interpretation which should be given to the mathematical formalism of the quantum theory. In February 1927 Heisenberg, remaining alone in Copenhagen while Bohr was skiing in Norway, had a new idea destined to clarify the compatibility between a *wave* description and a *corpuscular* description for a single quantum particle (such as an electron). As he himself recalled[20], the memory of his conversation with Einstein one year before played a crucial role in his thought-process:

---

[18]For historical references on Einstein's "ghost field" (*Gespensterfeld*) and on its influence on the probabilistic interpretation of the wave amplitude $\mathcal{A}$, see the biographies of Abraham Pais on Einstein (see Bibliography) and of Bohr (*Niels Bohr's Times*, Oxford, Clarendon Press, 1991).

[19]$\mathcal{A}$ being a complex number, $\mathcal{A} = \mathcal{A}_1 + i\mathcal{A}_2$, the "square" we speak of here is the squared modulus of $\mathcal{A}$: $|\mathcal{A}|^2 = (\mathcal{A}_1)^2 + (\mathcal{A}_2)^2$.

[20]See chapter VI of his book: *Physics and Beyond, op. cit.*

"That night, it was perhaps around midnight that I suddenly recalled my discussion with Einstein, and that I remembered his phrase: 'Only the theory decides what one can observe.' I realized immediately that it was within this remark that one must look for the key to the enigma which had so occupied [Bohr and me]. I then went for a nocturnal walk through the Fälledpark to reflect on the import of Einstein's comment."

It is in the course of this night-time walk, reflecting on the import of Einstein's phrase, that Heisenberg discovered his very famous "uncertainty relations[21]", saying that the product of the "uncertainty" in the position of a particle and the "uncertainty" in its momentum[22] must necessarily be greater than "Planck's constant" h[23].

Heisenberg understood that the uncertainty relations permitted a clarification of the conditions in which one could use the idea that a quantum particle is *simultaneously* described by a wave *and* by a corpuscle. For example, it seemed that the observation of rectilinear tracks, visible at the macroscopic level, left by particles in certain detectors implied that a particle must necessarily be described as a localized corpuscle. The uncertainty relations showed that the finite width of the tracks was compatible with a wave behavior on distance scales which were small compared with this width.

When Bohr returned from his vacation in Norway, Heisenberg enthusiastically explained to him what he had found by following Einstein's philosophy ("The theory alone decides what is observable"). In the interval, Bohr had continued his own reflections and had convinced himself that it was necessary to base the interpretation of quantum mechanics not on a logical derivation dictated by the theory itself (as Einstein had suggested) but on a new epistemological concept, introduced in *ad hoc* fashion for the interpretation of the quantum theory, called "complementarity". As Heisenberg said, in Bohr's mind "complementarity should describe a situation where we could grasp a single and identical phenomenon by two different modes of interpretation [for example, wave and corpuscle]. These two modes must both mutually exclude and complete each other; and it is only the juxtaposition of these contradictory modes which allows one to completely exhaust the visual content of the phenomenon."

The discussion between the young Heisenberg (who was then twenty-six years old) and Bohr (whose 1913 work had played a crucial role in the development of the quantum theory) was rather stormy. Heisenberg admired Bohr as a scientist, and also venerated him like a father. He had expected that Bohr would appreciate the innovative conceptual advance represented by the discovery of the uncertainty relations. In place of this, Bohr expressed some reservations and offered some detailed technical criticisms. Above all, he only considered his own idea of *com-*

[21]Also known as *indeterminacy relations* or *dispersion relations*.

[22]Recall that the (relativistic) *momentum* of a particle is given by $p = mv/\sqrt{(1 - v^2/c^2)}$, where m is the particle's mass (at rest), and v its speed.

[23]Depending on the precise technical definition of "uncertainty", the minimum of their product may differ from h by a numerical factor.

plementarity to be general enough to serve as a basis for a coherent interpretation of the quantum theory. The tension between the two men was great, and led to permanent damage of their relationship. Confronted with Bohr's stubbornness, Heisenberg gave up on convincing him of the soundness of the general epistemological attitude suggested by Einstein, and reluctantly accepted the necessity of using an *ad hoc* interpretive language based on *complementarity*. Heisenberg published his discovery of the uncertainty relations, and their consequences for the interpretation of quantum reality, by himself, and left Bohr to prepare a detailed article on the idea of complementarity, which Bohr presented some months later at the Solvay council in the Autumn of 1927.

10 A Watershed Moment

The fifth Solvay Council, held in Brussels in the Autumn of 1927, was a very important event. It was a *watershed moment*, both for the international community of theoretical physicists[24], and for Einstein's scientific career. It is at this meeting that Einstein was first confronted with the "interpretation" of the new quantum theory proposed by Bohr, starting from ideas of Born (the probabilistic interpretation of the amplitude $\mathcal{A}$), and Heisenberg (the uncertainty relations), and from the concept of complementarity. Each of the theoretical physicists waited with a passionate interest to see Einstein's reaction. For everyone, Einstein was not only the greatest living physicist, but also the one whose revolutionary ideas had been crucial for the discovery and comprehension of quantum reality. The physicists of the younger generation (Heisenberg, Jordan, Pauli, etc.) worshiped Einstein, and considered themselves to be his modest successors. Was the pope of theoretical physics going to bless, on the baptismal font of complementarity, the new quantum child that everyone considered as his intellectual "grandson"? Well ... no! Einstein was not convinced by the interpretation of quantum theory defended by Bohr.

For many, the disappointment was great. And some (like Paul Ehrenfest) went so far as to compare Einstein's attitude *vis-à-vis* the new quantum mechanics to those of the opponents of the theory of relativity, who had been disconcerted by the novelty of Einstein's ideas and had refused to change "their old habits". I think that the traditional image of Einstein as an aging revolutionary, refusing the new quantum ideas because they went against his prejudices about what reality must, *a priori*, be, is *inexact*. This does not mean that I think the attitude of Bohr, and of the majority of physicists who followed him by adopting what is called the "Copenhagen interpretation" of quantum theory, had been an error. Far from it! From a *practical* point of view, the consensus which crystallized at the Solvay council of 1927 around the "Copenhagen interpretation" helped the development

[24]In the sense that certain physicists followed Einstein in his doubts concerning the definitive and/or complete character of the quantum theory, while the majority rallied around the "Copenhagen interpretation".

of the new quantum ideas, and has permitted their application in an ever-growing domain of physics. A large part of the physics and technology of the twentieth century is based on the application of quantum theory (to the physics of solids, to atomic physics, to high energy physics, etc.). The interpretive scheme proposed by Bohr at the 1927 Solvay council helped to "put aside" the obscure epistemological aspects of quantum theory, and enabled the exploration of the new world which was opened up by its mathematical formalism. However, having said that, I think that it is time (above all on the occasion of the centenary of the revolutionary ideas proposed by Einstein in 1905) to give a description of Einstein's attitude *vis-à-vis* the quantum theory that is not a crude caricature, and at the same time to recognize both the fundamental soundness of his epistemological objections, and the visionary character of the works he undertook after 1927.

Fundamentally, I think that Einstein was not convinced by Bohr because the idea of *complementarity* was only a conceptually obscure and technically ill-defined cloak. In May 1928, in a letter to Schrödinger (who shared his doubts) Einstein compared the "Copenhagen interpretation" to a "soft pillow", on which one could fall asleep without asking oneself questions about quantum reality:

"The tranquilizing philosophy (or, dare I say, the religion?) of Heisenberg-Bohr is so delicately put together that, for the moment, it furnishes to the true believer a soft pillow that he has a hard time leaving."

Later (in 1939), when Bohr had ossified into his posture as the apostle of complementarity, now a panacea for all of the problems of interpretation mentioned by Einstein, Schrödinger, and others, Einstein described Bohr (in a letter to Schrödinger) as a "mystic, forbidding any questioning about whatever might exist independently of the observer ...".

In a more precise fashion, I think that Einstein's dissatisfaction came from the fact that the "Copenhagen interpretation" was not in agreement with the idea which Einstein had expressed to Heisenberg, and which had led the latter to the discovery of the uncertainty relations: "It is the theory which decides what is observable." Bohr was adding an entire interpretive superstructure to the mathematical formalism of quantum theory, founded on the utilization of a special language, and having recourse to another scientific theory ("classical" Newtonian physics) which was supposed to apply to macroscopic objects (like the measurement instruments). It is because Einstein had very high standards of conceptual clarity that he could not be satisfied with the "tranquilizing philosophy (or religion?) of Heisenberg-Bohr". The clearest formulation that Einstein gave of his conceptual dissatisfaction is probably that which he expressed in 1932 in a letter to Wolfgang Pauli. We quote it such as it is, even if the Latin it uses is awkward:

"Incidentally, I do not say *probabilitatem esse delendam*, but *probabilitatem esse deducendam*, which is not the same thing."

In other words, Einstein does not say that one must *get rid of* (*delendam*) the probabilities [which appear, according to Max Born, in the quantum theory], but that one must *deduce* (*deducendam*) the appearance of these probabilities [from the mathematical formalism which defines the quantum theory]. Recall that

Einstein was indeed an expert in the utilization of probabilities in classical physics (thermodynamics, Brownian motion), and it is he who introduced probabilities into quantum physics (in 1916, in his work on the absorption and emission of light by atoms). During the twenty or so years in which he had been (nearly) alone in believing in the quanta of light, he had spent countless hours trying to render the (deterministic) wave-like and (random) corpuscular descriptions of light compatible. He was not a man to resign himself to an abandonment of the logical, deductive character of science in favor of what the American physicist Bryce DeWitt recently called a "fuzzy metaphysics".

11 Adventurers in Entangled Reality

But we shall here concentrate on another work from the Princeton phase of Einstein's career, that which he completed in 1935, in collaboration with Boris Podolsky and Nathan Rosen. This work illustrates well the visionary profundity of Einstein's approach towards physics. We have remarked previously on Einstein's refusal, in 1927, to accept the "soft pillow" of the Copenhagen interpretation of quantum theory. For several years, Einstein hoped to find a technical fault in this interpretation, for example in the form of a subtle violation of the uncertainty relations. Rapidly enough he convinced himself of the absence of such faults. He then made an effort to more finely characterize his dissatisfaction *vis-à-vis* the Copenhagen interpretation, and his feeling that either this interpretation, or the quantum theory itself, was *incomplete*. The article by Einstein, Podolsky, and Rosen (EPR for short) marks a very important stage in the understanding of the deep structure of quantum theory. Indeed, this article brought to attention a paradoxical aspect of the formalism of quantum theory: the "entanglement" of two physical systems which have interacted (quantum mechanically) in the past, before separating.

Let us give an example of such an "EPR situation". Consider a system of two particles. For simplicity, we shall suppose that the masses of the particles are equal to each other. Heisenberg's uncertainty relations say that one cannot measure, with infinite precision, both the position and speed of the first particle at the same time. Likewise, they forbid a precise simultaneous measurement of the position and speed of the second particle. Nevertheless, it can be shown that nothing forbids the specification (or measurement), with infinite precision, of both the position of the *midpoint* (the center of mass) between the two particles and their *relative speed.* Because of this, one may initially prepare the system of two particles in a quantum state where the midpoint between the two particles is a well-localized point, that we can take as the origin of coordinates, and where, moreover, the relative speed is zero. We let this system evolve freely from this initial state. Then, at a certain moment, we make observations (very far from the origin of coordinates) on one of the two particles, let's say the first. Heisenberg's relations forbid the *simultaneous* measurement of the position and speed of the first particle but nothing, in quantum mechanics, forbids the measurement, with infinite precision, of one or the other. Imagine first that we were measuring the position of the first particle and found it to be equal to a certain value x_1. As we know that the midpoint of the particles is fixed at the origin of coordinates, we deduce from this measurement that the position of the second particle is well-determined, and takes the value $x_2 = -x_1$. However, imagine that we had decided to measure not the position of the first particle, but its speed, and that we had found a certain value v_1 for this speed. Since we know that the relative speed $(v_1 - v_2)$ between these particles is zero, we deduce from this measurement that the speed of the second particle is well-determined, and takes the value $v_2 = v_1$.

Thus, according to the arbitrary choice that one makes on the fashion in which one observes the first particle, one can determine, with certainty, the position or the speed of the second particle *without directly observing it* and thus without disturbing it in any way. Einstein, Podolsky, and Rosen assumed that every certain prediction that one could make for a system, without perturbing it in any way, must correspond to something "real". They thus deduced from the thought experiment which we have just described that *both* the position and the speed of the second particle were "real" quantities, since they could both be precisely determined in indirect fashion, without disturbing the second particle. This conclusion seemed to be in conflict with the uncertainty relations associated to the position and speed of the second particle, unless there is something "magical" in quantum theory, that is to say an intimate "link" between systems separated by very large distances, causing every observation performed on a system to instantaneously effect the other system, and thus making it capable of changing its "real state". Einstein, Podolsky and Rosen thought that the existence of distant *links* between spatially separated systems was not physically acceptable, and deduced from their reasoning that there was something *incomplete* in the quantum description of a system through the probability amplitude $\mathcal{A}(x_1, x_2)$ [which was the basis for their reasoning].

When it first appeared, the EPR article did not have a great impact on the community of physicists. Most of them rested their minds on the "soft pillow" of Copenhagen and took no pains to reflect on the new perspectives opened up by the EPR article. Only Niels Bohr and Erwin Schrödinger took a lively interest in this paper. Niels Bohr responded to the "EPR paradox" by publishing an article which essentially consisted in reaffirming the "dogma" of complementarity[25]. He thus justified what Einstein had written about him, just after the publication of the EPR article and before Bohr's response, in a letter to Schrödinger:

"As for the Talmudic philosopher, he doesn't give a hoot for "reality", that hobgoblin capable only of scaring naive souls. He explains that the two points of

[25] We note that there is nothing "incorrect" in Bohr's response, and that moreover it would not be "incorrect" to say that recent experiments on the EPR system have "vindicated" Bohr. The author, however, thinks that Einstein's approach, translating conceptual questions into thought experiments (which were subsequently realized) reflected a better sense of physics than that of an *a priori* rejection of any need for experimental verification through a quasi-religious belief in the metaphysically fuzzy concept of complementarity.

view differ only by their mode of expression".

Here the expression "Talmudic philosopher" refers to Bohr, thus comparing him to a commentator on the divine revelation (here understood as complementarity).

As for Schrödinger, he understood that Einstein had put his finger on an important structure in the quantum formalism. In the months following the publication of the EPR article, Einstein and Schrödinger held a discussion through the mail. In this exchange, Einstein suggested the consideration of an unstable system, like a gunpowder barrel which has a 50% chance of catching fire within a certain time. Einstein noted that after this interval of time the quantum theoretical representation of the gunpowder barrel by a probability amplitude "then describes a sort of mixture containing the system which has not yet exploded and the system which has already exploded." This suggestion by Einstein (to consider a macroscopic system whose state depends crucially on a random process) was soon taken up and improved by Schrödinger in his famous example of *Schrödinger's cat.* This is a living cat placed into a box with a diabolical mechanism which will either kill or not kill the cat within one hour, according to whether a single radioactive atom has decayed or not. At the end of an hour, quantum theory describes the cat by a "probability amplitude" $\mathcal{A}$ which corresponds to a *superposition*, with equal weight, of the amplitude for a living cat and the amplitude for a dead cat. How is this quantum description to be reconciled with the fact that we never observe such superpositions of half living and half dead cats, but only a living cat *or* a dead cat?

The heritage of the EPR argument did not stop there. In 1964, Nearly thirty years after the publication of the original article by Einstein, Podolsky, and Rosen, the Irish theoretical physicist John S. Bell took the EPR dilemma seriously, between a structure of reality called "separable", where spatially separated systems do not influence each other at a distance, and a "non-separable" structure where some spatially separated systems remain *linked* between themselves, or as is also said, are *entangled*, if they have had the opportunity to interact in the past. Bell understood that these two possibilities could be distinguished by certain types of measurements performed on quantum systems which have interacted in the past. More precisely, he showed that the quantum entanglement, *à la* EPR, of the "quantities of internal rotation", also called *spins* or *polarizations*, of two particles issuing from an initial state with zero spin, must lead to *correlations* between measurements of the polarizations of the two particles which are strictly greater in the case of a non-separable, quantum reality than in the case of a separable, "classical" reality.

Bell's theoretical discovery invoked great interest in entangled situations *à la* Einstein-Podolsky-Rosen and prodded several experimental teams to test the *inequalities* that Bell had deduced for the correlations between the polarizations of separate particles, issued from an initially correlated system. The most convincing experimental results were realized in 1982, at the University of Orsay, France, by a group led by Alain Aspect. These results were in full agreement with the predictions of quantum theory, that is to say with a non-separable structure of reality where two systems which have interacted in the past remain *entangled* in the future, even if they are spatially separated. The experiments at Orsay verified the reality of this EPR entanglement for the polarization of photons separated by a dozen meters. Some more recent experiments, completed near Geneva, Switzerland by the group of Nicolas Gisin, have verified the reality of EPR entanglement for the polarization of two photons separated by more than ten kilometers!

Experiments performed on systems of the Einstein-Podolsky-Rosen type have thus shown that two systems which have interacted in the past continue to behave like an inseparable whole in spite of the spatial distance between them. This shows that "quantum mechanical reality" is very different from "classical reality". In addition to leading to progress in our comprehension of quantum theory, the entangled EPR states are presently the object of numerous studies, for it is thought that they might have very important applications within the domains of quantum cryptography and quantum computation[26].

12 The Mouse and the Universe

Princeton University, United States, April 14th 1954

When the old man entered, silence fell suddenly upon the sixty or so students assembled in room 307 of the Palmer Physical Laboratory, on that 14th of April, 1954. The students were emotional and excited. Everyone knew that it was an exceptional event. Without a doubt the only time in their life that they would see, in flesh and blood, and hear the speech of, the greatest physicist of all time, the living legend of twentieth century science: Albert Einstein. They were going to attend the great man's final lecture.

Some of them had had the privilege, the preceding year, of being invited to take tea in Einstein's house, at 112 Mercer Street, and were able to hear the master's direct answers on all the questions they posed: from the nature of electricity and the foundations of the unified field theory, to the expansion of the universe and Einstein's position on quantum theory. Einstein had joined in the game with grace and good humor, and had responded in detail to all of their questions. He was not even offended when one student, bolder than the others, dared to ask him: "Professor Einstein, what will become of this house when you are no longer living?" A large smile lit up the old man's face. He replied, without becoming disconcerted, in good English with a melodious German accent: "This house will never become a place of pilgrimage where the pilgrims come to look at the bones of the saint."

The American theoretical physicist John Archibald Wheeler had begun teaching relativity (special and general) in the physics department of Princeton University starting in the fall of 1952. It had been his idea to invite the students of his

[26] See chapter 5 of Alain Aspect *et al.*, *Demain, la physique*, Paris, Éditions Odile Jacob, 2004.

course on relativity to take tea at Einstein's house, in May of 1953, to help motivate them to study this theory deeply. It was he as well who convinced Einstein, in the spring of 1954, to come give a seminar before a select group of students from the physics department. Of course, the grapevine had done its work, and a fair share of students from neighboring disciplines, especially mathematics, had come to hear him. Some professors also slipped in amongst the group of students which filled the small seminar room.

The central theme of this lecture – which was effectively Einstein's last seminar, given one year, nearly to the day, before his death – was quantum theory[27]. Einstein explained why he thought that this theory was not the last word on the question. He reviewed the process of an atomic transition to a state of higher energy under the influence of electromagnetic radiation. By continually lowering the intensity of radiation, this transition process becomes more and more rare. This led to the introduction of a probabilistic description of the process of transition. Thus probability was introduced into quantum theory[28]. "I am a heretic. If radiation causes jumps [between atomic states], it must have a granular character like matter," Einstein exclaimed. Then, he came to his crucial point: what is the real meaning of the probability amplitude $\mathcal{A}$? Does it give a complete description of the physical situation? "I knew in constructing special relativity that it was not complete. So is everything that we do in our time: with one hand we believe; with the other, we doubt." Then Einstein gave as an example the quantum description of a macroscopic object (a sphere of one millimeter diameter moving in a box). The description of the motion, for fixed energy, of the tiny ball given by the probability amplitude seems paradoxical for an object which one can see with the naked eye. The probability amplitude gives a fuzzy description of the ball's position, while everyday experience shows that the ball is always seen at a well defined location.

"It is difficult to believe that this description is complete. It seems to make the world quite nebulous unless somebody, like a mouse, is looking at it ... When a person such as a mouse observes the universe, does that change the state of the universe?"

Many of the attendees were struck by Einstein's evocative image. Einstein then mentioned that he believed that logical simplicity could, sometimes, be a good guide, for it is thus that he had constructed the theory of general relativity. He explained how he had found this theory, and why he thought it was incomplete: the description of matter by means of the distribution of energy and stress seemed to him to be something provisional, "a wooden nose in a snow man". He regretted that most physicists took quantum theory and the theory of special relativity as their

---

[27]The content of this lecture is known to us through notes taken by John A. Wheeler during the seminar, and by the memories reported by some of the attendees. See p. 201–211 of the book edited by Peter C. Aichelburg and Roman U. Sexl, *Albert Einstein, His Influence on Physics, Philosophy, and Politics*, Braunschweig/Wiesbaden, Vieweg, 1979.

[28]Recall that it is Einstein himself who introduced probability into quantum theory in the 1916 article where he described precisely this transition process between atomic levels under the influence of electromagnetic radiation.

starting point, while neglecting gravity as being unimportant. On the contrary, he thought that gravitation, or more generally the structure of space-time, must be taken into consideration from the beginning. He finished by indicating that: "There is much reason to be attracted to a theory with no space, no time. But nobody has any idea how to build it up."

Among Einstein's audience, on April 14th 1954, was an emaciated, nervous young man with an eagle's profile and an intense gaze: Hugh Everett III[29]. He was only twenty-three years old, and had come with his friend Charles Misner, who was taking Wheeler's course in Relativity. Hugh Everett would not have missed this opportunity to hear his idol for anything in the world. At twelve years old, he had written to Einstein to ask him whether the universe was based on a structure that was random or unified. And he had had the great surprise of receiving a friendly reply from Einstein himself. After having studied chemical engineering for the first two years of university in Washington, he had spent the last six months (since September 1953) at Princeton University, where he was affiliated with the mathematics department. However, he was in fact interested primarily in theoretical physics. Since classes had commenced in September 1953, he had followed in particular the course on introductory quantum mechanics given by Robert Dicke.

Hugh Everett was struck by Einstein's remarks on the apparently incomplete character of quantum theory, which offered a "nebulous" description of the universe, and which seemed to need the presence of living beings, even if it only be one mouse, to trigger what the partisans of the Copenhagen dogma called the "collapse of the wave-packet": the passage from a fuzzy world to the sharply defined world that we see around us. He began to seriously reflect on the physical meaning of the formalism of quantum theory.

Some months later, during an evening party soaked with sherry, an animated discussion took place at the Graduate College between Hugh Everett, Charlie Misner, and Aage Peterson, who was an assistant of Niels Bohr, and who was passionately interested in the problems posed by the interpretation of quantum theory. In the heat of conversation, Hugh sketched a new conceptual scheme for the interpretation of quantum theory in such a way as to avoid both the paradoxes raised by Einstein (and Schrödinger), and the necessity of assuming a mysterious random process of wave-packet collapse. This idea of genius, obtained when he was about twenty-four years old, was the seed of Hugh Everett's doctoral thesis, in which he developed a revolutionary interpretation of quantum theory.

Everett went to see John Wheeler (who had been a disciple and collaborator of Niels Bohr, and who was very interested in the meaning of quantum theory) and asked him to supervise his doctoral thesis. Wheeler accepted. This created some

[29]I thank Charles W. Misner for having confirmed to me the presence of Hugh Everett at this lecture. For a detailed biography of Hugh Everett III, see the text by Eugene Shikhovtsev (edited by Kenneth Ford) on Max Tegmark's internet site: http://space.mit.edu/home/tegmark/everett/index.html. We have pulled the greater part of the facts concerning Everett cited in the text from this biography.

problems for Everett. On the one hand, Wheeler was quite open to new ideas, and he encouraged his students to think for themselves. On the other hand, Wheeler had unconditional admiration for Bohr and his principle of complementarity. Because of this, while recognizing the innovative character of Everett's ideas, Wheeler presented many objections to the way in which they were expressed. For example, in a note to Everett from September 1955, Wheeler wrote that he would be "frankly bashful about showing it to Bohr in its present form" since it could be " subject to mystical misinterpretations by too many unskilled readers". Finally, after insistent advice from Wheeler, Everett summarized the long text in which he developed his ideas in detail into a much shorter text which he defended, as a doctoral thesis, in 1957, and which was published the same year, accompanied with an assessment by Wheeler.

Everett's interpretation of quantum theory is one of the great conceptual advances of twentieth century physics. The author of this book thinks that it would have pleased Einstein (who died in April 1955, when Everett had just begun to develop his idea). Indeed, not only did it supply a new response to the paradox of the mouse looking at the universe, mentioned by Einstein in his final lecture, but above all it fits perfectly with Einstein's scientific philosophy, such as we have previously outlined it. Let us recall Einstein's statement to Heisenberg, "The theory itself defines what is observable," which put Heisenberg on the path of one of the first conceptual advances in quantum theory: the "uncertainty relations". As we shall see, Everett's interpretation is the first to take Einstein's statement seriously[30].

Nevertheless, in spite of – or, perhaps, because of – its novelty, Everett's interpretation raised no interest. Before it was revived, through the efforts of the theoretical physicist Bryce DeWitt in the 1970s, it was completely ignored, even by the recognized experts on the history of quantum mechanics (like Max Jammer). This rejection is doubtless due in part to the total lack of interest in Everett's ideas shown by Niels Bohr himself. Bohr read the long version of Everett's thesis, and raised some objections. In the spring of 1959, at Wheeler's insistence, Everett visited Copenhagen for six weeks in order to meet Bohr and discuss his interpretation with him. Everett kept a very bad memory of this meeting. Bohr was not interested, and he gave Everett no opportunity to explain his ideas in detail[31]. Today, according to a recent poll conducted by e-mail, the majority of theoretical physicists interested in understanding cosmology within a quantum framework use Everett's interpretation. In fact, they have no choice. As written recently by Bryce DeWitt, who rescued Everett's interpretation from obscurity:

"Everett's interpretation has been adopted by the author [Bryce DeWitt] out of practical necessity: he knows of no other. At least he knows of no other

[30] I do not know if Everett had explicitly heard this phrase. He could have heard of its existence through John Wheeler, who must have known it. This phrase figures prominently in the book of John Archibald Wheeler and Wojciech Hubert Zurek, *Quantum Theory and Measurement*, Princeton, Princeton University Press, 1983.

[31] See the Everett biography by Eugene Shikhovtsev (edited by Kenneth Ford), *op. cit.*

that imposes no artificial limitations or fuzzy metaphysics while remaining able to serve the varied needs of quantum cosmology, mesoscopic quantum physics, and the looming discipline of quantum computation[32]."

13 The Multiple World

What is the essential idea of Everett's interpretation? To introduce it, let us recall the central paradox of quantum theory, such as was highlighted by the arguments of *Einstein's gunpowder barrel* (half-exploded, half-intact) and *Schrödinger's cat* (half-living, half-dead). Quantum theory describes the system consisting of the cat and its environment (the box enclosing it, the air it breathes, the lethal mechanism triggered by a radioactive atom, etc.) by a function of the *configuration* of the system. To each configuration q of the system is associated a (complex) number $\mathcal{A}(q)$ that we shall simply call the *amplitude* of configuration q. What is a configuration q, considered at a fixed time t, and how is it described? For example, one could describe each possible instantaneous *configuration* of the cat and its environment by specifying the position in space of each of the system's atoms[33] (the atoms making up the cat, those in the air, those in the lethal mechanism, etc.). The position of each atom is specified by giving its three coordinates in space (length, width, and height). Let N be the number of atoms in the system. The number N is gigantic. Indeed, we recall that a gram of matter contains around six hundred thousand billion billion (6×10^{23}) atoms. A configuration of the total system is thus specified by giving a (gigantic) list of $3N$ numbers. The notation q denotes such a list[34].

Dear reader, maybe you take fright at the thought of considering a quantity $\mathcal{A}$ which depends on such a gigantic number of variables. All the more so since, as we have briefly indicated, the amplitude $\mathcal{A}$ is not a regular "real" number (like 2.5 or 3.1416) but a complex number, which is to say essentially an arrow within a plane, which requires two "real" numbers for its description (for example the length of the arrow, and the angle that it makes with an arrow pointing east). To visualize what such an amplitude $\mathcal{A}$ means, we can use a representation introduced by the author in a previous book[35]. It consists of (mentally) using the techniques of film-making.

First, each configuration q of the system is represented by a (holographic[36])

[32] Bryce DeWitt, *The Global Approach to Quantum Field Theory*, Oxford, Clarendon Press, 2003; volume 1, page 144.

[33] To be more precise, we must consider all of the stable elementary particles of the system (electrons, quarks) and include as well a description of the various interaction fields (electromagnetic, weak and strong nuclear, and gravitational).

[34] In other words $q = (x_1, y_1, z_1; x_2, y_2, z_2; \ldots; x_N, y_N, z_N)$. The amplitude $\mathcal{A}$ is a complex function of the time t (at which the configuration is considered) and of the $3N$ real variables q.

[35] Thibault Damour and Jean-Claude Carrière, *Entretiens sur la multitude du monde*, Paris, Éditions Odile Jacob, 2002.

[36] An ordinary photographic image is an imperfect representation since it projects a three-dimensional configuration onto a flat, two-dimensional film. The reader must imagine that either

photographic image of the system at the instant considered. To each q, that is, to each photographic image of the system, we want to associate a certain amplitude $\mathcal{A}$, determined by an arrow in a plane, having a certain length and pointing in a certain direction. To each direction of the arrow, one may associate a *particular hue* of color on the "color wheel": for example we associate to the direction east (on a map) the color orange, then, as one rotates the direction in a clock-wise direction, one changes the corresponding color by passing successively from orange (east) to red (south-east) to violet (south) to indigo (south-west) to blue (west) to blue-green (north-west) to green (north), and finally to yellow (north-east). As we continue to rotate from north-east to east, the hue evolves continually from yellow back to orange, in such a way that we land again solidly on our feet, having spread a full spectrum of hues around the circle. We have already mentioned that each amplitude $\mathcal{A}$ corresponds to *a length and a direction.* To the *length*, we can associate an *intensity of light* (a weak intensity if the length is short and a strong one if the length is long), and to the *direction* a *hue* of color (for example orange). Thus we can fix each complex amplitude $\mathcal{A}$ by a *color*, having both a particular intensity and a particular hue: for example a high-intensity orange, or a medium-intensity red, or a weak-intensity green, etc.

Let us then combine these two representations: that of the spatial configuration of the system by a *photographic image* (initially in black and white), and that of the "amplitude" $\mathcal{A}$ associated to this configuration by a *color* (intensity and hue). This gives us a *photographic image having a certain intensity and a certain hue.* For example, at a given instant, the living cat with his environment is represented by an intense blue image, and the dead cat with his environment by a red image of the same intensity. We may now *superpose* these two images, by the film-making technique of *double exposure.* That is to say, we print onto the same frame the two preceding images. This multiple exposure of images of the system, colored more or less intensely, gives a fairly faithful representation of the mathematical notion of a complex amplitude $\mathcal{A}$ depending on a spatial configuration q. To complete this representation, it suffices to vary the *instant* t at which we consider the system. Thus, to each instant t there corresponds a frame, multiply exposed to several colored images with more or less intensity. By considering all the successive instants, we thus obtain a (continuous) series of (colored and multiply exposed) images, that is to say a *film, in color and with multiple exposures.* Finally, we must imagine that the hue of each configuration changes extremely quickly, moving rapidly around the color wheel, as soon as the configuration is modified, even in an infinitesimal fashion (for example as soon as a single atom of the configuration moves). Moreover, even for an empty "still frame", where the configuration does not move at all, we must imagine that its hue changes very rapidly in the course of time, rotating at top speed around the color wheel (while the intensity of the light remains constant)[37].

we are speaking of three-dimensional photographs or of two-dimensional holograms containing all of the spatial information of the configuration.

[37]More precisely, the frequency f with which the hue of a physical system turns on the color

Let us now explain Everett's idea. It consists in taking seriously Einstein's statement: "The theory itself defines what is observable." Let us first take quantum theory seriously and ask it to define "what is real". Each configuration q will have "more or less reality" according to the value of the amplitude $\mathcal{A}(q)$. In other words, we interpret $\mathcal{A}$ as an *existence amplitude*, and not (like in the Born-Heisenberg-Bohr interpretation) as a *probability amplitude*. Indeed, the notion of probability amplitude for a certain configuration q suggests, from the very beginning, a random process by which only one configuration, among an ensemble of possible configurations, is realized, passing from the possible to the actual. By contrast, the notion of existence amplitude suggests the simultaneous existence (within a multiply-exposed frame) of all possible configurations, each actually "existing", but with more or less intensity (with the color encoding the "orientation" of the amplitude $\mathcal{A}$, which in physics is called its "phase").

Using the film-making analogy explained above, let us now describe the two basic elements making up the Everett interpretation. The first consists in saying that "quantum reality" is a color film with multiple exposures. At each instant, all of the individual images in the exposure "exist" with an intensity given by the length of the complex amplitude $\mathcal{A}$. The only configurations q which "do not exist" are those with a null amplitude $\mathcal{A}(q) = 0$. Having arrived at this stage, the reader may say to him- or her-self that the film obtained by successively projecting all these multiply-exposed images will be effectively invisible. It will only offer an infinite jumble of confused images. We seem to thus recover the "nebulous" or fuzzy description which Einstein and Schrödinger complained of, while we actually see, around us, reality "existing" in one well-defined configuration, like in a unique film with sharp images and no double exposures.

It is here that the second element of Everett's interpretation comes into play. To completely explain this second element, we must first have recourse to certain mathematical characterizations measuring the fact that certain images (or certain successions of images, that is to say certain films) are so different from each other that, when we superimpose them, they "don't interfere" with each other, with the effect that we can "focus" on one image or the other. We allude here to a mathematical phenomenon similar to what is known as [38] the "cocktail party effect": the possibility for two people to have a conversation between themselves, in the middle of the brouhaha formed by the intersecting conversations of other people. Another analogy, helpful for radio owners, would be that of changing the reception frequency to be able to listen, without "interference" from the other

---

wheel is given by the Planck-Einstein relation ($E = hf$). That is, it takes the value $f = E/h$ where E is the total energy of the system and h is Planck's constant. This link between the energy of the system and the frequency of rotation around the circle of the complex amplitude $\mathcal{A}$ essentially constitutes the famous "Schrödinger's Equation". Because of the extremely small numerical value of Planck's constant h, the frequency f is extremely large for any macroscopic energy E.

[38] For the "cocktail party effect" and more generally for a more detailed explanation of the notion of existence amplitude and of Everett's interpretation, see T. Damour and J.-C. Carrière, *op. cit.*

channels on the dial, to one particular station.

In other words, to return to our cinematic analogy, Everett tells us that among the hodge-podge of the total multiply-exposed film, there exist sub-films with (more or less) sharp images, which evolve in time according to (more or less) logical scenarios. The important point here is that the characters who evolve within such a sub-film act, at each instant, under the influence (almost) exclusively of those things which they have seen or felt in the previous images of the same sub-film.

Let us give a cinematic example of this idea. In the middle of Frank Capra's beautiful film *It's a Wonderful Life*, the hero, George Bailey, played by Jimmy Stewart, wants to commit suicide on Christmas Eve, because he believes himself to be a useless failure. Clarence the angel then plays out before his (and our) eyes, from the beginning, the film of what would have happened if George had never existed. This second film also develops in a coherent manner, and progressively becomes quite different from the first, which is to say the first half of Capra's film. Everett's idea is essentially that, in the total quantum reality, the two halves of the film (with or without George Bailey) are superimposed on each other, and thus play out simultaneously. Nevertheless, within each sub-film each character only has knowledge of what has happened and is happening in their own layer of the film, and thus has no "consciousness" of the "existence" of the other sub-film, playing out on a neighboring layer.

Let us finally note that Everett did not completely establish the necessity of what he proposed. By making the hypothesis of the existence of sub-films, which do not interfere with each other, he realized an essential desideratum of Einstein ("Probabilitatem esse deducendam"), that of justifying the connection between the existence amplitude $\mathcal{A}(q)$ and the *probability* for an observer to see the corresponding configuration q[39]. Later, other physicists justified the (apparent) existence of sub-films which do not interfere with each other by studying what is now called the *decoherence* between two possible sub-films[40].

We further note that Everett's interpretation was called, by he who brought it back from obscurity, Bryce DeWitt, the "Many-Worlds Interpretation". This name refers to the existence of numerous non-interfering sub-films in the midst of the total, multiply-exposed film. One then says that the world "splits" at every moment into multiple, slightly different versions which, in their turn, split in the following instant, etc. This leads to an image of a world which continually "branches out" in a multiplicity of separate worlds. This image has been used by excellent physicists who well understood Everett's interpretation: notably Bryce DeWitt and David

[39]Later, other physicists, notably Bryce DeWitt, would improve the proof sketched by Everett.

[40]One of the first physicists to understand the role of decoherence in quantum theory was Hans Dieter Zeh (1970). The first rigorous result on decoherence, and on its role in justifying the "quantum theory of measurement", is due to the Swiss mathematical physicist Klaus Hepp (1972). Decoherence is presently the object of many experimental studies (notably by the group led by the French physicist Serge Haroche). It is indeed essential to understand and master decoherence in order to envisage utilizing all the possibilities offered by quantum theory in cryptography and computation.

Deutsch[41]. I nevertheless find this image inappropriate, as it suggests a complete splitting between separate classical worlds, similar to the splitting from one cell into two, and so on to an irreversible multiplication. I prefer to remain closer to the formalism of the theory itself and to speak of *a multiple world*, that is, *one* film, multiply-exposed.

We mention finally that by calling reality a *multiple world* one could (and in fact one should) understand the word "world" in the sense used by Minkowski, which is to say *a space-time*. *Classical* (in the sense of pre-quantum) relativistic reality is identified with a unique space-time, that is to say with a four-dimensional world. In our cinematic analogy, such a world corresponds to *one film*: a sequence (or a "stack") of three-dimensional images. *Quantum* reality corresponds to a multiply-exposed film, which is to say a stack of superimposed images. Note that, starting from this stack, one can distinguish *a priori* a very large number of sub-films, many more than the number of layers of exposure within one instantaneous image. Indeed, if one considers a mini-film of three successive images, each of which has two layers of exposure, one can assemble $2 \times 2 \times 2 = 2^3$ sub-films of three successive images, each of which is taken at random from the two possible images at each of the three instants of the total film. Everett nevertheless tells us that most of these sub-films only "exist" with an amplitude too weak to be perceived. Only certain "quasi-classical" sub-films, whose amplitudes are reinforced by a process of *constructive interference*, will "exist" with an amplitude strong enough to be perceived.

14 The Kantian Quantum

The reader may be saying to him- or her-self that Everett, and those who adopt his point of view, have truly passed outside the boundaries of the "reasonable", and that this idea of a multiple, phantasmagorical world is too "absurd" to be taken seriously. It is indeed because of the revolutionarily "absurd" character of Everett's idea that it was ignored (particularly by Bohr), rejected or considered taboo for nearly thirty years. Even today, some experts in the interpretation of quantum mechanics reject, with an incredulous and disapproving sneer, Everett's interpretation by arguing that it shamefully violates the principle of logical economy proposed by William of Ockham: "One should not increase, without necessity, the number of entities required to explain something."

On the contrary, we would like to point out that Everett's interpretation is characterized by its logical economy. It is the only interpretation of quantum theory which does not add foreign (physical or metaphysical) elements to the theory. As for us, we consider that it is the only possible interpretation (see also the above quote by Bryce DeWitt) and that it finds its justification in the most rigorous and most rational epistemology, notably that of the German philosopher Immanuel

[41] See the stimulating book by David Deutsch, *The Fabric of Reality: The Science of Parallel Universes and Its Implications*, New York, Penguin Books, 1997.

Kant.

One of Kant's goals was to clarify the "nature" of the objects (space, time, force, and matter) which science speaks of, and to understand to what extent science makes "true" assertions about these objects. For example, is the absolute "space" evoked by Newton something "real", which "exists" by itself independently of things? Is the Euclidean geometry that is attributed to space "true" in an *a priori* fashion, before making measurements to verify it? This is not the place to discuss in any detail the responses Kant gave to these questions[42]. Let us simply say that while Kant of course recognizes the essential role of *experimentation* in the progress of physics, he strongly insists on the fact that experimentation is only truly fruitful if *reason* "takes the forefront" by posing a logical and mathematical framework permitting the interpretation of experimental results, and giving them meaning. This concept upends the very notion of "reality", that is to say the meaning of what is "an object" or "a thing" for the rational investigator. As Kant wrote:

"Hitherto it has been assumed that all our knowledge must conform to objects.[...] Let us try to see whether we may not have more success in the tasks of metaphysics, by assuming that objects must conform to our knowledge."

Let's apply this philosophy to the interpretation of quantum theory. This will lead us to what I like to call "the Kantian Quantum[43]", where the word *Kantian* makes reference to a perfectly rational approach. The Kantian Quantum is thus an approach in which we must adjust our understanding of "objects", that is to say the very notion of reality (the word "reality" derives from the Latin *res* = thing), to "our knowledge", or, more specifically, to the quantum theory itself.

Indeed, quantum theory has been verified by a huge number of experiments which have, in particular, confirmed the validity of its most "bizarre" consequences, like the entanglement predicted by Einstein-Podolsky-Rosen between separated systems, and the superposition of different macroscopic states, of the type of Einstein's gun-powder barrel or Schrödinger's cat[44]. One can, and in fact should until getting contrary information, consider quantum theory as firmly established *knowledge*. Then, starting from this knowledge, that is, from the mathematical formalism of quantum theory, if we ask this formalism to define the nature of "quantum objects", or "quantum reality", one necessarily falls back on Everett's point of view, since it is the only "interpretation" which is founded uniquely on the formalism of the theory, adding neither "fuzzy metaphysics", nor verbal incantations, nor new, non-verified hypotheses.

I have stated several times that Einstein himself had asserted his adhesion to

[42]See Immanuel Kant, *Critique of Pure Reason*, New York, Palgrave Macmillan (2003). See also the previously cited book of Martin Heidegger, *What Is a Thing?*.

[43]In French, "Le Kantique du Quantique", which is a (non-translatable) play on the French "Le Cantique des Cantiques", which refers to the "Song of Songs" in the Bible.

[44]Some recent experiments, due notably to the group of the physicist Serge Haroche, have permitted detailed observation of situations of the "Schrödinger's Cat" type for *mesoscopic* systems (intermediate between the microscopic level and the macroscopic level).

a point of view close to Kant's ("Only the theory decides what is observable"). It is of interest to note that he expressed himself in a letter to Schrödinger (written just after the appearance of the EPR article) in a manner very close to Kantian views, and this precisely in regard to the mysterious character of quantum reality:

"The true difficulty lies in the fact that physics is a sort of metaphysics: physics describes 'reality'. However, we do not know what 'reality' is, we only know it through the description given by physics!"

References

[1] Michel Paty, *Einstein philosophe*, Presses Universitaires de France, 1993.

[2] Paul Arthur Schilpp, Albert Einstein: philosopher-scientist, The library of living philosophers, Lassalle (Ill.), open Court, 1949.

[3] Albert Einstein, *Sur la méthodologie de la physique théorique* (Herbert Spencer Lecture, 1933) in Œuvres choisies, under the direction of Françoise Balibar, Editions du Seuil and Editions du CNRS, tome 5, p. 102.

[4] For a recent review of Poincaré's contributions to Relativity, see T. Damour, *Poincaré, Relativity, Billiards and Symmetry* (conference given at the Symposium Solvay on Henri Poincaré, Brussels, 8–9 octobre 2004), hep-th/0501168.

[5] Albert Einstein, *Remarks on Bertrand Russell's Theory of Knowledge,* in "The Philosophy of Bertrand Russell, Vol. V of 'The Library of Living Philosophers,'" edited by Paul Arthur Schilpp, Tudor Publishers, 1971, reprinted in "Ideas and Opinions," Three Rivers Press, 1954/82.

[6] Albert Einstein, lettre du 7 mai 1952 à Maurice Solovine, in Œuvres choisies, tome 4, pp. 310–311 ; see also pp. 410–411 of [1].

[7] Maurice de Broglie, "Poincaré et la philosophie", *in* Le livre du centenaire de la naissance de Henri Poincaré, 1854–1954, *in* Œuvres de Poincaré, XI (deuxième partie du tome XI) pp. 71–77 (Gauthier-Villars, Paris, 1956) ; cité in [1].

[8] Thibault Damour, *Si Einstein m'était conté*, Le Cherche-Midi, Paris, 2005; *Once Upon Einstein* (English translation by Eric Novak), A.K.Peters Ltd, Boston, 2005, in press.

Thibault Damour
Institut des Hautes Études Scientifiques
35, route de Chartres
91440 Bures-sur-Yvette
email: `damour@ihes.fr`

Einstein, 183– 199
© Birkhäuser Verlag, Basel, 2005

Poincaré Seminar 2005

On Boltzmann's Principle and Some Immediate Consequences Thereof

Albert Einstein

Abstract. This is a previously unpublished manuscript of the lecture given by A. EINSTEIN on November 2, 1910 before the Zürich Physical Society. The original document can be found in the bequest of Heinrich ZANGGER, at the Manuscripts Department of the Central Library of Zürich. Transcribed with the permission of the Central Library by W. HUNZIKER and A. SCHULTZE (Institute for Theoretical Physics, ETH Zürich), kindly transmitted by N. STRAUMANN (Institute for Theoretical Physics, University of Zürich-Irchel).[1] The notes are by B. DUPLANTIER (Department of Theoretical Physics, Saclay). Copyright: EINSTEIN, ALBERT: THE COLLECTED PAPERS OF ALBERT EINSTEIN, ©1987–2005 Hebrew University and Princeton University Press. Reprinted by permission of Princeton University Press.

Translation by:
Bertrand DUPLANTIER and Emily PARKS
from the original German text into French and English [2]

All natural science is based upon the hypothesis of the complete causal connection of all events. Let us suppose that Galileo, during his research on the pendulum, had found that the latter oscillates in such a way that the duration of one oscillation varies in an irregular way. Let us moreover suppose that this variation could not be put in relation with the variation of any other observable circumstances. It would have been then impossible for Galileo to reunify his observations under one law. Had all phenomena accessible to us a character as irregular as we have just imagined in this fictitious case, it would certainly never have befallen mankind to pursue the natural sciences.

Where do we stand now with our present-day knowledge of the complete causal connection of events? One should not answer the question before having made it more precise. We shall do this right now by making use of an example. Here sits a copper cube of a given size. By external influence we set in the cube

[1]We are especially grateful to Norbert STRAUMANN for having mentioned the existence of this unpublished manuscript by Einstein, and for having kindly transmitted a copy of it and its transcription.

[2]We thank Simone WARZEL (Princeton University) for kindly helping with the resulting translation and checking its faithfulness to Einstein's beautiful but complex style in German, and Thomas C. HALSEY (ExxonMobil R&E, Annandale) for a final reading of the manuscript.

a well-defined temperature distribution and we leave it, after having embedded it inside a thermal-insulating wrap at a given time, by itself. We know that by the process of heat conduction a temperature equilibrium is then reached in the course of time. The flow of temperature at any point of the cube will appear as "entirely determined" by the initial state only; by the expression "entirely determined" we mean that we will always observe the same diffusion of temperature each time we repeat the experiment, i.e., each time we set the temperature distribution at the beginning and then leave the cube by itself. Such an undeniable certainty of evolution, such a complete causal connection of events, do they actually exist? To answer a readily apparent, although uninteresting, objection, we prefer to put the question in the following way : do we observe a complete causal chain of events always with a precision which is the greater the more exactly we realize the initial state and the more exactly we pursue the measure of the course of time?

The physicist's point of view on this question has considerably changed during the last century. If for a moment we put aside Brownian motion, radioactive fluctuations, and a small number of other phenomena on which scientific interest has focused in recent years only, we then firmly and definitively arrive at the judgement that a complete causal connection (with the meaning explained above) exists by experience. However the physicists, and quite especially the kinetic theorists, came to deny the existence of a complete causal connection of events, more precisely events insofar as the latter can be subjected to observation. Let us turn our attention for a fleeting moment to this development ! From the simple representation of gases as made up of material points (molecules), that essentially act mechanically upon each other just by contact (collision), Clausius was able to deduce a relation between the specific heats and the constant of the state equation for monoatomic gases, as well as a relation between heat conduction, internal friction and diffusion in gases, for which the magnitude and the phenomena, respectively, stayed without any nexus in the absence of Clausius's theory. This great success led physicists to impute thermal phenomena to the irregular movement of molecules. But this kinetic theory of heat conveyed the idea that the laws of heat conduction, etc., had to be considered as approximately valid laws only; from this theory, there cannot generally exist an exactly valid law of heat conduction, except as a law valid for mean values. That the deviations from such mean value laws are usually very small is, in principle, irrelevant.

The kinetic theory of heat, experimentally verified in such an overwhelming way, is not only incompatible with the hypothesis that observable events are connected to one another in a completely causal way. The investigations led by Maxwell, Boltzmann, and Gibbs also show that deviations from these mean value laws, of any magnitude and accessible to observation, must appear, even if that occurs according to the theory so rarely in most systems that we would not be able to actually ascertain those deviations.

The following well-known consideration most strikingly shows that heat conduction laws, as well as all other laws concerning non-reversible processes, cannot be exact. According to the kinetic theory of heat, the temporal inversion of each molecular motion is an equally possible motion; there is therefore in general no thermal process that could not run backwards. So one should consider as possible, from the point of view of the molecular theory of heat, that by simple heat conduction, some heat flows over from a colder to a warmer body. Why do we not observe that? Does not this consideration show that the kinetic theory of heat should be abandoned?

This question was answered by Boltzmann and, to be more precise, in the following way : Consider any isolated physical system the energy of which has a fixed given value. We label by $Z_1, Z_2 \cdots Z_l$ all the observable states of the system with the same given energy value. In the example of the copper cube, each Z_ν would represent a determined temperature distribution, where a total of l different temperature distributions are possible. But let us suppose now that these states Z have rather distinct probabilities, such that among all states close to a given state Z_a, one (Z_b) is much more likely than all the others, at least in the case where Z_a differs considerably from the so-called thermodynamic equilibrium state. When the system is brought into the state Z_a and next left by itself, it evolves much more probably into the state Z_b than into any of the other neighbouring states of Z_a. The probability for this to happen can be as close to unity as one wishes (i.e., to certainty); whereas it remains ruled out in principle that this transformation is absolutely certain. This means that when we prepare many times the system in the state Z_a, the state Z_b will follow from state Z_a in the large majority of cases, but by no means always; a transformation into any of the other neighbouring states of Z_a will occur occasionally, although extremely rarely. What was said about the transformation of the state Z_a into the neighbouring state Z_b, applies again to the evolution of the system from the state Z_b at the following instant. One reaches thus a conception of the (apparently) irreversible processes.

Such a sketch of Boltzmann's conception is incomplete. One still needs to answer the following questions : "What is to be understood by the probability of individual states $Z_1, Z_2 \cdots$" and "Why is the transformation of the state Z_a into the most probable neighbouring state, more probable than a transformation towards the remaining neighbouring states?"

Regarding the first question we make the following remark. According to the kinetic theory of heat, there cannot be a temperature equilibrium in a strict sense. The state we call of thermal equilibrium is the one most frequently occupied by a system left alone for an extremely long time. But a consequence of the kinetic theory is that in the course of long periods of time the system takes on by itself all possible states, and the further away a state is from thermodynamic equilibrium, the more rarely the system takes it on. The copper cube left on its own for an infinite time changes its temperature distribution endlessly, while extremely rarely taking temperature distributions that differ significantly from the thermal equilibrium temperature distribution. If we imagine a system to be observed for an enormously long time T, then there will be for most of the states Z_ν a very small part, τ, of this time duration T during which the system takes on exactly the state Z_ν. The proportion $\frac{\tau}{T}$ we call the probability W of the state concerned.

If one lays down this definition of the probability of a state, then one can see generally that from a state Z_a a system changes on average so that this state is followed by the neighbouring state Z_b of largest probability. I can only mention this, without entering into its justification.

It is essential that one can define the probability of a state independently of its kinetic picture; the probability W is a quantity in principle accessible to observation, even if its direct observation is excluded in most cases because of the shortness of time at our disposal.

If we leave alone a system in a state considerably far from thermodynamic equilibrium, it takes on successive states with growing W. This property is common to the probability W of a state and the entropy S of the system, and it was Boltzmann who found that between W and S exists the relation

$$S = k \ln W,$$

where k is a universal constant, i.e., independent of the choice of the system.

This equation of Boltzmann can be applied in two different ways. First, a more or less complete picture of molecular theory can be present, on which one can ground a calculation of the probability W. Boltzmann's equation then yields the entropy S. So was Boltzmann's equation applied most of the time until now.

Example. In a volume V_0 let $\mathcal{N}$ molecules of a certain kind be present, i.e., one gram-molecule. Let the volume be sufficiently large with respect to the proper volume of the $\mathcal{N}$ molecules, and —if present—let matter other than the $\mathcal{N}$ molecules be uniformly distributed in V_0, so that the different points of V_0 are equivalent for each of the $\mathcal{N}$ molecules. This is a partial expression of the picture we make for ourselves of an ideal gas or of a diluted solution. What is the probability W for all $\mathcal{N}$ molecules to be in the sub-volume V of volume V_0, at a randomly chosen instant?

A simple consideration gives

$$W = \left(\frac{V}{V_0}\right)^{\mathcal{N}}.$$

By using Boltzmann's equation we deduce

$$S = k\mathcal{N} \ln\left(\frac{V}{V_0}\right) = k\mathcal{N} \ln V + \text{const.},$$

where the constant "const." can well depend on the temperature, but not on the volume. We immediately deduce the force that $\mathcal{N}$ molecules exert on a surface constraining them to stay in the volume V. In fact the energy of the system is independent of V, and by calling dG the work received in a reversible way during an infinitely small expansion of the volume V, one obtains

$$p\,dV = +dG = +T\,dS = +k\mathcal{N}T\frac{dV}{V},$$

so

$$pV = k\mathcal{N}T.$$

We have therefore obtained the equation of perfect gases and of the osmotic pressure. At the same time, it turns out that the universal constant $k\mathcal{N}$ of this equation is equal to the constant R of the equation for perfect gases.

In my opinion, the main meaning of Boltzmann's equation does not however lie in the fact that, given a known molecular picture, we can calculate with its help the entropy. The most important way to use it consists much more in that, from the empirically determined entropy function, we can with the help of Boltzmann's equation in the opposite sense determine the statistical probability of the individual states. We thus obtain the possibility of judging how much the behaviour of the system differs from the behaviour required by thermodynamics.

Example. A particle suspended in a fluid, and which is somewhat heavier than the displaced fluid.
Such a particle should, according to thermodynamics, sink to the bottom of the container and stay there. According to Boltzmann's equation, however, each height z above the bottom will get a probability W; the particle changes its height constantly and in an irregular way. We want to estimate S and from it W. Let μ be the mass of the particle, and μ_0 be that of the fluid displaced by the latter; one thus must perform the work $A = (\mu - \mu_0)g\,z$ to lift the particle to a height z from the bottom. For the system's energy to remain meanwhile constant, one must extract from the system the quantity of heat $G = A$, thereby the entropy decreases by $\frac{G}{T} = \frac{A}{T}$. There is then

$$S = \text{const} - \frac{1}{T}(\mu - \mu_0)g\,z.$$

From Boltzmann's equation it follows, when one substitutes k with the value $\frac{R}{\mathcal{N}}$, that

$$W = \text{const}\; e^{-\frac{\mathcal{N}}{RT}(\mu-\mu_0)g\,z}.$$

If many identical particles are present in the fluid instead of a single one, then the right side of the equation gives the density of the particles as a function of the depth. This relation was tested and confirmed by Perrin.

From this relation the law of Brownian motion can be very easily deduced. It follows indeed immediately from it firstly that the average height $\bar{z}$ of a particle above the bottom of the container is equal to

$$\frac{\int z\, e^{-\frac{\mathcal{N}}{RT}(\mu-\mu_0)g\,z} dz}{\int e^{-\frac{\mathcal{N}}{RT}(\mu-\mu_0)g\,z} dz} = \frac{RT}{\mathcal{N}} \frac{1}{g(\mu-\mu_0)}.$$

But now, because of its larger density, the particle falls according to Stokes' law by $D = \frac{g(\mu-\mu_0)}{6\pi\eta P}\tau$ in a time τ, where η represents the viscosity coefficient of the fluid and P is the radius of the (spherically shaped) particle. But there will be besides, in the same time τ, because of the irregularity of heat phenomena, an up or down displacement by a distance Δ, where positive and negative values of Δ occur equally frequently, and therefore $\bar{\Delta} = 0$.

A particle that at the beginning of time τ is at height z, will be at the end of time τ at height $z - D + \Delta = z'$. For the distribution law of any particle not to depend on time, the average value of z^2 must be equal to that of ${z'}^2$, therefore

$$\overline{(z - D + \Delta)^2} = \overline{z^2},$$

or, for τ small enough to neglect D^2, and $\overline{z\Delta} = \overline{D\Delta} = 0$

$$\overline{\Delta^2} = 2\,\bar{z}\,D = \frac{RT}{\mathcal{N}} \frac{1}{3\pi\eta P}\tau. \tag{1}$$

This is the known law of Brownian motion, which was also confirmed by experiments.[3]—

[3] *Note by the translator:* This peculiar, and quite elegant derivation by Einstein of the law of Brownian motion for a suspended particle does not seem to have appeared elsewhere than in the context of this lecture. Notice the particular roles played in the above derivation, on the one hand by gravity, and on the second by the second moment of position z. One can thus think of generalizing Einstein's demonstration to the case of any potential, and to any moment of the position; see the Commentary after Einstein's text.

The demonstration by Einstein for a suspended Brownian particle is actually related to similar works of his in the same period, in the different context of thermal radiation.

In a previous publication in 1909 ["Zum gegenwärtigen Stand des Strahlungsproblem", *Physikalische Zeitschrift* **10**, 185-193 (1909)], Einstein studied the Brownian motion of a mirror immersed in a thermal radiation bath. He made use of the statistical approach introduced in his original article on Brownian motion; relativity theory and Lorentz transformations were also involved in the description of the radiative damping ("Brehmstrahlung") of the mirror. Einstein calculated exactly in this way the momentum fluctuations of the mirror.

In a later article from 1910, written a little before the present Zürich lecture of November 1910, Einstein and his student and collaborator Ludwig Hopf then addressed the similar question of the momentum fluctuations of a charged particle ("resonator"), in the presence of an electromagnetic radiation field. [A. EINSTEIN and L. HOPF, "Statistische Untersuchung der Bewegung eines Resonators in einem Strahlungsfeld", *Ann. d. Physik* **33**, 1105-1115 (1910). See also "Statistical investigation of a resonator's motion in a radiation field," *The Collected papers of Albert Einstein*, J. STACHEL ed., Vol. **3**, Princeton University Press (1993)].

They use an astute method very similar to the one described in the present text, namely the stationarity of the second moment of the momentum of the charged particle :

$$\overline{(mv)^2_{t=0}} = \overline{(mv)^2_{t=\tau}} = \overline{(mv_{t=0} + \Delta - \mathcal{P}v\tau)^2}.$$

The example just considered of the particle suspended in a fluid gives a perfect illustration of Boltzmann's point of view on irreversible phenomena. Let us indeed consider a particle in suspension, which is in a container of a certain height, and which is so much heavier than the displaced fluid that the expression for the probability W to be at a minute height z above the bottom of the container

---

Here v is the speed of the particle, Δ the magnitude of the fluctuating electromagnetic impulse on the particle, and $-\mathcal{P}v\tau$ the reduction of particle's momentum due to the viscous electromagnetic drag by a damping force $\mathcal{K} = -\mathcal{P}v$ during time τ. This equation is of course very similar to the one used by Einstein in the lecture presented above, for the second moment of the position of a suspended particle.

By arguments of stochastic independence of the various terms, similar to the ones given there for arriving at result (1) , one arrives immediately at

$$\overline{\Delta^2} = 2m\overline{v^2}\mathcal{P}\tau.$$

Equipartition of kinetic energy for the particle, $m\overline{v^2} = \frac{RT}{\mathcal{N}}$, coming from the by then well-established kinetic theory of gases, then implied :

$$\overline{\Delta^2} = 2\frac{RT}{\mathcal{N}}\mathcal{P}\tau, \tag{2}$$

a result entirely analogue to the famous Sutherland-Einstein relation (1) for standard Brownian motion.

Einstein and Hopf then follow with a detailed calculation of the friction force $\mathcal{K} = -\mathcal{P}v$ (the analogue of Stokes' force), using Relativity theory, and of the electomagnetic fluctuations of impulse, $\overline{\Delta^2}$. The results are expressed in terms of the density of thermal radiation $u(\nu)$ at the characteristic frequency ν of the resonator.

By substituting into (2), Einstein and Hopf obtain a differential equation for the thermal radiation density $u(\nu)$ per frequency interval $d\nu$, which has for solution :

$$u(\nu) = \frac{8\pi}{c^3}\frac{RT}{\mathcal{N}}\nu^2.$$

Therefore only Rayleigh-Jeans black-body law, and not Planck's one, can be recovered from the thermal equilibrium of the radiation field with a *classical* charged resonator.

The classical oscillator obeys indeed the classical law of equipartition of energy, $\langle E\rangle_{\rm cl} = \frac{RT}{\mathcal{N}} = kT$ for its mean total energy, whence the factor kT in the aforementioned law. Planck's black-body quantum radiation law,

$$u_{\rm qu}(\nu) = \frac{8\pi h}{c^3}\frac{\nu^3}{e^{h\nu/kT}-1},$$

does reduce to the Rayleigh-Jeans law for high temperature T or low frequency ν, while in the complementary domain of minute overall radiation intensity (i.e., low T), or high frequency ν, it recovers Wien's law : $u(\nu) = \frac{8\pi\nu^2}{c^3}h\nu e^{-h\nu/kT}$.

One notes that the substitution of the mean energy of the *quantum* oscillator,

$$\langle E\rangle_{\rm qu} = \frac{h\nu}{e^{h\nu/kT}-1},$$

to the classical equipartition value kT into the Rayleigh-Jeans law, actually allows the exact recovery of Planck's formula. This is of course in complete agreement with Planck's original result, derived from Maxwell theory, that the black-body spectral density $u(\nu)$ and the (temporal) average energy U of the resonator interacting with it, are related by the necessary identity

$$u(\nu) = \frac{8\pi\nu^2}{c^3}U.$$

is already very small compared to the value W_0 for $z = 0$, such that the particle will very rarely rise notably above the bottom, once it will have touched the bottom (thermodynamic equilibrium). If we lift the particle up to a noticeable height z, then obviously with the largest probability it will sink back (non-reversible process) down to the bottom, next dancing up and down as before within the neighbourhood of the latter. If this sinking back did not take place in an overwhelming number of cases, a probability distribution with the assumed character could just not be valid.

Before arriving at other applications of Boltzmann's equation, I will derive a universal consequence of the latter, concerning the mean fluctuations of the system parameters around the ideal values at thermodynamic equilibrium. Let $\lambda_1 \cdots \lambda_n$ be the parameters that determine the state of a system. Let the zero values of λ's be chosen such that at thermodynamic equilibrium one has $\lambda_1 = \lambda_2 \cdots = 0$. Let A be the work one should perform, according to thermodynamics, to bring the system from the thermodynamic equilibrium state towards a state very close to it, characterized by the values $\lambda_1 \cdots \lambda_n$:

$$A = \sum A_\nu = \sum_1^n \frac{a_\nu}{2} \lambda_\nu^2.$$

Once this state is established, for the energy of the system to be the same as before, a quantity of heat $G = A$ must be extracted, which means a decrease of the entropy of the system by $\frac{G}{T} = \frac{A}{T}$. Therefore if the system has reached by itself the state considered, its entropy is

$$S = \text{const} - \frac{1}{T} \sum_1^n \frac{a_\nu}{2} \lambda_\nu^2.$$

When substituting this into Boltzmann's equation, one then obtains

$$W = \text{const} \exp\left(- \frac{\mathcal{N}}{RT} \sum_1^n \frac{a_\nu}{2} \lambda_\nu^2 \right).$$

In this case therefore Gauss' error law applies to the fluctuations of the single parameters around their thermodynamic equilibrium values. For the average value $\overline{A_\nu}$ of the work that one should perform according to thermodynamics, to bring by a reversible process the parameter λ_ν from an equilibrium value to the time average $\sqrt{\overline{\lambda_\nu^2}}$, one obtains the value

$$\overline{A_\nu} = \frac{RT}{2\mathcal{N}}.$$

One can express this result as follows. In the case where close to thermodynamic equilibrium A lends itself to be represented in the way given above, deviations from the state of ideal thermodynamic equilibrium arise spontaneously; for each parameter, the size of these deviations is on average such that the work necessary for the arbitrary production of the deviation is equal to one third of the average kinetic energy of the ballistic motion of a gas molecule at the same temperature. Perceptible deviations from the ideal thermodynamic equilibrium appear therefore everywhere a perceptible effect can be attained by performing just as small an amount of work. The measurement of each deviation of this type brings us to a determination of the energy of a monoatomic gas molecule, therefore also a determination of the absolute size of atoms.

A very interesting application of this general result was given by Smoluchowski. According to classic thermodynamics, at thermodynamic equilibrium the independent components of a phase are uniformly distributed throughout phase volume. On the contrary, according to what was said above, irregularities must appear in the spatial distribution of matter, that are the larger, the smaller the forces that oppose any change of the uniform distribution of the matter, or of the single independent components, respectively. The phase is therefore in reality inhomogenous, which manifests itself by optical turbidity (opalescence). Such an opalescence is particularly strong near critical points (in homogeneous substances and in solutions), because here only minute forces oppose a change of density, or concentration, respectively. I recently showed that, on the ground of the conception sketched by Smoluchowski, an exact calculation of the light diffused by opalescence is possible.

I would not like to leave unmentioned that with the help of Boltzmann's equation the statistical properties of thermal radiation can be deduced in a simple way from the law of thermal radiation, and certainly without the help of electromagnetism and of the theory of heat. The problem is the following. In a cavity, which is surrounded by opaque bodies at temperature T, there is a radiation whose properties are determined uniquely by the temperature. Across a surface σ, that is thought of as situated anywhere in the cavity, a determined radiation energy $\mathcal{E}$ passes during the time τ, whose direction is given within the elementary cone $d\Omega$ and whose frequency domain is $d\nu$.

If one were to measure this energy repeatedly, and truly very precisely, one will not always find the same value $\mathcal{E}$, but a value $\mathcal{E} = \mathcal{E}_0 + \varepsilon$ somewhat fluctuating around a mean value $\mathcal{E}_0$. One asks about the quadratic mean value $\overline{\varepsilon^2}$ of the variable ε. This problem has a crucial interest for the very reason that its solution contains an expression of the structure of thermal radiation.[4]

[4] *Note by the translator:* A. Einstein alludes very likely here to the results of his articles "Zum gegenwärtigen Stand des Strahlungsproblem", *Physikalische Zeitschrift* **10**, 185-193 (1909), cited above, and "Über die Entwicklung unserer Anschauungen über das Wesen und die Konstitution der Strahlung", *Physikalische Zeitschrift* **10**, 817-825 (1909). In a work of 1904, Einstein had derived the famous relation between the fluctuations of the energy E of a system about its mean, $\overline{E}$, and its specific heat, namely

$$\overline{(E - \overline{E})^2} = kT^2 \frac{d\overline{E}}{dT}.$$

I will only outline the way this problem can be solved. When any body K is in thermal contact with a similar one of a relatively infinite thermal capacity, K will then, according to thermodynamics, reach the temperature of this second body and keep it permanently. But from Boltzmann's principle, the temperature of K will unceasingly change, although it moves only rarely in a noticeable manner away from the thermal equilibrium temperature; Boltzmann's equation gives the mean value of such temperature fluctuations. The temperature fluctuations thus experienced are absolutely independent of the way the thermal exchange occurs between K and the body of relatively infinite size; the temperature fluctuation is therefore then also of the size calculated when such a thermal exchange occurs exclusively through radiation. There remains therefore then only to examine the question : What the statistical properties of radiation must be, so that the calculated temperature fluctuations actually happen? When pursuing the outlined study, one then obtains the result that the temporal fluctuations of thermal radiation are much larger, for minute radiation intensity and large frequency, than was expected from our present-day theory.[5] —

Applying this to the elementary energy $\mathcal{E}$ of a radiation field of frequency range ν within $d\nu$, distributed according to Planck's law, he finds

$$\frac{\overline{(\mathcal{E}-\overline{\mathcal{E}})^2}}{\overline{\mathcal{E}}^2} = \left(\frac{8\pi\nu^2}{c^3}\mathcal{V}d\nu\right)^{-1} + \left(\frac{\overline{\mathcal{E}}}{h\nu}\right)^{-1}.$$

This is exactly the law of large numbers, but with two *(complementary)* terms. The first fluctuation term is the inverse of the elementary number of electromagnetic modes, $\frac{8\pi\nu^2}{c^3}\mathcal{V}d\nu$, in frequency domain ν within $d\nu$, and in volume $\mathcal{V}$. The second is, in modern language, the inverse of the mean number of photons, $\frac{\overline{\mathcal{E}}}{h\nu}$, present in the electromagnetic field.

In the setting described by Einstein in this lecture, the volume delimited by the surface area σ, the elementary cone angle $d\Omega$, and light rays covering a distance $c\tau$ in time τ, is simply $\mathcal{V} = \sigma c\tau d\Omega$. The corresponding fluctuation formula alluded to by Einstein should therefore be

$$\frac{\overline{\varepsilon^2}}{\mathcal{E}_0^2} = \left(\frac{8\pi\nu^2}{c^2}\sigma\tau d\Omega d\nu\right)^{-1} + \left(\frac{\mathcal{E}_0}{h\nu}\right)^{-1}, \tag{3}$$

where, still in Einstein's notations, the elementary energy $\mathcal{E} = \mathcal{E}_0 + \varepsilon$ fluctuates about its mean value $\overline{\mathcal{E}} = \mathcal{E}_0$.

[5]The difficulty alluded to by Einstein : "*. . . the temporal fluctuations of thermal radiation are much larger, for minute radiation intensity and large frequency, than was expected from our present-day theory*", corresponds in Eq. (3) to the fact that at low temperature, or equivalently, high frequency, the second fluctuation term dominates the first. This term, which manifests the corpuscular character of radiation, could not be understood at that time from Maxwell theory. The latter can explain only the first fluctuation term, as resulting from the constructive interference of electromagnetic waves which leads to a definite number of stationary modes.

This difficulty in reconciling both points of views is best illustrated in the Nobel lecture by Wilhelm Wien in 1910. He writes very interestingly :

Einstein investigated the fluctuations to which radiation is continuously subjected even in the state of equilibrium as a result of the irregularities of the thermal processes. If we imagine a small plate in a cavity filled with radiation, this plate will be subjected to a radiation pressure which is the same on average on both sides of the plate. Since the radiation must contain irregularities, the pressure will alternately be greater on one or the other side so that the plate will execute small irregular movements, similar to the Brownian movement of a dust particle suspended

If to conclude we ask once more the question : "Are the observable physical facts correlated one to another in an entirely causal way?", we must surely answer this question in the negative. The positions of a particle engaged in a Brownian motion at two instants separated by one second must always appear, even to the most consciencious observer, as independent of each other, and the greatest mathematician would never succeed in any determined case to compute in advance, even approximately, the path covered in a second by such a particle. According to the theory, to be able to do so one should know the position and speed of each molecule exactly, which appears in principle excluded. However, the laws of mean values, which proved themselves everywhere, as well as the statistical laws of fluctuations, valid in these domains of finest effects, lead us to the conviction that in theory we must firmly hold onto the hypothesis of a complete causal connection of events, even if we should not hope to ever obtain by improved observations of Nature the direct confirmation of such a concept.

---

in a liquid. These fluctuations can be derived from probability calculations. According to the Boltzmann theorem there is a simple relationship between entropy and probability. The entropy of radiation is known from the radiation law, so that the probability of a state is also known, from which the fluctuations can be calculated. The mathematical expression for these fluctuations consists, in a peculiar manner, of two members. The first is readily understandable : it is due to irregularities which arise as a result of the mutual interference of the many independent beams which meet in one point. Where the density of radiation energy is high, this term alone predominates; it corresponds to the radiation range that obeys Rayleigh's law.

The other term, which cannot be directly explained by the undulation theory, predominates at low density of radiation energy, where the radiation obeys the law formulated by me [Wien's law]. *It would be understandable if the radiation consisted of the Planck energy elements which would be localized even in an empty space. We cannot shake the undulation theory of light, which is one of the most firmly established constructions in the whole of physics. Moreover, the term to be explained by localized energy elements, is not present by itself, and it is a priori impossible to introduce a dualistic approach into optics, e.g., to assume simultaneously Huyghens' wave theory and Newton's emanation theory. All we can do is to relinquish the Boltzmann method of applying probability calculations to this type of fluctuations, or to assume that a new irregularity is introduced into radiation with the process of reflection.*

In view of the magnitude of the difficulties it is natural that opinions about the path to be pursued should differ greatly. Some are of the opinion that the fundamental principles of electrodynamics must be changed. And yet, previous theory embraces a vast range of facts, it accounts for events even in the most rapid movements of the X-rays, it has proved itself in the most precise optical measurements. In my view, all the signs suggest that the deviations from current theory are due to events within the atom. None of the processes in which the interior of the atom participates are amenable to current theory.

Indeed Quantum Mechanics was still to be born. Nevertheless, Einstein's calculations were entirely correct, and already announced fully the dual nature of light. The complete quantization of radiation was achieved in 1925 by Pascual Jordan and 1926 by Paul Dirac, thereby giving a formal confirmation of the existence of the light quanta, introduced by Einstein twenty years before in his 1905 article "Über einen die Erzeugung und Vervandlung des Lichtes betreffenden heuristischen Gesichtspunkt" [*Annalen der Physik* **17**, 132-148 (1905)]. It was indeed a *truly revolutionary* point of view, as he wrote to his friend Conrad Habicht in the Spring 1905 ! As is well-known, Einstein received in 1922 the Nobel prize in Physics for this work.

Commentary

Bertrand Duplantier

Abstract. The argument given above by Einstein for deriving the Brownian diffusion constant of a suspended particle is generalized to arbitrary potentials. We also check its consistency by considering moments of the particle's position of any order n.

The demonstration made here by Einstein concerns Boltzmann's distribution in a uniform gravitational field of strength g, and the second moment, $\overline{z^2}$, of the position z of a particle in suspension. The Sutherland-Einstein result[6] about the mean quadratic fluctuations of position in time τ,

$$\overline{\Delta^2} = \frac{RT}{\mathcal{N}} \frac{1}{3\pi\eta P} \tau, \tag{4}$$

does not depend on the field g, and should apply in all generality. It is then natural to verify the generality of this demonstration in the case of a generic force field. In a second section, we are also considering the case of nth-order moments, $\overline{z^n}$, of the fluctuating position z of the suspended particle. The consistency of the approach by Einstein requires that their stationarity in the course of time leads to the same result (4) for the mean quadratic fluctuations.

General potential

Let us thus consider a potential energy $V(z)$, where z is (for simplicity) a one-dimensional coordinate, which is a generalization of the height coordinate. The associated force is :

$$F = -\frac{\partial V}{\partial z}.$$

The contribution of the potential energy to the entropy is then

$$S = \text{const.} - \frac{1}{T} V(z).$$

[6]It seems to be unsufficiently known that William Sutherland from Melbourne, Australia, arrived at formula (4) before Einstein. He announced it in a conference in Dunedin (New-Zealand) in January 1904 [*The measurement of large molecular masses*, Report of the 1Oth Meeting of the Australasian Association for the Advancement of Science, Dunedin, 1904), 117-121 (1904)], and published it also in 1905 [*A dynamical theory of diffusion for non-electrolytes and the molecular mass of albumin*, Phil. Mag., S.6, **9**, 781-785 (1905)]. For a more detailed history, see, e.g., A. Pais, *"Subtle is the Lord..."*, *The Science and Life of Albert Einstein*, Oxford University Press (1982); B. Duplantier, *Brownian Motion, "Diverse and Undulating"*, Poincaré Seminar, Birkhäuser Verlag, Basel, 2005.

According to Boltzmann's equation, there exists a probability W for a particle to be at a height z above the bottom

$$W = \text{const.}\ e^{-\frac{\mathcal{N}}{RT}V(z)}.$$

In the force field F and according to Stokes' law, the particle follows the law of displacement during a time τ

$$D = \frac{F}{6\pi\eta P}\tau = -\frac{\tau}{6\pi\eta P}\frac{\partial V}{\partial z}, \tag{5}$$

where η is the coefficient of fluid viscosity and P is the radius of the particle, following Einstein's notations. For the sake of notational simplicity, we set in the following :

$$a \equiv \frac{\tau}{6\pi\eta P}.$$

By still calling Δ the irregular displacement caused by heat, such that $\bar{\Delta} = 0$, we state with Einstein that a particle being at height z at the origin of time interval τ, will be at height $z + D + \Delta = z'$ at the end of this time interval, where now the displacement D (5) is counted algebraically. Because of the time-invariance of the distribution law for a particle, the average value of z^2 must be equal to that of z'^2, i.e.,

$$\overline{(z + D + \Delta)^2} = \overline{z^2}.$$

For τ small enough to be able to neglect D^2, and by using $\overline{z\Delta} = \overline{D\Delta} = 0$

$$\overline{\Delta^2} = -2\,\overline{z\,D},$$

where this time we compute the average by taking into account the non-uniformity of the force field F, and therefore of the displacement D. This is in contradistinction to Einstein's case of a uniform gravitational field, for which the mean-square displacement (1) factorized as $\overline{\Delta^2} = -2\,\overline{z}\,D$.

What is left to evaluate is the average of $\overline{z\,D}$. By definition this can be written as :

$$\overline{z\,D} = -a\frac{\int z\frac{\partial V}{\partial z}\,e^{-\frac{\mathcal{N}}{RT}V(z)}dz}{\int e^{-\frac{\mathcal{N}}{RT}V(z)}dz} = -\frac{a}{\mathcal{Z}}\int z\,\frac{\partial V}{\partial z}e^{-\beta V(z)}dz, \tag{6}$$

where we have introduced the partition function

$$\mathcal{Z} = \int e^{-\beta V(z)}dz, \tag{7}$$

and we used the well-known notation $\beta = \frac{\mathcal{N}}{RT} = \frac{1}{k_B T}$, where k_B is Boltzmann's constant.

One must notice here that the boundaries have not been specified yet, and that they can be imagined to be taken at infinity, as follows. The potential $V(z)$

is infinite (positive) for z smaller than some value z_0 marking the bottom of the box, and again infinite (positive) at infinity, or for z above some value $z_1 (z_1 > z_0)$ marking the top of the box (and possibly pushed to infinity). We thus take at infinity in both directions

$$V(z = \pm\infty) = +\infty. \tag{8}$$

The probability density $W(z)$, which is proportional to $e^{-\beta V(z)}$, vanishes outside the box interval $[z_0, z_1]$ and therefore at infinity too.

Integrating by parts in the expression (6) of $\overline{z\,D}$, one finds[7]

$$\begin{aligned} \int z \frac{\partial V}{\partial z} e^{-\beta V(z)} dz &= -\int z \frac{1}{\beta} \frac{\partial}{\partial z} e^{-\beta V(z)} dz \\ &= -z \frac{1}{\beta} e^{-\beta V(z)} |_{-\infty}^{+\infty} + \frac{1}{\beta} \int e^{-\beta V(z)} dz = \frac{1}{\beta} \mathcal{Z}. \end{aligned} \tag{9}$$

The partition function $\mathcal{Z}$ thus cancels out from (6), and we get therefore a result which is independent of the potential :

$$\overline{z\,D} = -\frac{a}{\beta},$$

from which

$$\overline{\Delta^2} = -2\,\overline{z\,D} = \frac{2a}{\beta} = \frac{k_B T}{3\pi\eta P}\tau \tag{10}$$

follows, which is of course the result of Sutherland-Einstein.

This argument for calculating Brownian diffusion, given by Einstein at the Zürich lecture in 1910, appears to be remarkably simple even when generalized. It is enough to compare the quadratic averages $\overline{z^2}$ and $\overline{z'^2}$ by applying the hypothesis of stationarity at thermal equilibrium. It is rather remarkable that the consideration of quadratic averages plays such an important role. Nevertheless, the miracle (*à la* Einstein) : we obtain the correct result, in all generality !

Let us pursue the analysis a little further : why not simply apply the same hypothesis of stationarity to the average linear displacements $\overline{z}$ and $\overline{z'}$ themselves? And what about the comparison of moments of higher order, $\overline{z^n}$ and $\overline{z'^n}$?

---

[7]In the case considered by Einstein of the linear gravitational potential, the potential was infinite positive neither at the bottom, nor at $-\infty$. Nevertheless, it happened that the height variable z was bounded below by 0, which insured implicitly the vanishing at $z = 0$ of the term integrated by parts (owing to the presence of the z factor in the integrand of (9)). In fact, the simplest physical way is to consider, as done here, a box limited by potential walls, i.e., to confine the system.

Moments of any order

We consider then the generic moment

$$\overline{z'^n} = \overline{(z + D + \Delta)^n}.$$

First, according to Leibniz' formula we write :

$$(z + D + \Delta)^n = \sum_{n_1 \geq 0,\ n_2 \geq 0,\ n_3 \geq 0;\ n_1+n_2+n_3=n} \frac{n!}{n_1!\ n_2!\ n_3!} z^{n_1} D^{n_2} \Delta^{n_3}.$$

We then should evaluate

$$\overline{(z + D + \Delta)^n} = \sum_{n_1 \geq 0,\ n_2 \geq 0,\ n_3 \geq 0;\ n_1+n_2+n_3=n} \frac{n!}{n_1!\ n_2!\ n_3!} \overline{z^{n_1} D^{n_2} \Delta^{n_3}}. \tag{11}$$

Here the average has a double significance, being at the same time an average over the space coordinates z and a local average over the stochastic variable Δ. We can assume a hypothesis, which will be useful in the following, that these two averages are independent. This independence corresponds to the fact that the local fluctuations of Δ are created by thermal agitation, independently from the value of (the gradient of) the local potential $V(z)$. We have then

$$\overline{z^{n_1} D^{n_2} \Delta^{n_3}} = \overline{z^{n_1} D^{n_2}} \times \overline{\Delta^{n_3}}. \tag{12}$$

Another related hypothesis is that of the local symmetry of thermal fluctuations, and therefore of Brownian motion, $\Delta \leftrightarrow -\Delta$, already used by Einstein. We deduce from it, for odd values $n_3 = 2n'_3 + 1$,

$$\overline{z^{n_1} D^{n_2} \Delta^{2n'_3+1}} = \overline{z^{n_1} D^{n_2}} \times \overline{\Delta^{2n'_3+1}} = 0.$$

This identity generalizes those used by Einstein at first order : $\overline{z\Delta} = \overline{D\Delta} = 0$.

Let us recall now that

$$D = -a\frac{\partial V}{\partial z},\ a = \frac{\tau}{6\pi\eta P},$$

and that we are working in the limit of an infinitesimal time $\tau \to 0$, in which $\mathcal{O}(a) = \mathcal{O}(\tau)$. Within this limit, we have the orders of magnitude $\mathcal{O}(D) = \mathcal{O}(\Delta^2) = \mathcal{O}(\tau)$, as can be anticipated from the result of Sutherland-Einstein $\overline{\Delta^2} = 2a/\beta$, while of course $\mathcal{O}(z) = 1$. We find then

$$\mathcal{O}\left(\overline{z^{n_1} D^{n_2} \Delta^{n_3}}\right) = \mathcal{O}\left(\tau^{n_2 + \frac{1}{2}n_3}\right).$$

Moreover, as n_3 must be even so that the average moment does not vanish identically, we find that the two dominant moments, with the constraint $n_1+n_2+n_3 = n$, are obtained for

$$(n_1, n_2, n_3) = (n-1, 1, 0) \text{ or } (n-2, 0, 2),$$

for which $n_2 + \frac{1}{2}n_3 = 1$, giving

$$\mathcal{O}\left(\overline{z^{n-1}D}\right) = \mathcal{O}\left(\overline{z^{n-2}\Delta^2}\right) = \mathcal{O}(\tau).$$

From equation (11) finally follows that

$$\overline{z'^n} = \overline{(z + D + \Delta)^n} = \overline{z^n} + n\,\overline{z^{n-1}D} + \frac{n(n-1)}{2}\,\overline{z^{n-2}\Delta^2} + \mathcal{O}(\tau^2).$$

From the hypothesis of stationarity $\overline{z'^n} = \overline{z^n}$, and owing to (12), we then deduce at first order in τ the identity

$$\overline{z^{n-1}D} + \frac{(n-1)}{2}\,\overline{z^{n-2}}\;\overline{\Delta^2} = 0. \tag{13}$$

Let us then express the two associated moments. We have by definition

$$\overline{z^{n-1}D} = -\frac{a}{\mathcal{Z}}\int z^{n-1}\frac{\partial V}{\partial z}e^{-\beta V(z)}dz,$$

$$\overline{z^{n-2}\Delta^2} = \overline{z^{n-2}} \times \overline{\Delta^2} = \frac{1}{\mathcal{Z}}\int z^{n-2}\,e^{-\beta V(z)}dz \times \overline{\Delta^2}.$$

By integrating by parts, the first moment is written as

$$\begin{aligned}\overline{z^{n-1}D} &= \frac{a}{\beta}\frac{1}{\mathcal{Z}}\left[z^{n-1}e^{-\beta V(z)}|_{-\infty}^{+\infty} - (n-1)\int z^{n-2}\,e^{-\beta V(z)}dz\right]\\ &= -(n-1)\frac{a}{\beta}\,\overline{z^{n-2}},\end{aligned}$$

where we used the boundary conditions (8) for the confining potential.

The identity (13) is therefore satisfied *for all* n, $n > 1$, when

$$\overline{\Delta^2} = 2\frac{a}{\beta} = \frac{2}{\beta}\frac{\tau}{6\pi\eta P},$$

which indeed coincides with the result (10) obtained by Einstein in the $n = 2$ case.

Finally we notice that the $n = 1$ case corresponds to the condition

$$\overline{z + D + \Delta} = \overline{z},$$

which is identically satisfied, because we have separately

$$\overline{D} = -\frac{a}{\mathcal{Z}}\int\frac{\partial V}{\partial z}e^{-\beta V(z)}dz = \frac{a}{\beta}\frac{1}{\mathcal{Z}}e^{-\beta V(z)}|_{-\infty}^{+\infty} = 0,\ \ \overline{\Delta} = 0,$$

where we again used (8), as well as the symmetry of Brownian fluctuations.

The simple method used by Einstein to obtain the equation of Brownian diffusion, starting from the dynamics of the quadratic moment of the position of a particle in suspension in a gravitational field, can thus be generalized to any potential and moments of any order, as it should. The result, the Sutherland-Einstein relation (4), is *universal.*

Bertrand Duplantier
Service de Physique Théorique
Orme des Merisiers
CEA/Saclay
F-91191 Gif-sur-Yvette Cedex
France
email: Bertrand.Duplantier@cea.fr

Einstein, 201 – 293
© Birkhäuser Verlag, Basel, 2005

Brownian Motion, "Diverse and Undulating"

Bertrand Duplantier

Translation by Emily PARKS *from the original French text*

Abstract. *Truly man is a marvelously vain, diverse, and undulating object. It is hard to found any constant and uniform judgment on him.* MICHEL DE MONTAIGNE, Les Essais, Book I, Chapter 1: "By diverse means we arrive at the same end"; in *The Complete Essays of Montaigne*, Donald M. Frame *transl.*, Stanford University Press (1958).

Pour distinguer les choses les plus simples de celles qui sont compliquées et pour les chercher avec ordre, il faut, dans chaque série de choses où nous avons déduit directement quelques vérités d'autres vérités, voir quelle est la chose la plus simple, et comment toutes les autres en sont plus, ou moins, ou également éloignées. RENÉ DESCARTES, Règles pour la direction de l'esprit, Règle VI.

In order to distinguish what is most simple from what is complex, and to deal with things in an orderly way, what we must do, whenever we have a series in which we have directly deduced a number of truths one from another, is to observe which one is most simple, and how far all the others are removed from this – whether more, or less, or equally. RENÉ DESCARTES, Rules for the Direction of the Mind, Rule VI.

Car, supposons, par exemple que quelqu'un fasse quantité de points sur le papier à tout hasard, comme font ceux qui exercent l'art ridicule de la géomance. Je dis qu'il est possible de trouver une ligne géométrique dont la notion soit constante et uniforme suivant une certaine règle, en sorte que cette ligne passe par tous ces points, et dans le même ordre que la main les avaient marqués.

... Mais quand une règle est fort composée, ce qui luy est conforme, passe pour irrégulier.

G. W. LEIBNIZ, Discours de métaphysique, H. Lestienne *ed.*, FÉLIX ALCAN, Paris (1907).

Thus, let us assume, for example, that someone jots down a number of points at random on a piece of paper, as do those who practice the ridiculous art of geomancy.[1] *I maintain that it is possible to find a geometric line whose notion is constant and uniform, following a certain rule, such that this line passes through all the points in the same order in which the hand jotted them down.*

... But, when the rule is extremely complex, what is in conformity with it passes for irregular.

G. W. LEIBNIZ, Discourse on Metaphysics.

Mens agitat molem. VIRGIL, AEneid. lib. VI.

Un coup de dés jamais n'abolira le hasard. STÉPHANE MALLARMÉ, Cosmopolis, 1897.

A throw of the dice never will abolish chance.

[1]Note: From *géomance*, a way to foretell the future; a form of divination.

L'antimodernisme, c'est la liberté des modernes. ANTOINE COMPAGNON, about his book "Les antimodernes: de Joseph de Maistre à Roland Barthes", Bibliothèque des Idées, Gallimard, March 2005.

Antimodernism is the liberty of modern men.

Here we briefly describe the history of Brownian motion, as well as the contributions of Einstein, Sutherland, Smoluchowski, Bachelier, Perrin and Langevin to its theory. The always topical importance in physics of the theory of Brownian motion is illustrated by recent biophysical experiments, where it serves, for instance, for the measurement of the pulling force on a single DNA molecule.

In the second part, we stress the mathematical importance of the theory of Brownian motion, illustrated by two chosen examples. The by-now classic representation of the Newtonian potential by Brownian motion is explained in an elementary way. We conclude with the description of recent progress seen in the geometry of the planar Brownian curve. At its heart lie the concepts of conformal invariance and multifractality, associated with the potential theory of the Brownian curve itself.

1 A brief history of Brownian motion

Several great classic works give a historical view of Brownian motion. Amongst them, we cite those of Brush,[2] Nelson,[3] Nye,[4] Pais[5], Stachel[6] and Wax.[7] We also cite a number of essays in mathematics,[8] physics,[9] especially those which have appeared very recently for the centenary of Einstein's 1905 articles,[10] and in biology.[11]

[2]Stephen G. Brush, *The Kind of Motion We Call Heat*, Book 2, p. 688, North Holland (1976).

[3]E. Nelson, *Dynamical Theories of Brownian motion*, Princeton University Press (1967), second ed., August 2001, http://www.math.princeton.edu/~nelson/books.html/.

[4]Mary Jo Nye, *Molecular Reality: A Perspective on the Scientific Work of Jean Perrin*, New York: American Elsevier (1972).

[5]Abraham Pais, *"Subtle is the Lord...", The Science and Life of Albert Einstein*, Oxford University Press (1982).

[6]John Stachel, *Einstein's Miraculous Year* (Princeton University Press, Princeton, New Jersey, 1998); *Einstein from 'B' to 'Z'*, Birkhäuser, Boston, Basel, Berlin (2002).

[7]N. Wax, *Selected Papers on Noise and Stochastic Processes*, New York, Dover (1954). It contains articles by Chandrasekhar, Uhlenbeck and Ornstein, Wang and Uhlenbeck, Rice, Kac, Doob.

[8]J. P. Kahane, Le mouvement brownien : un essai sur les origines de la théorie mathématique, in *Matériaux pour l'histoire des mathématiques au XXème siècle, Actes du colloque à la mémoire de Jean Dieudonné (Nice, 1996)*, volume 3 of *Séminaires et congrès*, pages 123–155, French Mathematical Society (1998).

[9]M. D. Haw, *J. Phys. C* **14**, 7769 (2002).

[10]B. Derrida and É. Brunet in *Einstein aujourd'hui*, edited by M. Leduc and M. Le Bellac, Savoirs actuels, EDP Sciences/CNRS Editions (2005); P. Hänggi *et al.*, *New J. Phys.* **7** (2005); J. Renn, *Einstein's invention of Brownian motion*, *Ann. Phys.* (Leipzig) **14**, Supplement, pp. 23–37 (2005); D. Giulini & N. Straumann, *Einstein's Impact on the Physics of the Twentieth Century*, arXiv:physics/0507107; N. Straumann, *On Einstein's Doctoral Thesis*, arXiv:physics/0504201.

[11]E. Frey and K. Krey, arXiv:cond-mat/0502602.

Figure 1: *Brownian motion described by the center of gravity of a pollen particle in suspension.*

1.1 Robert Brown and his precursors

In an article published in the *Edinburgh Journal of Science* in 1828, and republished multiple times elsewhere,[12] entitled *"A Brief Account of Microscopical Observations Made in the Months of June, July and August,* 1827, *on the Particles Contained in the Pollen of Plants; and on the General Existence of Active Molecules in Organic and Inorganic Bodies"*, the botanist Robert Brown reported on the random movement of different particles that are small enough to be in suspension in water. It is an extremely erratic motion, apparently without end (see figure 1)[13].

Brown may not have been the first, however, to observe Brownian motion. The universal and irregular motion of small grains suspended in a fluid may have been observed soon after the discovery of the microscope.[14] Brown made his observations just after the introduction of the first achromatic objectives for microscopes. In fact, it is nowadays sufficient to look in a microscope to see these small objects dancing.

The story begins with Anthony van Leeuwenhoek (1632–1723), a famous constructor of microscopes in Delft, who in 1676 was also designated executor of the estate of the no-less-famous painter Johannes Vermeer, who was apparently a

[12]R. Brown, *Edinburgh New Phil. J.* **5**, 358 (1828); *Ann. Sci. Naturelles*, (Paris) **14**, 341 (1828); *Phil. Mag.* **4**, 161 (1828); *Ann. d. Phys. u. Chem.* **14**, 294 (1828).

[13]One can find examples of real Brownian motion at the web site: www.lpthe.jussieu.fr/poincare/.

[14]S. Gray, *Phil. Trans.* **19**, 280 (1696).

personal friend.[15] Leeuwenhoek built several hundred simple "microscopes", with which he went as far as to observe living bacteria.

Next, one meets Buffon and Spallanzani, the two 18th-century protagonists of the debate on spontaneous generation,[16] and lastly Bywater, cited by Brown in his second article, who published in 1819 the conclusion that "not only organic tissues, but also inorganic substances, consist of animated or irritable particles", and therefore are subject to Brownian motion. In fact, in 1827 similar observations to those of Brown were alluded to in France by Adolphe Brongniart,[17] one year before the publication by Brown.

Robert Brown (1773–1858) was one of the greatest botanists of his time in England. He is known for his discovery of the nucleus of plants, and for the classification of several exotic plants he brought back to England from a trip to Australia in 1801–1805. Indeed, he was the botanist on the *Investigator* during Flinder's circumnavigation of Australia.[18] His first publication on the erratic motion of pollen garnered much attention, but the use of the ambiguous terms *"active molecules"* by Brown brought him criticisms based on some misunderstanding. Indeed, under the influence of Buffon, the similar expression *"organic molecules"* represented hypothetical entities, elementary bricks all living beings would be made of. This type of theory was still around at the beginning of the 19th century, so much that one thinks that Brown's opinion was that the particles themselves were animated. Faraday himself had to defend him during a Friday night lesson he gave at the Royal Society on February 21, 1829, about Brownian motion![19]

Brown's merit was rather in emancipating himself from this misconception and in making a systematic study of the movement named after him, with grains of pollen, dust and soot, pulverized rock, and even a fragment from the Great Sphinx. This served to eliminate as well the "vital force" hypothesis, where the movement was reserved to organic particles. As for the nature of Brownian motion, even if he could not explain it, he eliminated easy explanations, like those linked

[15]Although no document exists testifying a relationship between Vermeer and van Leeuwenhoek, it seems impossible that they did not know one another. The two men were born in Delft the same year, their respective families were involved in the textile business and they were both fascinated by science and optics. A commonly accepted and probable hypothesis is that Anthony van Leeuwenhoek was in fact a model for Vermeer, and perhaps also the source of his scientific information, for the two famous scientific portraits, *The Astronomer*, 1668, (Louvre Museum, Paris), and *The Geographer*, 1668–69, (Städelsches Kunstinstitut am Main, Frankfurt). (See *Johannes Vermeer*, B. Broos et al., National Gallery of Art, Washington, Mauritshuis, The Hague, Waanders Publishers, Zwolle (1995).)

[16]Jean Perrin, in his book *Les Atomes* (*Atoms*, translated by D. Ll. Hammick, Ox Bow Press, Woodbridge (1990)), writes: "Buffon and Spallanzani knew of the phenomenon but, possibly owing to the lack of good microscopes, they did not grasp its nature and regarded the "dancing particles" as rudimentary animalculae (Ramsey: Bristol Naturalists' Society, 1881)".

[17]A. Brongniart, *Ann. Sci. Naturelles* (Paris) **12**, 41 (1827).

[18]A rivulet south of Hobart is named after him (as mentioned by Bruce H. J. McKellar, in *Einstein, Sutherland, Atoms, and Brownian Motion*, Einstein International Year of Physics 2005, Melbourne AAPPS Conference, July 2005, http://www.ph.unimelb.edu.au/.

[19]S. G. Brush, *The Kind of Motion We Call Heat*, Book 2, p. 688, North Holland (1976).

to convection currents or to evaporation, by showing that the Brownian motion of a *simple* particle stayed "tireless" even in a isolated drop of water in oil! On the same occasion he eliminated as well the hypothesis of movements created by interactions between Brownian particles, a hypothesis that would nevertheless reappear later. The theoretical picture made perhaps by Brown, which he however always carefully avoided presenting as the conclusion of his studies, could be that the particles of matter were animated into a rapid and irregular movement whose source was in the particles themselves and not in the surrounding fluid. Before leaving Robert Brown, one cannot refrain from citing Charles Darwin's recollection from the 1830s:

> "I saw a good deal of Robert Brown, *"facile Princeps Botanicorum"*, as he was called by Humboldt. He seemed to me to be chiefly remarkable by the minuteness of his observations and their perfect accuracy. His knowledge was extraordinarily great, and much died with him, owing to his excessive fear of ever making a mistake. He poured out his knowledge to me in the most unreserved manner, yet was strangely jealous on some points. I called on him two or three times before the voyage of the *Beagle* [1831], and on one occasion he asked me to look through a microscope and describe what I saw. This I did, and believe now that it was the marvelous currents of protoplasm in some vegetable cell. I then asked him what I had seen; but he answered me, *"That is my little secret."*[20]

1.2 The period before Einstein

Between 1831 and 1857 it seems that one can no longer find references to Brown's observations, but from the 1860s forward his work began to draw large interest. It was noticed soon thereafter in literary circles, if we are to judge by a passage of *"Middlemarch"* published by George Eliot in 1872, where the surgeon Lydgate offered to Rev. M. Farebrother, in exchange for marine specimens, "the latest of Robert Brown's discoveries, *Microscopic Observations on the Pollen of Plants*, if you don't already have it."

Jean Perrin wrote in his famous 1909 memoir *Brownian Motion and Molecular Reality*:[21]

> "The singular phenomenon discovered by Brown did not attract much attention. It remained, moreover, for a long time ignored by the majority of physicists, and it may be supposed that those who had heard of it thought it analogous to the movement of the dust particles, which can be seen dancing in a ray of sunlight, under the influence of feeble currents of air which set up small differences of pressure or temperature. When we reflect that this apparent explanation was able to satisfy even thoughtful minds, we ought the more to admire the acuteness of those physicists, who have recognised in this, supposed insignificant, phenomenon a fundamental property of matter."

[20] *Charles Darwin: His Life told in an autobiographical Chapter, and in a selected series of his published letters*, ed. by his son, Francis Darwin, London (1892); New York: Schuman (1950), p. 46; quoted by S. G. Brush in *The Kind of Motion We Call Heat, op. cit.*

[21] J. Perrin, *Mouvement brownien et réalité moléculaire, Ann. de Chim. et de Phys.* **18**, 1 (1909). Translated by Frederick Soddy in *Brownian Motion and Molecular Reality*, Taylor and Francis, London (1910); facsimile reprint in David M. Knight, ed., *Classical scientific papers: chemistry*, American Elsevier, New York (1968).

1.2.1 Brownian motion and the kinetic theory of gases

It became clear from experiments made in various laboratories that Brownian motion increases when the size of the suspended particles decreases (one essentially ceases to observe it when the radius is above several microns), when the viscosity of the fluid decreases, or when the temperature increases. In the 1860's, the idea emerged that the cause of the Brownian motion has to be found in the internal motion of the fluid, namely that the zigzag motion of suspended particles is due to collisions with the molecules of the fluid.

The first name worth citing in this regard is probably that of Christian Wiener, holder of the Chair of Descriptive Geometry at Karlsruhe, who in 1863 reaffirmed in the conclusions to his observations that the motion could be due neither to the interactions between particles, nor to differences in temperature, nor to evaporation or convection currents, but that the cause must be found in the liquid itself.[22] That being so, his theory on atomic motion anticipated those of Clausius and Maxwell, implicating not only the motion of molecules but also the motion of "ether atoms". The Brownian motion was thus bound to the vibrations of the ether, to the wavelength corresponding to that of red light and to the size of the smallest group of molecules moving together in the liquid. Such an explanation was criticized by R. Mead Bache, who showed that the motion was insensitive to the color of light, whether it was violet or red.[23] Christian Wiener is nevertheless credited by some authors as the first to discover that molecular motion could explain the phenomenon.[24]

At least three other people proposed the same idea: Giovanni Cantoni of Pavia, and two Belgian Jesuits, Joseph Delsaulx and Ignace Carbonelle. The Italian physicist attributed Brownian movement to thermal motions in the liquid, and considered that this phenomenon provides a "beautiful and direct demonstration of the fundamental principles of the mechanical theory of heat".[25] The Belgian physicists published in the *Royal Microscopical Society* and in the *Revue des Questions scientifiques*, from 1877 to 1880, various Notes on the *Thermodynamical Origin of the Brownian Movement.* In a Note by Father Delsaulx, for example, one may read:[26]

"The agitation of small corpuscles in suspension in liquids truly constitutes a general phenomenon", that it is "henceforth natural to ascribe a phenomenon having this universality to some property of matter", and that "in this train of ideas the internal movements of translation

---

[22] Chr. Wiener, *Erklärung des atomischen Wessens des flüssigen Körperzustandes und Bestätigung desselben durch die sogennanten Molekularbewegungen, Ann. d. Physik* **118**, 79 (1863).

[23] R. Mead Bache, *Proc. Am. Phil. Soc.* **33**, 163 (1894).

[24] J. Perrin, *Mouvement brownien et réalité moléculaire, op. cit.*

[25] See the reprint with notes by J. Thirion in *Revue des Questions Scientifiques* **15**, 251 (1909).

[26] "See for this bibliography an article which appeared in the *Revue des Questions Scientifiques*, January 1909, [*op. cit.*], where M. Thirion very properly calls attention to the ideas of these *savants*, with whom he collaborated." [original citation and note by J. Perrin in *Brownian Motion and Molecular Reality, op. cit.*]

which constitute the calorific state of gases, vapours and liquids, can very well account for the facts established by experiment".

Such a point of view, parallel to that of the kinetic theory of gases, faced strong opposition. One opponent, cytologist Karl von Nägeli of Switzerland, familiar with the kinetic theory of gases and the orders of magnitude involved, likewise the British chemist William Ramsey (the future Nobel laureate in Chemistry), commented that the particles in suspension have a mass several hundreds of millions of times larger than that of the molecules in the fluid. Each random collision with a molecule of the surrounding fluid produces therefore an effect far too small to displace the suspended particle. Nägeli wrote for example about a supposedly similar motion of micro-organisms in the air:

"The motion which a sun-mote, and on the whole any particle found in the air, can acquire by the collisions of an individual gas molecule or a multitude of such molecules is therefore so extraordinarily small, and the number of simultaneous collisions against the particle from all sides so extraordinarily large, that the particle behaves as if it were completely at rest."

He believed instead that the cause of the motion was not the thermal molecular motion but some attractive or repulsive forces.

Nevertheless, the second part of his proposition about the frequency of such collisions held the principle of the solution. Because it is a collective statistical effect, as described in perspicacious manner by Father Carbonelle:

"In the case of a surface having a certain area, the molecular collisions of the liquid, which cause the pressure, would not produce any perturbation of the suspended particles, because these, as a whole, urge the particles equally in all directions. But if the surface is of area less than necessary to insure the compensation of irregularities, there is no longer any ground for considering the mean pressure; the inequal pressure, continually varying from place to place, must be recognised, as the law of large numbers no longer leads to uniformity; and the resultant will not now be zero but will change continually in intensity and direction. Further, the inequalities will become more and more apparent the smaller the body is supposed to be, and in consequence the oscillations will at the same time become more and more brisk..."

Perrin mentions these authors to conclude:

"These remarkable reflections unfortunately remained as little known as those of Wiener. Besides it does not appear that they were accompanied by an experimental trial sufficient to dispel the superficial explanation indicated a moment ago; in consequence, the proposed theory did not impress itself on those who had become acquainted with it."

He continues:

"On the contrary, it was established by the work of M. *Gouy* (1888), not only that the hypothesis of molecular agitation gave an admissible explanation of the Brownian movement, but that no other cause of the movement could be imagined, which especially increased the significance of the hypothesis.[27] This work immediately evoked a considerable response, and it is only from this time that the Brownian movement took a place among the important problems of general physics."

Indeed in 1888 the French physicist Louis-Georges Gouy made the best observations on Brownian motion, from which he drew the following conclusions:

[27] L.-G. Gouy, *J. de Physique* **7**, 561 (1888); *C. R. Acad. Sc. Paris*, **109**, 102 (1889); *Revue générale des Sciences*, **1** (1895).

- The motion is extremely irregular, and the trajectory seems not to have a tangent.
- Two Brownian particles, even close, have independent motion from one another.
- The smaller the particles, the livelier their motion.
- The nature and the density of the particles have no influence.
- The motion is most active in less viscous liquids.
- The motion is most active at higher temperatures.
- The motion never stops.

Gouy seemed, however, to claim again that one cannot explain Brownian motion by disordered molecular motion, but only by the partially organized movements over the order of a micron within the liquid.

But somehow he became known as the "discoverer" of the cause of Brownian motion, as Jean Perrin wrote about his experimental conclusions:

> "Thus comes into evidence, in what is termed a *fluid in equilibrium*, a property eternal and profound. This equilibrium only exists as an average and for large masses; it is a statistical equilibrium. In reality the whole fluid is agitated indefinitely and *spontaneously* by motions the more violent and rapid the smaller the portion taken into account; the statical notion of equilibrium is completely illusory."[28]

1.2.2 Brownian motion and Carnot's principle

Brownian agitation continues indefinitely. It does not contradict the principle of energy conservation, because any increase in the velocity of a grain, for instance, is accompanied by a local cooling of the surrounding fluid, and the thermal equilibrium is statistical.

Gouy was the first to note the apparent contradiction between Brownian motion and *Carnot's principle*. The latter states that one cannot extract work from a simple source of heat. However, it really seems that some work is made, in a *fluctuating* manner, by the thermal motion of the molecules of the fluid. Gouy mentioned the theoretical possibility to extract work by a mechanism attached to a Brownian particle, and he concluded that Carnot's principle perhaps was no longer valid for dimensions of the size of a micron, in that echoing Helmholtz's reservations about the validity of such principle for living tissues.

These questions sparked the interest of Poincaré, who announced at the following lecture of the Congress of Arts and Sciences in St. Louis in 1904, about the "Present Crisis of Mathematical Physics"[29]:

[28] J. Perrin, *Mouvement brownien et réalité moléculaire, op. cit.*

[29] Henri Poincaré, *La valeur de la science*, Bibliothèque de philosophie scientifique, Flammarion, Paris (1905); in *Congress of Arts and Sciences, Universal Exposition, St. Louis, 1904*, Houghton, Mifflin and Co., Boston and New York (1905).

"But here the stage changes. Long ago the biologist, armed with his microscope, noticed in his specimens disorganized movements of small particles in suspension; that is the Brownian motion. He believed at first that it is a vital phenomenon, but soon he saw that inanimate bodies did not dance with less fervor than the others, so he handed over to physicists. Unfortunately, physicists have been uninterested for a long while in this question; one concentrates light to enlighten the microscopic specimens, they thought; light does not go without heat, from which inhomogeneities of temperature, and then internal currents in the liquid that produce the motion we are speaking about.

M. Gouy had the idea to look closer and he saw, or believed he saw that this explanation is unsustainable, that the motion becomes more lively the smaller the particles, but that they are not influenced by light. So, if the motion never stops, or more exactly is continually reborn without end, without an external source of energy, what are we to believe? We must not, without any doubt, renounce the conservation of energy because of this, but we see before our own eyes both motion transform into heat by friction, and inversely heat transform into motion; and all that while nothing is lost, as the motion lasts forever. This is the opposite of Carnot's principle. If this is the case, to see the world develop in reverse, we no longer have need of the infinitely subtle eye of Maxwell's demon, a microscope will suffice. The largest of bodies, those that have for example, a tenth of a millimeter, collide with atoms in motion from all sides, but they do not move at all as the shocks are so numerous that the law of chance says they compensate one another; however the smallest particles do not receive enough shocks for the compensation to be exact and they are unendingly tossed around. And *voilà*, one of our principles already in danger."

It is rather subtle to prove that the Brownian phenomenon does not infringe on the impossibility of creating perpetual motion (called of the second kind), where work is extracted in a coherent manner by the observer (recalling Maxwell's famous demon). One had to wait for Leo Szilard, who hinted in 1929 that, because of the amount of information required by such an attempt, the total produced entropy would compensate the apparent entropy reduction due to the coherent use of fluctuations. We shall briefly return to this question later, after having described Smoluchowski's contributions.

1.2.3 The kinetic molecular "hypothesis"

Nowadays it seems evident to us that the world is made up of particles, of atoms and of molecules. However, it was not always the case, and the hypothesis of a continuous structure of matter was relentlessly defended until the end of the nineteenth century by famous names such as Duhem, Ostwald, and Mach.

The intuition or the idea that gases are composed of individual molecules was already present in the eighteenth century, and in 1738 David Bernoulli was probably the first to affirm that the pressure of a gas on its container is due to collisions of molecules with the walls. Avogadro made the radical affirmation in 1811 that two gases at the same pressure and same temperature contain the same number of molecules. When such conditions are of one atmosphere and of 25° Celsius, the number contained in a volume of one liter is noted as $\mathcal{N}$, and called *Avogadro's number.*

To understand the stakes surrounding the determination of Avogadro's number, one must recall that the constant R in the perfect gas law has been experimentally accessible since the eighteenth century, thanks to the work of Boyle, Mariotte, Charles, and later Gay-Lussac. It is in fact associated to the number of moles, $N/\mathcal{N}$, which is an experimental macroscopic parameter, contrary to the total number of particles N, and of Avogadro's number $\mathcal{N}$, that are microscopic quantities.

The study of Brownian motion played an essential role in establishing the "molecular hypothesis" definitively. As Jean Perrin observed, the "hypothesis" that bodies, despite their homogeneous appearance, are made up of distinct molecules, in unending agitation which increases with temperature, is logically suggested by the phenomenon of Brownian motion alone, even before providing an explanation.

In fact, according to Perrin, what is really strange and new in Brownian motion, is, precisely, that it never stops, contrary to our every-day experience with friction phenomena. If, for example, we pour a bucket of water into a tub, the initial coherent motion possessed by the liquid mass disappears, de-coordinated by the multiple rebounds on the boundaries of the tub, until an apparent equilibrium settles within the fluid at rest. Does such a de-coordination of the motion of the particles proceed indefinitely, as it would in an ideal continuous medium? The answer by Perrin is exceptionnally convincing:[30]

> "To have information on this point and to follow this de-coordination as far as possible after having ceased to observe it with the naked eye, a microscope will be of assistance, and microscopic powders will be taken as indicators of the movement. Now these are precisely the conditions under which the Brownian motion is perceived: we are therefore *assured* that the de-coordination of motion, so evident on the ordinary scale of our observations, does not proceed indefinitely, and on the scale of microscopic observation, we *establish* an equilibrium between coordination and de-coordination. If, that is to say, at each instant, certain of the indicating granules stop, there are some in other regions at the same instant, the movement of which is re-coordinated automatically by their being given the speed of granules which have come to rest. So that it does not seem possible to escape the following conclusion:
>
> Since the distribution of motion in a fluid does not progress indefinitely, and is limited by a spontaneous re-coordination, it follows that the fluids are themselves composed of granules or *molecules*, which can assume all possible motions relative to one another, but in the interior of which dissemination of motion is impossible. If such molecules had no existence it is not apparent how there would be any limit to the de-coordination of motion [...] In brief, the examination of Brownian movement alone suffices to suggest that every fluid is formed of elastic molecules, animated by perpetual motion."

In 1905 Albert Einstein was the first, actually along with (but independently from) William Sutherland from Melbourne, to propose a quantitative theory of Brownian motion. This theory will allow Perrin to determine the precise value of Avogadro's number $\mathcal{N}$, in his famous experiments of 1908–1909. Sutherland and Einstein succeeded where many others failed, because they used an ingenious and global reasoning of statistical mechanics, that we will explain here. Marian von Smoluchowski made at the same time an analysis according to a different "Gedankenweg", more probabilistic, which led him to similar conclusions. We will came back to this point later in the paper.

[30]Translation by Frederick Soddy, *op. cit.*

1.3 William Sutherland, 1904–05

In his famous biography of Einstein, *Subtle is the Lord...* (1982), Abraham Pais noted, while describing Einstein's route to his well-known diffusion relation, that the same relation had been discovered "at pratically the same time" by the Melbourne theoretical physicist William Sutherland, following similar reasoning to Einstein's, and had been submitted for publication to the *Philosophical Magazine* in March 1905, shortly before Einstein completed the doctoral thesis in which he first announced the relation. Pais, therefore, proposed that the relation be called "the Sutherland–Einstein relation".

We follow here the introduction of the essay, *Speculating about Atoms in Early 20th-century Melbourne: William Sutherland and the 'Sutherland–Einstein' Diffusion Relation*, written recently by the Australian historian of science Rod W. Home.[31] In this section we shall briefly discuss Sutherland's work, and the factors that may have led to his work being over-shadowed by Einstein's, and soon forgotten. When the Einstein International Year of Physics commemorates the hundredth anniversary of the *Annus Mirabilis* papers' release, focusing also on W. Sutherland's achievements seems to be just fair!

1.3.1 Sutherland's papers

Sutherland's paper to which Pais refers was actually published in June 1905,[32] after Einstein completed his thesis, but shortly before he submitted it for examination. We seem to be looking here at a perfect example of effectively simultaneous discovery. However, as Rod Home notes, the story is still a little more complicated, for Sutherland had already reported his derivation over a year earlier, at the congress of the Australasian Association for the Advancement of Science held in Dunedin, New Zealand, in January 1904, and his paper had been published in the congress proceedings in early 1905![33] Unfortunately, there was a *misprint* in the crucial equation giving the diffusion coefficient of a large molecular mass in terms of physical parameters: Avogadro's constant was missing![34]

The correct and extended equation, finally published in the *Philosophical Magazine*, is

$$D = \frac{RT}{\mathcal{N}} \frac{1}{6\pi\eta\, a} \frac{1 + 3\eta/\beta a}{1 + 2\eta/\beta a}, \tag{1}$$

[31] Most of the material presented in this section originates from the 2005 essay by R. W. Home, *Speculating about Atoms in Early 20th-century Melbourne: William Sutherland and the 'Sutherland–Einstein' Diffusion Relation*, Sutherland Lecture, 16th National Congress, Australian Institute of Physics, Canberra, January 2005. See also the interesting note by Bruce H. J. McKellar, *The Sutherland–Einstein Equation*, *AAPPS Bulletin*, February 2005, 35.

[32] W. Sutherland, *A Dynamical Theory for Non-Electrolytes and the Molecular Mass of Albumin*, *Phil. Mag.* S.6, **9**, 781–785 (1905).

[33] W. Sutherland, *The Measurement of Large Molecular Masses*, Report of the 10th Meeting of the Australasian Association for the Advancement of Science, Dunedin, pp. 117–21 (1904).

[34] As R. W. Home remarks, it is clear that one is looking at a genuine misprint in the proceedings, since the preceding line was given correctly.

where R is the perfect gas constant, T the absolute temperature, $\mathcal{N}$ Avogadro's number, η the fluid viscosity, a the radius of the (spherical) diffusing molecule, and β the coefficient of sliding friction if there is slip between the diffusing molecule and the solution.[35] To deal with the available empirical data, Sutherland had indeed to allow for a varying coefficient of sliding friction between the diffusing molecule and the solution. By taking β to infinity, so there is no slip at the boundary, one recovers the usual form of the equation:

$$D = \frac{RT}{\mathcal{N}} \frac{1}{6\pi\eta\, a}. \tag{2}$$

Since in a fluid the molecules are close packed the molecular radius a should be proportional to the cube root of the molar volume $\mathcal{B}$, the volume occupied by Avogadro's number of particles. Hence, from the constancy of the product aD in relation (2), should follow that of $\mathcal{B}^{1/3}D$. After having estimated this constant from experimental data on the diffusion of various dissolved substances, Sutherland could obtain the molar volume of albumin, and got an estimate of its atomic mass[36] as 32814 Da.[37]

1.3.2 Sutherland, Einstein and Besso

In 1903, Einstein and his friend Michele Besso discussed a theory of dissociation that required the assumption of molecular aggregates in combination with water, the "hypothesis of ionic aggregates", as Besso called it. This assumption opens the way to a simple calculation of the sizes of ions in solution, based on hydrodynamical considerations. In 1902, Sutherland had considered in *Ionization, Ionic Velocities, and Atomic Sizes*[38] a calculation of the sizes of ions on the basis of Stokes' law, but criticized it as in disagreement with experimental data.[39] The very same idea of determining sizes of ions by means of classical hydrodynamics occurred to Einstein in his letter of 17 March 1903 to Besso,[40] where he proposed what appears to be just the calculation that Sutherland had performed:

[35] Sutherland uses the version of Stokes' law, $F = 6\pi\eta\, a \frac{1+2\eta/\beta a}{1+3\eta/\beta a} V$, relating the viscous friction force F to the velocity of the particle. This relation is generalized here to the case where slip occurs at the boundary between the fluid and the moving sphere. For a derivation, see H. Lamb, *Hydrodynamics*, Cambridge University Press (1932).

[36] The dalton (Da) is the atomic mass unit; it honors the English chemist John Dalton (1766–1844), who revived the atomic theory of matter in 1803.

[37] The present-day value is 43 kDa for ovalbumin.

[38] W. Sutherland, *Ionization, Ionic Velocities, and Atomic Sizes*, *Phil. Mag.* S.6, **4**, 625–645 (1902).

[39] He wrote: *"Now this simple theory must have been written down by many a physicist and found to be wanting, for it makes the ionic velocities of the different atoms at infinite dilution stand to one another inversely as their radii, a result which a brief study of data as to ionic velocities and relative atomic sizes shows to be not verified"*. Sutherland did not use the assumption of ionic hydrates, which can avoid such disagreement by permitting ionic sizes to vary with temperature and concentration.

[40] Albert Einstein, Michele Besso, *Correspondance 1903–1955*, translation, notes and introduction by Pierre Speziali, Herrmann, Paris (1979).

"Have you already calculated the absolute magnitude of ions on the assumption that they are spheres and so large that the hydrodynamical equations for viscous fluids are applicable? With our knowledge of the absolute magnitude of the electron [charge] this would be a simple matter indeed. I would have done it myself but lack the reference material and time; you could also bring in diffusion in order to obtain information about neutral salt molecules in solution."

As the editors of Einstein's Collected Papers remark, "This passage is remarkable, because both key elements of Einstein's method for the determination of molecular dimensions, the theories of hydrodynamics and diffusion, are already mentioned, although the reference to hydrodynamics probably covers only Stokes' law".[41]

It is also striking that a former letter of 11–17 February 1903, this time from Besso to Einstein, clearly indicates that they had been discussing Sutherland's work together. This letter contains two parts. The first deals with experimental data in connection to the dissociation of bi-ionic molecules. The second discusses what Besso calls "Sutherland's hypothesis", in connection to dissociation or dissolution. He states that the theory of "ionic hydrates", as he calls it, rescues temporarily this hypothesis with regard to Ostwald's dilution law. Since Besso also discusses the role of imperfect semi-permeable membranes as a possible experimental test of Sutherland's hypothesis, P. Speziali, in the French edition of the Einstein–Besso correspondence, has indicated that Besso would have been discussing in this letter another of Sutherland's papers, entitled *"Causes of osmotic pressure and of the simplicity of the laws of dilute solutions"*.[42]

However, upon reading these letters of 1903, one cannot refrain from wondering whether Besso and Einstein were not also acquainted with and discussing Sutherland's 1902 paper on ionic sizes. In that case, Sutherland suggestion to use hydrodynamic Stokes' law to determine the size of molecules would have been a direct inspiration to Einstein's dissertation and subsequent work on Brownian motion!

1.3.3 Sutherland's legacy

That Sutherland, in spite of his isolation in Melbourne, was well-known in physics circles is also evidenced by the fact that he was invited to contribute to the Boltzmann Festschrift in 1904 – the only other non-European contributor being J. Willard Gibbs! – If so, why did Einstein and not Sutherland become famous?

Sutherland had assumed the existence of atoms, and attacked a practical question, the measurement of large molecular masses. He was interested in these masses because of their role in the chemical analysis of organic substances. While that is what everyone now uses the Sutherland–Einstein equation for, it was perhaps not of so widespread interest at the time. However, we have just seen from the

[41] *The Collected Papers of Albert Einstein*, volume 2, The Swiss Years: Writings, 1900–1909, John Stachel ed., pp. 170–182, Princeton University Press (1989).

[42] *Causes of Osmotic Pressure and of the Simplicity of the Laws of Dilute Solutions*, *Phil. Mag.*, S.5, **44**, 52–55 (1897).

Einstein–Besso correspondence how extremely important Sutherland's idea was of determining the sizes of ions or molecules by means of classical hydrodynamics.

On the other hand, as stressed by the editors of *The Collected Papers*:

"In developing in his dissertation a new method for the determination of molecular dimensions, Einstein was concerned with several problems on different levels of generality. An outstanding current problem of the theory of solutions was whether molecules of the solvent are attached to the molecules or ions of the solute. Einstein's dissertation contributed to the solution of this problem. He recalled in 1909:

"At the time I used the viscosity of the solution to determine the volume of sugar dissolved in water because in this way I hoped to take into account the volume of any attached water molecule."

The results obtained in his dissertation indicate that such an attachment does occur. Einstein's concerns extended beyond this particular question to more general problems of the foundations of the theory of radiation and the existence of atoms. He later emphasized:

"A precise determination of the size of the molecules seems to me of the highest importance because Planck's radiation formula can be tested more precisely through such a determination than through measurements on radiation."

The dissertation also marked the first major success in Einstein's effort to find further evidence for the atomic hypothesis, an effort that culminated in his explanation of Brownian motion."

To conclude, it is probably most appropriate to cite R. W. Home:

"Of course, the diffusion-viscosity relation is generally known as the Einstein relation, not the Sutherland–Einstein relation. Why? In part, I think, this happened because in the early 20th century, theoretical physics was a largely German affair. In so far as the relation was taken up, and initially it was not taken up much at all, it was taken up by Continental researchers who had read Einstein's work but failed to notice that the relation was also buried in a paper in the Philosophical Magazine entitled *"A dynamical theory for non-electrolytes and the molecular mass of albumin"*. In the English-speaking world, where the Philosophical Magazine was one of the leading journals in the field, there were very few people pursuing theoretical physics in the German style. There is plenty of testimony that experimentally orientated British physicists were at something of a loss as how to assess Sutherland's work. His obituary in Nature makes the point very clearly:[43]

"His papers are well known to the scientific world. They are distinguished by great width of reading in the latest phases of the subjects he treated, combined with very bold speculation always brought into ample comparison with experimental knowledge. His generalisations were, indeed, so numerous that it was often a difficult task to try to estimate their value."

So in Britain, Surtherland didn't have a readership likely to be alert to the significance of his announcement of a relationship between diffusion and viscosity, in the way some Continental readers of Einstein's work were. And, finally, Sutherland's own presentation surely would not have helped, with the relation itself being almost submerged by his lengthy computations relating to the molecular mass of albumin. He would have done much better to highlight the relation, alone, in a paper to itself. But that was not his style! His mind was firmly fixed on the problem of determining molecular masses of large molecules, and he clearly saw the diffusion-viscosity relation as an incidental result arrived at on the way to achieving that larger goal, not as something of particular value in its own right."

In this year 2005, it is definitely time, I think, for the physics community to finally recognize Sutherland's achievements, and following Pais' suggestion, to re-baptize the famous relation (2) with a double name!

---

[43] "*Nature*, 23 November 1911, p. 116. The obituary is signed "J. L." [Joseph Larmor?]." [original note]

1.4 Albert Einstein, 1905

Mens agitat molem

1.4.1 Einstein's Dissertation

One finds nowadays in the literature excellent descriptions of Einstein's dissertation. An outstanding presentation is given in the Editorial Notes of the *Collected Papers of Albert Einstein.*[44] Their presentation is closely followed in this section, which incorporates some material of the editorial notes of the chapter entitled *"Einstein's dissertation on the determination of molecular dimensions"*.[45] The interested reader can also find a detailed scientific study of Einstein's doctoral thesis in a recent article by Norbert Straumann.[46]

Einstein completed his dissertation on *"A New Determination of Molecular Dimensions"* on 30 April 1905, and submitted it to the University of Zürich on 20 July.[47] Shortly after being accepted there, the manuscript was sent for publication to the *Annalen der Physik*, where it would be published in 1906.[48] On 11 May 1905, eleven days after finishing his thesis, Einstein had also sent the manuscript of his first paper on Brownian motion to the *Annalen*, which would publish it on 18 July 1905.

Einstein's central assumption is the validity of using classical hydrodynamics to calculate the effect of solute molecules, treated as rigid spheres, on the viscosity of the solvent in a dilute solution. His method is well suited to determine the size of solute molecules that are large compared to those of the solvent, and he applied it to solute sugar molecules. As we have seen above, Sutherland published in 1905 a method for determining the masses of large molecules, with which Einstein's method shares many important elements. Both methods make use of the molecular theory of diffusion that Nernst[49] developed on the basis of van 't Hoff's analogy between solutions and gases, and of Stokes' law of hydrodynamic friction.

The first of the results in the dissertation is a relation between the coefficients of viscosity of a liquid with and without suspended molecules (η^* and η, respectively),

$$\eta^* = \eta\,(1+\varphi)\,, \tag{3}$$

[44]Editorial notes of the chapter *"Einstein's dissertation on the determination of molecular dimensions"*, in *The Collected Papers of Albert Einstein*, volume 2, *op. cit.*, pp. 170–182; see also John Stachel, *Einstein's Miraculous Year*, *op. cit.*, pp. 31–43.

[45]With kind permission of John Stachel, EDITOR.

[46]Norbert Straumann, *On Einstein's Doctoral Thesis*, arXiv:physics/0504201.

[47]Einstein had already submitted a dissertation in 1901, on "a topic in the kinematic theory of gases". By February 1902, he had withdrawn the dissertation, possibly at his advisor's suggestion to avoid a controversy with Boltzmann. (For a detailed analysis, see the Editorial Notes of *The Collected Papers of Albert Einstein*, volume 2, *op. cit.*, pp. 174–175). Nevertheless, there is no doubt that Einstein was a great admirer of Boltzmann. (For a biography of the latter, see C. Cercignani, *Ludwig Boltzmann, The Man Who Trusted Atoms*, Oxford University Press (1998).)

[48]*Eine neue Bestimmung der Moleküldimensionen*, *Ann. d. Phys.* **19**, pp. 289–305 (1906).

[49]W. Nernst, *Z. Phys. Chem. Stöchiometrie Verwandschaftslehre*, **2**, 613–639 (1888).

where φ is the fraction of the volume occupied by the solute molecules.

The second result is the famous expression (2) for the coefficient of diffusion D of the solute molecules. Like Loschmidt's method based on the kinetic theory of gases, the expressions obtained by Einstein give two equations for two unknowns, Avogadro's number $\mathcal{N}$, and the molecular radius a of the suspended particles, hence providing a possible determination of molecular dimensions!

The derivation of eq. (3) represents the technically difficult part of Einstein's dissertation. It rests on the assumption that the motion of the fluid can be described by the hydrodynamical equations for stationary flow of an incompressible homogeneous fluid, even in the presence of solute molecules; that the inertia of these molecules can be neglected; that they do not interact; and that they can be treated as rigid spheres moving in the liquid without slipping, under the sole influence of hydrodynamical stress.

Eq. (2) follows from the conditions of dynamical and thermodynamical equilibrium in the fluid. Its derivation, as does Sutherland's paper, requires the identification of the force on a single large molecule, which appears in Stokes' law, with the apparent force due to the osmotic pressure. We shall return to this derivation in detail in the next section, when describing the content of Einstein's first paper on Brownian motion. In the dissertation, Einstein's derivation of eq. (2) does not involve yet the theoretical tools he developed in his work on the statistical foundations of thermodynamics in the preceding years. Here he simply states the osmotic pressure law, while in his first paper on Brownian motion, he will instead derive from first principles the validity of van 't Hoff's law for large suspended particles.

In 1909, Einstein drew Perrin's attention to his method for determining the size of solute molecules, which allows one to take into account the volume of any water molecule attached to the latter, and he suggested its application to the suspensions studied by Perrin in relation to Brownian motion. In the following year, an experimental study of formula (3) for the viscosity coefficient was performed by a pupil of Perrin, Jacques Bancelin. Using the same aqueous emulsions of gum-resin ("gamboge"), he confirmed that the increase in viscosity does not depend on the size of the solute molecules, but only on their volume fraction. However, the coefficient of φ in eq. (3) was found to be close to 3.9, instead of the predicted value 1. That prompted Einstein, after an unsuccessful attempt to find an error, to ask his student and collaborator Ludwig Hopf to check his calculations and arguments:

> "I have checked my previous calculations and arguments and found no error in them. You would be doing a great service in this matter if you would carefully recheck my investigation. Either there is an error in the work, or the volume of Perrin's substance in the suspended state is greater than Perrin believes."[50]

Hopf did find an error in the dissertation, namely in the derivatives of some velocity components, and obtained for φ a corrected coefficient 2.5. The remaining

[50] *The Collected Papers of Albert Einstein*, volume 2, *op. cit.*, pp. 180–181.

discrepancy between this corrected theoretical factor and the experimental one led Einstein to suspect that there might be also an experimental error.[51]

In early 1911 Einstein submitted his correction for publication, and recalculated Avogadro's number. He obtained a value of 6.56×10^{23} per mole, a value that is close to those derived from kinetic theory and Planck's black-body radiation theory.

The paper published in 1911 by Bancelin in the *Comptes rendus de l'Académie des Sciences* gave an experimental value of 2.9 as the coefficient of φ in eq. (3). Extrapolating his results to sugar solutions, Bancelin recalculated Avogadro's number, and found a value of 7.0×10^{23} per mole.

Einstein's dissertation was at first overshadowed by his more spectacular work on Brownian motion, and it required an initiative by Einstein to bring it to the attention of the scientists of his time. The paper on Brownian motion, the first of several on this subject that Einstein published over the course of the next couple of years, actually included his first published statement of the famous relationship linking diffusion with viscosity, that he had derived in his thesis.

As Abraham Pais points out in *Subtle is the Lord...*, this equation has found widespread applications, as a result of which Einstein's January 1906 paper in the *Annalen der Physik*, the published version of his dissertation, later became his most frequently cited paper![52] As stressed by R. H. Home in his essay on Sutherland, Pais also goes on to argue that the thesis was also one of Einstein's *"most fundamental papers"*, of comparable intrinsic significance to the other papers Einstein wrote in that year of 1905. *"In my opinion"*, Pais writes, *"the thesis is on a par with [Einstein's] Brownian motion article"*: indeed, *"in some if not all respects, his results are by-products of his thesis work"*.

It is now time to turn to this famous 1905 Brownian motion article.

1.4.2 The 1905 article on Brownian motion

The 1905 article is entitled: *"On the Motion of Small Particles Suspended in Liquids at Rest, Required by the Molecular-Kinetic Theory of Heat."*[53] There, Einstein tried to establish the existence and the size of molecules, and to determine a theoretical method for computing Avogadro's number precisely, by using the molecular kinetic theory of heat. In fact, he concluded:

---

[51] He asked Perrin: *"Wouldn't it be possible that your mastic particles, like colloids, are in a swollen state? The influence of such a swelling 3.9/2.5 would be of rather slight influence on Brownian motion, so that it might possibly have escaped you."*, Einstein to Perrin, 12 January 1911, in *The Collected Papers of Albert Einstein*, volume 2, *op. cit.*, p. 181.

[52] According to R. W. Home, *op. cit.*, it became the paper most widely cited in the period 1961–75, the period surveyed for the citation analysis of any scientific article published by any author before 1912. According to B. H. J. McKellar, *op. cit.*, the 1905 citation count is as follows (from World of Science, Dec. 2004): *Ann. d. Phys.* **17**, 132 (1905): **325** (photoelectric effect); *Ann. d. Phys.* **17**, 549 (1905): **1368** (Brownian motion); *Ann. d. Phys.* **17**, 891 (1905): **664** (special relativity); *Ann. d. Phys.* **18**, 639 (1905): **91** ($E = mc^2$); *Ann. d. Phys.* **19**, 289 (1906): **1447** (molecular dimensions, Einstein's thesis).

[53] A. Einstein, *Ann. d. Physik* **17**, 549–560 (1905).

"Möge es bald einem Forscher gelingen, die hier aufgeworfene, für die Theorie der Wärme wichtige Frage zu entscheiden !"[54]

Astonishingly enough, he was not yet certain that one could apply it to Brownian motion. In fact, his introduction opens with: *"In this paper it will be shown that, according to the molecular-kinetic theory of heat, bodies of a microscopically visible size suspended in liquids must, as a result of thermal molecular motions, perform motions of such magnitude that they can be easily observed with a microscope. It is possible that the motions to be discussed here are identical with so-called Brownian molecular motion; however the data available to me on the latter are so imprecise that I could not form a judgement on the question."*

Einstein relied on the results of his thesis, that he completed eleven days before submitting his famous article on the suspensions of particles. Only later would his predictions be progressively confirmed by refined experimental data on Brownian motion.[55]

1.4.3 The Einstein–Sutherland derivation

The demonstration is based on two distinct elements from apparently contradicting domains.

It seemed initially natural to use a *hydrodynamic* representation for particles in suspensions with size much greater than that of the liquid's molecules. A substantial amount of knowledge on the subject was available, in particular the famous "Stokes' formula", which gives the force of friction opposing to a sphere moving in the liquid.

But at the same time it was necessary for Einstein to exploit the kinetic theory of heat, pulling it away from the original context of the theory of gases and bringing it closer to the context of liquids, where the state of the theory was much less advanced. It was the crucial notion of *osmotic pressure*, developed by van 't Hoff, that made the passage possible. It is based on the concept of kinetic molecular disorder, where solute molecules, with a size comparable to that of the liquid's molecules, participate to the general motion like in a dilute gas.

Einstein was in possession of two theories about particles in a fluid. The first: Stokes' hydrodynamic theory, based on the hypothesis that a liquid is a continuous medium which adheres to a large solid surface moving through it, without any turbulence, and where the molecular agitation does not seem to play any role. The other: van 't Hoff's osmotic theory, based on the hypothesis that a particle in solution is similar to any other fluid molecule, and therefore is subjected to the same laws of molecular agitation.

One needed Einstein's perspicacity and his profound knowledge of statistical

[54] "Let us hope that a researcher will soon succeed in solving the problem presented here, which is so important for the theory of heat!"

[55] This led J. Renn, *op. cit.*, to speak of "Einstein's invention of Brownian motion".

mechanics to understand and to prove that the two points of view were simultaneously valid for particles as big as Brownian particles.

Einstein first studied the osmotic pressure created in the solution by solute molecules. This notion was developed by J. H. van 't Hoff[56] who, for dilute solutions, showed the identity between the pressure exerted on semi-permeable walls by molecules in solution and the partial pressure exerted by a gas. For sufficiently dilute solutions, this additional pressure p due to the molecules in solution satisfies the law of perfect gases

$$p = \frac{n}{\mathcal{N}} RT, \tag{4}$$

where R is the ideal gas constant, T is the absolute temperature, and n is the number of solute particles per unit volume, or particle density.

In his thesis, Einstein considered the effect of the density of such molecules on the viscosity, such as in the case of sugar in water. This time the particles in suspension are much larger so as to be observable under a microscope. Einstein right away affirms that the difference between solute molecules and particles in suspension is only a matter of size, and that van 't Hoff's law must be applied as well to particles in suspension. Next, he proves this fact and formula (4), by determining the free energy of an ensemble of such particles in suspension. In fact, he calculates the associated partition function by the phase space method.

Einstein then imagines that the numerous particles of the suspension are subjected to an external force F, which may depend on their positions but not on time.[57] This force, acting along the x axis for instance, moves each particle of the solute, and generates a gradient of concentration. Let $n(x, y, z; t)$ be the number of particles in suspension per unit volume around the point x, y, z at the instant t. From (4), a non-uniform osmotic pressure corresponds to a gradient of concentration of particles in suspension. By considering the resultant of all pressure forces on an elementary interval $\mathrm{d}x$, one also obtains the force of the osmotic pressure *per unit volume*:

$$\Pi = -\frac{\partial p}{\partial x} = -\mathrm{grad}\, p = -\frac{R}{\mathcal{N}} T \mathrm{grad}\, n(x, y, z; t), \tag{5}$$

where here the gradient is the spatial derivative in the direction x of the force.

In addition, the quantity $\Pi_F = n\,F$ represents the total external force per unit volume acting on the Brownian particles in suspension. From both a *hydrostatic* and *thermodynamic* point of view, one imagines *a priori* that the equilibrium of a unit of volume of the suspension is established when the force Π_F is balanced by the osmotic pressure force Π. In fact, by using arguments of equilibrium invariance

[56] J. H. van 't Hoff, *Kongliga Svenska Vetenskaps-Academiens Handlingar*, Stockholm, **21**, 1 (1884).

[57] This force can be, for example gravitational, as in the sedimentation experiments by Jean Perrin, but the beauty of the argument is that the result does not depend on the nature of the force, that can even be virtual, as in the notion of "virtual work" of the eighteenth century Mechanics.

of the free energy with respect to virtual displacements, Einstein demonstrates that actually the sum of the external and osmotic forces per unit volume cancels:

$$\Pi_F + \Pi = 0, \tag{6}$$

$$n\,F = \frac{R}{\mathcal{N}} T \operatorname{grad} n. \tag{7}$$

One notices that he directly obtained the explicit formula (7) from the free energy of the particles in suspension, without relying on the result (4), which shows the two results come from the same approach.

The second part of this argument focuses on the *dynamics of the flux equilibrium*. Equilibrium in the fluid is actually just an apparent effect: while the force F moves the particles in suspension, these are also subjected to Brownian motion which reflects the kinetic nature of heat.

By moving in the liquid under the force F, each particle in suspension experiences an opposing force of viscous friction. This brings the particle to a limit velocity $V = F/\mu$, where μ is the coefficient of viscous friction for each particle in suspension. The result is a flux of particles

$$\Phi_F = n\,V = n\,F/\mu, \tag{8}$$

that is the number of particles crossing a unit surface perpendicular to the direction x of the force.

The particle density $n(x, y, z; t)$ satisfies the local diffusion equation

$$\frac{\partial n}{\partial t} = D\Delta n, \tag{9}$$

where Δ is the Laplacian $\Delta = \frac{\partial^2}{\partial x^2} + \frac{\partial^2}{\partial y^2} + \frac{\partial^2}{\partial z^2}$, and where D is a coefficient, called the coefficient of diffusion, measured in square meters per second units. To this equation is naturally associated a diffusion flux Φ_D, which is the number of particles diffusing across a unit surface per unit of time. This flux is directly connected to the concentration gradient by[58]

$$\Phi_D = -D \operatorname{grad} n. \tag{10}$$

At equilibrium, here both local and dynamic, the force-driven flux Φ_F (8) and the flux of diffusion Φ_D (10), cancel:

$$\Phi_F + \Phi_D = 0, \tag{11}$$

$$n\,F/\mu = D \operatorname{grad} n. \tag{12}$$

[58] Einstein, like Sutherland, writes this equation directly, without passing through the diffusion equation he will prove farther along. This is indeed the celebrated Fick's law (A. Fick, *Über Diffusion*, *Ann. Phys. Chem.* **4**, 59–86 (1855)). For mathematically inclined readers, let us recall that the Laplacian is also $\Delta = \operatorname{div}(\operatorname{grad})$, where the divergence is the operator of derivation of a vector $\vec{A}$: $\operatorname{div}\vec{A} = \vec{\nabla}.\vec{A} = \frac{\partial A_x}{\partial x} + \frac{\partial A_y}{\partial y} + \frac{\partial A_z}{\partial z}$, and where the gradient is the vector operator of derivation $\operatorname{grad} = \left(\frac{\partial}{\partial x}, \frac{\partial}{\partial y}, \frac{\partial}{\partial z}\right)$. From the diffusion equation, $\frac{\partial n}{\partial t} = D\Delta n$, by counting the number of particles crossing an arbitrary closed surface and by applying the Green–Ostrogradski theorem, one immediately finds the existence across the surface of a diffusion flux $\Phi_D = -D \operatorname{grad} n$.

By comparing the static equation (7) and the dynamic equation (12), one sees that they have identical structures for the dependence on n and its gradient, from which we obtain the required identity between the coefficients:

$$D = \frac{1}{\mu}\frac{RT}{\mathcal{N}}. \tag{13}$$

By supposing that the particles in suspension are all spheres of radius a, Einstein uses at last Stokes' relation which gives the coefficient of friction μ of a sphere immersed in a (continuous) fluid with viscosity η:

$$\mu = 6\pi\eta\, a, \tag{14}$$

from which he finally deduced:

$$D = \frac{RT}{\mathcal{N}}\frac{1}{6\pi\eta\, a}. \tag{15}$$

This is Einstein's famous relation, which is already in his thesis. In fact, as mentioned above, the same relation was discovered earlier in Australia and, by a remarkable coincidence, published at practically the same moment as Einstein was working on his thesis. William Sutherland published his Philosphical Magazine article in March of 1905. One should therefore definitely call this relation the Sutherland–Einstein relation.

In the 1905 article, Einstein completes these results by means of mathematical and probabilistic considerations. Let $P(x, y, z; t)$ be the probability density of finding a Brownian particle at a point x, y, z at the time t. This density satisfies the diffusion equation:

$$\frac{\partial P}{\partial t} = D\,\Delta P. \tag{16}$$

Let us follow Einstein in his demonstration.

He starts by introducing a time interval τ, small compared to the duration of the observation, but large enough for the motions made by a particle during two consecutive intervals of time τ to be considered as independent events. Let us suppose then that in a liquid suspension there is a total number of particles N. During the time interval τ, the coordinates of each particle along the x axis will change by an amount Δ, where Δ takes a different value (positive or negative) for each particle. A probability distribution governs Δ: the number $\mathrm{d}N$ of particles with a displacement between Δ and $\Delta + \mathrm{d}\Delta$ is:

$$\mathrm{d}N = N\varphi_\tau(\Delta)\mathrm{d}\Delta,$$

where

$$\int_{-\infty}^{+\infty} \varphi_\tau(\Delta)\mathrm{d}\Delta = 1, \tag{17}$$

and where, for small τ, $\varphi_\tau(\Delta)$ differs from zero only for very small values of Δ. This function also satisfies the symmetry condition

$$\varphi_\tau(\Delta) = \varphi_\tau(-\Delta). \tag{18}$$

Einstein tries then to determine how the coefficient of diffusion depends on φ, once again by considering only the unidimensional case where the particle density n depends only on x and t. We can thus write $n = f(x,t)$ (the number of particles per unity volume) and we calculate the particle distribution at the time $t+\tau$ given the distribution at the time t. From the definition of the function $\varphi_\tau(\Delta)$, we obtain the number of particles between two planes in x and $x + \mathrm{d}x$ at the time $t + \tau$:

$$f(x, t+\tau)\mathrm{d}x = \mathrm{d}x \int_{-\infty}^{+\infty} f(x+\Delta, t)\varphi_\tau(\Delta)\mathrm{d}\Delta. \tag{19}$$

Since τ is very small, we can assume that

$$f(x, t+\tau) = f(x,t) + \tau\frac{\partial f}{\partial t}. \tag{20}$$

Moreover, expand $f(x+\Delta, t)$ in powers of Δ:

$$f(x+\Delta, t) = f(x,t) + \Delta\frac{\partial f(x,t)}{\partial x} + \frac{\Delta^2}{2}\frac{\partial^2 f(x,t)}{\partial x^2} + \cdots$$

We can then substitute such an expansion inside the integral in (19) as only very small values of Δ contribute to this last one. We obtain:

$$f+\tau\frac{\partial f}{\partial t} = f\int_{-\infty}^{+\infty} \varphi_\tau(\Delta)\mathrm{d}\Delta + \frac{\partial f}{\partial x}\int_{-\infty}^{+\infty} \Delta\varphi_\tau(\Delta)\mathrm{d}\Delta + \frac{\partial^2 f}{\partial x^2}\int_{-\infty}^{+\infty} \frac{\Delta^2}{2}\varphi_\tau(\Delta)\mathrm{d}\Delta + \cdots$$

On the right side, the second term, fourth term, etc., cancel out because of the parity property (18), while each of the other terms is very small in relation to the preceding one. From this equation, taking into account the conservation property (17), defining

$$\frac{1}{\tau}\int_{-\infty}^{+\infty} \frac{\Delta^2}{2}\varphi_\tau(\Delta)\mathrm{d}\Delta = D \tag{21}$$

and keeping only the first and the third terms on the right hand side, we obtain

$$\frac{\partial f}{\partial t} = D\frac{\partial^2 f}{\partial x^2}. \tag{22}$$

This is the famous diffusion equation, where the diffusion coefficient D is given by (21).

We comment now on the method of Einstein. The definition (21) of the diffusion coefficient D can be rewritten as

$$\langle\Delta^2\rangle_\tau \equiv \int_{-\infty}^{+\infty} \Delta^2\varphi_\tau(\Delta)\mathrm{d}\Delta = 2D\tau, \tag{23}$$

which is the average quadratic variation produced by the thermal agitation during the time τ. Formally identical to formula (28) (see below) which gives the law of the average quadratic displacement as a function of time, it somehow contains the latter tautologically. Moreover, as τ is assumed to be small, this definition implies the existence of the limit (21) for $\tau \to 0$, if one requires D to be independent of τ.

Einstein continues by noting that until then all particles have been considered with respect to a common origin on the x axis, but that their independence also allows us to consider each particle with respect to the position it occupied at the time $t = 0$. Therefore $f(x,t)\,\mathrm{d}x$ is also the number of particles (per unit area) whose abscissa x has changed by an amount comprised between x and $x+\mathrm{d}x$, over the time interval from 0 to t. The function f then obeys the diffusion equation (22). Einstein also says that evidently one must have, for $t = 0$,

$$f(x, t = 0) = 0, \forall x \neq 0\,; \quad \text{and} \int_{-\infty}^{+\infty} f(x,t)\mathrm{d}x = N.$$

The problem thus coincides with that of diffusion from a given point (neglecting the interactions between diffusing particles), and is now entirely determined mathematically; its solution is:

$$f(x;t) = \frac{N}{(4\pi Dt)^{1/2}} \exp\left(-\frac{x^2}{4Dt}\right). \tag{24}$$

The probability density $P(x,t) = f(x,t)/N$ for a Brownian particle to be within $\mathrm{d}x$ of x, assuming it was at $x = 0$ at the instant $t = 0$, is thus the normalized *Gaussian* distribution

$$P(x;t) = \frac{1}{(4\pi Dt)^{1/2}} \exp\left(-\frac{x^2}{4Dt}\right). \tag{25}$$

In three dimensions, if the Brownian particle is at $\vec{0}$ at the instant $t = 0$ then the solution of equation (16) is still Gaussian and written as:

$$P(x,y,z;t) = \frac{1}{(4\pi Dt)^{3/2}} \exp\left(-\frac{x^2+y^2+z^2}{4Dt}\right). \tag{26}$$

One clearly finds the previous density $P(x,t)$ by integrating over the variables y and z.

From these results one can evaluate the integral of the average quadratic displacement along, say, the x axis. One finds

$$\begin{aligned} \langle x^2 \rangle_t &= \int_{-\infty}^{+\infty} x^2 P(x;t)\,\mathrm{d}x = \frac{1}{(4\pi Dt)^{1/2}} \int_{-\infty}^{+\infty} x^2 \exp\left(-\frac{x^2}{4Dt}\right) \mathrm{d}x \\ &= 2Dt. \end{aligned} \tag{27}$$

As already pointed out above, this result for $\langle x^2 \rangle_t$ is absolutely identical to the result (23) for $\langle \Delta^2 \rangle_\tau$, which is just a reflection of the *scale invariance* of Brownian motion, a notion perhaps not yet completely mastered in 1905!

From the Sutherland–Einstein relation (15), one finally obtains the average Brownian displacement as a function of time

$$\langle x^2 \rangle_t = 2Dt = \frac{RT}{\mathcal{N}} \frac{1}{3\pi\eta\, a} t. \tag{28}$$

This is the first appearance of a *fluctuation-dissipation* relation, linking position fluctuations and a property of dissipation (the viscosity). In this fundamental equation for Brownian motion, $\langle x^2 \rangle$, t, a and η are measurable quantities and thus Avogadro's number can be determined. This is an astonishing result: first prepare a suspension of small spheres, but large however with respect to molecular dimensions, then take a chronometer and a microscope, and finally measure $\mathcal{N}$! Einstein gave this example: for water at 17°C,[59] $a \approx 0.001\,\mathrm{mm} = 1\,\mu\mathrm{m}$, $\mathcal{N} \approx 6 \times 10^{23}$, one finds a displacement of $\langle x^2 \rangle \approx 6\,\mu\mathrm{m}$ for $t = 1\,\mathrm{mn}$.

One can ask to what extent does the Sutherland–Einstein formula (13) or (15) prove the existence of molecules. In other words, what will be the limit of the diffusion coefficient $D = \frac{RT}{\mu \mathcal{N}}$ if nature was continuous, i.e., if Avogadro's number was infinite? Then D would cancel out, and the displacement of Brownian diffusion (28) would simply *disappear* in this limit, but one should verify, for the sake of rigour, the simultaneous existence of a finite continuous limit of the friction coefficient μ or of the viscosity η when $\mathcal{N} \to \infty$. We will come back to this point in section (1.7.4) where the study of a microscopic model allows for an explicit calculation of μ, and for concluding that Brownian motion is surely a manifestation of the existence of molecules!

1.4.4 Einstein, 1906, general theory of Brownian motion

In another article written in December 1905 and received on the 19th of the same month by *Annalen der Physik*,[60] this time entitled: *"On the Theory of Brownian Motion"*, Einstein mentions that *"Soon after the appearance of my paper on the movements of particles suspended in liquids required by the molecular theory of heat, Siedentopf (from Jena) informed me that he and other physicists – firstly, Prof. Gouy (of Lyons) – had been convinced by direct observation that the so-called Brownian motion is caused by the irregular thermal movements of the molecules of the liquid.*

[59] According to John Stachel in *Einstein's Miraculous Year* (Princeton University Press, New Jersey, 1998), the data Einstein uses on the viscosity of water is taken from his thesis, and in fact corresponds to the temperature 9.5°C.

[60] A. Einstein, *Ann. d. Physik* 19, 371 (1906); translated in A. Einstein, *Investigations on the Theory of the Brownian Movement*, R. Fürth *Ed.*, A. D. Cowper *Transl.*, Dover Publications, p. 19 (1956).

Not only the qualitative properties of Brownian motion, but also the order of magnitude of the paths described by the particles correspond completely with the results of the theory."

This time Einstein is convinced that Brownian motion is the phenomenon he just described. He then gives another, more general, theoretical approach. It can be applied not only to the translational, but also rotational diffusion motion of particles in suspension, or to charge fluctuations in an electric resistance. We briefly describe such a general and, from our standpoint, very enlightening approach. It shows the central role of the Boltzmann's distribution at thermodynamic equilibrium, and shows that its stationarity in time requires the existence of Brownian motion and its link to the molecular nature of heat.

Einstein considers a quantity α, which has a Boltzmann distribution

$$\mathrm{d}n = Ae^{-\frac{\mathcal{N}}{RT}\Phi(\alpha)}\mathrm{d}\alpha = F(\alpha)\mathrm{d}\alpha, \tag{29}$$

where A is a normalization coefficient and $\Phi(\alpha)$ is the potential energy associated to the parameter α. Here $\mathrm{d}n$ is proportional to the probability density of α and gives the number of systems (*à la* Gibbs) identical to the present system taken in the same state.

Einstein uses that relation for determining the irregular changes of the parameter α produced by thermal phenomena. He states that the function $F(\alpha)$ does not change during a time interval t under the combined effect of the force corresponding to the potential Φ and the irregular thermal phenomena; t is so small that all changes of the variable α can be considered as infinitesimally small in the arguments of the function $F(\alpha)$.

We consider the real line representing all α values and take an arbitrary point α_0 on it. During the time interval t, the same number of systems must pass through the point α_0 in one direction as in the other. The force $-\frac{\partial\Phi}{\partial\alpha}$ corresponding to the potential Φ induces a change of the parameter α per unit of time:

$$\frac{\mathrm{d}\alpha}{\mathrm{d}t} = -B\frac{\partial\Phi}{\partial\alpha}, \tag{30}$$

where B is, according to Einstein's words, the "mobility of the system with respect to α". This is an equation of viscous-friction type, like equation (8) with $B = 1/\mu$. According to (29), the variation of the number of systems passing through the point α_0 during the time interval t is:

$$n_1 = -B\left(\frac{\partial\Phi}{\partial\alpha}\right)_{\alpha=\alpha_0} \times tF(\alpha_0), \tag{31}$$

where the number of systems is counted algebraically (positive or negative) according to the side of α_0 they are moving from, i.e., according to the sign of the velocity (30).

Let us suppose that the probability that the parameter α changes of an amount between Δ and $\Delta + \mathrm{d}\Delta$, during the time t and under the effect of the irregular thermal processes, is equal to $\psi_t(\Delta)\mathrm{d}\Delta$, where $\psi_t(\Delta) = \psi_t(-\Delta)$ is independent of α. This last assumption reflects the intrinsic nature of thermal agitation. The number of systems passing through the point α_0 during the time t in the positive direction is given by

$$n_2 = \int_0^{+\infty} F(\alpha_0 - \Delta)\chi_t(\Delta)\mathrm{d}\Delta, \tag{32}$$

where $\chi_t(\Delta)$ is the cumulative probability that the system makes a jump to the right of size at least Δ during the time t:

$$\chi_t(\Delta) = \int_\Delta^{+\infty} \psi_t(\Delta')\mathrm{d}\Delta'. \tag{33}$$

Analogously, the number of systems that, under the effect of thermal fluctuations, pass through the value α_0 in the negative direction during the same time is (taking into account the algebraic sign),

$$n_3 = -\int_0^{+\infty} F(\alpha_0 + \Delta)\chi_t(\Delta)\mathrm{d}\Delta, \tag{34}$$

where we have used the symmetry property

$$\chi_t(\Delta) = \int_\Delta^{+\infty} \psi_t(-\Delta')\mathrm{d}\Delta'. \tag{35}$$

The equation which mathematically states the invariance of the equilibrium distribution $F(\alpha)$ is thus the law of algebraic conservation of the number of ensembles

$$n_1 + n_2 + n_3 = 0. \tag{36}$$

By substituting the expressions for n_1, n_2, and n_3, by remembering that t is infinitesimally small, as well as the values of Δ for which $\psi_t(\Delta)$ is different from 0, and by performing a first order expansion, one finds the essential equation[61]:

$$B\left(\frac{\partial\Phi}{\partial\alpha}\right)_{\alpha=\alpha_0} \times tF(\alpha_0) + \frac{1}{2}F'(\alpha_0)\langle\Delta^2\rangle_t = 0. \tag{37}$$

[61] In fact, we find that for the part concerning the thermal fluctuations

$$n_2 + n_3 = \int_0^{+\infty} \mathrm{d}\Delta\,[F(\alpha_0 - \Delta) - F(\alpha_0 + \Delta)]\,\chi_t(\Delta) = -2F'(\alpha_0)\int_0^{+\infty} \mathrm{d}\Delta\,\Delta\,\chi_t(\Delta),$$

where the integral is explicitly written

$$2\int_0^{+\infty} \Delta\,\mathrm{d}\Delta \int_\Delta^{+\infty} \psi_t(\Delta')\mathrm{d}\Delta' = \int_0^{+\infty} (\Delta')^2\psi_t(\Delta')\mathrm{d}\Delta' = \frac{1}{2}\langle\Delta^2\rangle_t,$$

after having exchanged the order of integrations or again integrated by parts.

Here

$$\langle\Delta^2\rangle_t = \int_{-\infty}^{+\infty} \Delta^2 \psi_t(\Delta)\mathrm{d}\Delta$$

represents the average quadratic variation of the quantity α due to thermal agitation during time t.

Then, by using Boltzmann's distribution $F(\alpha) \propto \exp\left[-\frac{\mathcal{N}}{RT}\Phi(\alpha)\right]$ which automatically satisfies equation (37) for any potential, Einstein obtains the average quadratic fluctuation

$$\langle\Delta^2\rangle_t = 2B\frac{RT}{\mathcal{N}}t. \tag{38}$$

Here, as before, R is the perfect gas constant, $\mathcal{N}$ is Avogadro's number, B is the system mobility with respect to the parameter α, T is the absolute temperature, and t is the time interval during which α changes due to thermal agitation.

Einstein's study shows that Boltzmann's equilibrium distribution, *dynamically* interpreted as in the conservation equation (36), implies the existence of Brownian diffusion for any physical quantity α for which the system possesses a mobility.

This idea is so rich that one can reverse the point of view and consider the equilibrium equation (37) as an equation for $F(\alpha)$, where $\langle\Delta^2\rangle_t$ is independent of α and where t is arbitrary. It is then remarkable that the solution of (37) necessarily has the exponential form of Boltzmann's distribution (29), where $\frac{RT}{\mathcal{N}}$ appears as a parameter connected with Brownian diffusion , according to the identity (38). In other words, Einstein's study of the general dynamics of Brownian motion implies equally well the particular form of the Boltzmann–Gibbs equilibrium distribution[62].

Einstein applies the result (38) to translational and rotational Brownian motions. For translational motions, the parameter α is any spatial coordinate x, and one needs to insert the corresponding value of the mobility B. For a sphere of radius a in suspension in a liquid of viscosity η, Stokes' formula, for which he cites Kirchhoff's course[63], gives

$$B = \frac{1}{\mu} = \frac{1}{6\pi\eta a},$$

[62]This strongly suggests introducing, in courses on Statistical Physics, Einstein's demonstration of Brownian motion, in order to clarify the statistical and dynamical nature of thermodynamic equilibrium. In fact, in the usual approach, Brownian motion is not taught at first, and even when it is, it appears more as a curiosity. The approach that one usually takes consists of introducing Boltzmann's distribution, either via the microcanonical ensemble and the associated Boltzmann entropy, and by evaluating the latter for a small system in contact with a thermostat, or via Shannon statistical entropy and the canonical ensemble. In these formal approaches, the emphasis is put on the probabilities and one does not see the necessity of the thermal agitation process for keeping the equilibrium distribution dynamically. After all, molecules or particles in suspension, even when initially distributed according to Boltzmann's statistics, will always fall to the bottom of the container under the effect of gravity in the absence of thermal agitation!

[63]G. Kirchhoff, *Vorlesungen über Mechanik*, 26. Vorl., 4.

and we find the famous formula (28) again:

$$\langle x^2 \rangle_t = \frac{RT}{\mathcal{N}} \frac{1}{3\pi\eta\, a} t. \tag{39}$$

Next, Einstein considers for the first time the Brownian motion of the *rotation* of a sphere suspended in a liquid, and he considers the squared fluctuations $\langle \vartheta^2 \rangle$ of any rotation angle ϑ resulting from the thermal agitation.

If one then defines $\Gamma = -\frac{\partial \Phi}{\partial \vartheta}$ the moment of the forces acting on a sphere suspended in a liquid with viscosity η, then the associated angular limit velocity is (again from Kirchhoff):

$$\frac{\mathrm{d}\vartheta}{\mathrm{d}t} = \frac{\Gamma}{8\pi\eta a^3}, \tag{40}$$

and in this case, one has:

$$B = \frac{1}{8\pi\eta a^3}.$$

One deduces

$$\langle \vartheta^2 \rangle_t = \frac{RT}{\mathcal{N}} \frac{1}{4\pi\eta\, a^3} t. \tag{41}$$

The angular motion produced by the molecular thermal agitation decreases with the radius of the sphere much faster than the translational motion does.

For $a = 0.5\,\mathrm{mm}$, and with water at 17° C, the formula gives, for $t = 1\,\mathrm{s}$, an angular shift of roughly 11 seconds of an arc, while for $a = 50\,\mu\mathrm{m}$ it gives for the same time duration roughly 100° of arc.

Finally Einstein mentions that the same formula (38) for $\langle \Delta^2 \rangle_t$ can be applied to other situations. For example, if B is chosen as the inverse of the electric resistance ρ of a closed circuit, the formula indicates the average squared total charge

$$\langle e^2 \rangle_t = 2\frac{RT}{\mathcal{N}} \frac{1}{\rho} t$$

which moves through any section of the circuit during time t.

Einstein concludes his article by assessing the limits of applicability of his formula at very short time scales, for which memory effects can occur. He arrives thereby at the estimate that the formula is valid for t large compared to a characteristic time $\tau' = m'B$, where m' is the mass of the fluid displaced by the sphere.

1.4.5 The problem of measuring the velocity

In subsequent articles, published in 1907 and 1908 in the *Zeitschrift für Elektrochemie*, Einstein tries to draw experimenters' attention to his results and to explain them in a simpler manner. He comes back to the average velocity of a particle in suspension, which must follow the equipartition law

$$\frac{1}{2}m\langle v^2\rangle = \frac{3}{2}\frac{RT}{\mathcal{N}}.$$

For Svedberg's colloid solutions of platinum, of mass $m \approx 2.5 \times 10^{-15}$ g, it gives an average velocity of 8.6 cm/s. However Einstein says that there is no possibility to observe such a velocity because of the effectiveness of viscous friction, which reduces the velocity to 1/16 of its initial value in 3.3×10^{-7} s. He continues:[64]

> "But, at the same time, we must assume that the particle gets new impulses to movement during this time by some process that is the inverse of viscosity, so that it retains a velocity which on average is equal to $\sqrt{\langle v^2\rangle}$. But since we must imagine that direction and magnitude of these impulses are (approximately) independent of the original direction of motion and velocity of the particle, we must conclude that the velocity and direction of motion of the particle will be already very greatly altered in the extraordinarily short time θ [$= 3.3 \times 10^{-7}$ s] and, indeed, in a totally irregular manner.
>
> It is therefore impossible – at least for ultra-microscopic particles – to ascertain $\sqrt{\langle v^2\rangle}$ by observation."

According to Einstein's result (28), the apparent velocity in a time interval τ is inversely proportional to $\sqrt{\tau}$ and therefore grows without limit when this time interval becomes shorter. Any attempt to measure the instantaneous velocity of a particle brings one to erratic results. This explains experimenters' repeated failures to obtain well defined conclusions for the velocity of particles in suspension. They simply were not measuring the correct quantity, and they had to wait for Einstein to show that only the ratio of the quadratic displacement over time has a theoretical limit for the experiments to connect to the theory.

As Brush remarked,[65] it was not the first time that the particular nature of a motion governed by a diffusion equation pointed out something right under one's nose. In 1854, William Thomson (who would go on to become Lord Kelvin) applied the diffusion equation (i.e., Fourier's equation for heat conduction) in his study of motion of electricity in telegraph lines. After having carried out almost exactly the same mathematical analysis that Einstein would do fifty years later, Thomson wrote:

> "We may infer that the signal delays are proportional to the squares of the distances, and not to the distances simply; and hence different observers, believing they have found a "velocity of electric propagation," may well have obtained widely discrepant results; and the apparent velocity would, *caetaris paribus*, be the less, the greater the length of wire used in the observation."

A better estimate of the very short time behavior of particles in suspension follows from subsequent work made by many physicists,[66] among which those of

[64] A. Einstein, *Zeit. f. Elektrochemie*, **13**, pp 41–42 (1907); translated in A. Einstein, *Investigations on the Theory of the Brownian Movement*, *op. cit.*

[65] S. G. Brush, *opus cit.*

[66] P. Langevin, *C. R. Ac. Sci. Paris* **146**, 530 (1908); L. S. Ornstein, *Proc. Amst.* **21**, 96 (1918); L. de Haas–Lorentz, *"The Brownian Mouvement and some Related Phenomena"*, *Sammlung Wissenschaft*, B. 52, Vieweg (1913); R. Fürth, *Zeit. f. Physik* **2**, 244 (1920).

Langevin, through his stochastic equation that we will see later, and that culminated with the Ornstein–Uhlenbeck process.[67]

A more complete formula is actually

$$\langle \Delta^2 \rangle_t = 2D\left[t - mB\left(1 - e^{-\frac{t}{mB}}\right)\right], \tag{42}$$

where $D = B\frac{RT}{\mathcal{N}}$ is the diffusion coefficient, and m this time is the mass of the particle. Therefore we clearly get the formula (38) for t large compared to the microscopic time

$$\tau = mB = \frac{m}{\mu}, \tag{43}$$

of the same order of magnitude as the time τ' estimated by Einstein.

For t smaller than τ, we find a *ballistic* regime

$$\langle \Delta^2 \rangle_t = D\frac{t^2}{mB} = \frac{RT}{\mathcal{N}}\frac{1}{m}t^2, \quad \tau \gg t, \tag{44}$$

independent of the viscosity of the medium, and which remarkably can be interpreted as corresponding to the energy equipartition theorem, this time in the form:

$$\frac{1}{2}m\frac{\langle \Delta^2 \rangle_t}{t^2} = \frac{1}{2}\frac{RT}{\mathcal{N}} \quad \tau \gg t.$$

1.4.6 Einstein's third derivation of Brownian motion

A third approach to Brownian motion was incidentally offered by Einstein in a lecture given in front of the Zürich Physical Society, on 2 November 1910, which was entitled: "On Boltzmann's Principle and Some Immediate Consequences Thereof".[68] This text seems not to have appeared in print before, so an English translation, followed by a commentary, is included in this volume.

In this fascinating lecture, Einstein describes his point of view on Statistical Physics at that time. He illustrates it by stressing the role of fluctuations, in relation to Boltzmann's formula for the entropy. This text is of particular importance, since Einstein asks more generally whether a complete causal connection can always be found between physical events; this epistemological interrogation takes place at the dawn of Quantum Mechanics.

Among other examples, Einstein considers the case of a suspended particle in a gravitational field, and performs a calculation of the mean square position of the particle. From the simple assumption of the stationarity of that average, he rederives the famous Sutherland–Einstein formula (15). This is perhaps the most direct and illuminating derivation of the Brownian diffusion formula!

---

[67] G. E. Uhlenbeck and L. S. Ornstein, *On the Theory of Brownian Motion*, *Phys. Rev.* **36**, 823–841 (1930).

[68] *Über das Boltzmann'sche Prinzip und einige unmittelbar aus demselben fliessende Folgerungen, Vorlesungen für die Physikalische Gesellschaft Zürich*, 2 November 1910, Zangger Nachlaß, Zentral Bibliothek Zürich.

1.5 Marian von Smoluchowski

"A throw of the dice never will abolish chance." (Stéphane Mallarmé, 1897)

1.5.1 Probabilities and stochasticity

Smoluchowski's name is closely attached to Brownian motion and the theory of diffusion, as we will find out here. However, as Marc Kac wrote about Smoluchowski[69], the latter showed through a true intellectual *tour de force*, that the notion of a game of chance lies at the heart of our comprehension of physical phenomena. We are indebted to him for his original and bold introduction to the calculus of probability in statistical physics, and he deserves a place beside the great names of Maxwell, Boltzmann, and Gibbs.

Marian von Smoluchowski was born in 1872, the same year as Paul Langevin, the year Boltzmann published the great memoir containing the equation that bears his name, as well as the famous "H theorem". There, Boltzmann derives the irreversible increase of entropy linked to the second principle of thermodynamic, in the area of classic Newtonian mechanics, with the help of a hypothesis of molecular chaos that Smoluchowski thought should have been instead a consequence in this framework. This brought about serious paradoxes (Loschmidt, Zermelo), because the equations of classical mechanics are reversible and have recurring cycles, called Poincaré recurrence cycles. So this forbade *a priori* the monotone growth of a function of positions and the momenta, as seen for Boltzmann's H function which is directly connected to entropy. On the defensive, Boltzmann had to introduce probabilistic and statistical arguments to justify his results, often by completely changing his point of view about the true nature of the probabilities involved. The situation became so confused that Paul and Tatiana Ehrenfest, for example, tried to clarify Boltzmann's ideas by banishing the term (but not the concept) "probability" from their famous 1912 Encyclopedia memoir!

As S. G. Brush noted[70], the research line of the kinetic theory of gases that Smoluchowski pursued was a continuation of that of Clausius, Maxwell, O.E. Meyer, Tait and Jeans, according to which one describes the effects of collisions on the trajectory of a molecule, and therefore on the properties of the gas. Einstein, on the contrary, followed the path opened by Boltzmann, Maxwell (in his subsequent articles) and Gibbs, where the objective was to obtain more general laws starting from statistical distributions postulated for molecular ensembles, without making any assumption about intramolecular forces and collision mechanisms. It is then extremely interesting to see these two "Gedankenwege", kinetic theory and statistical mechanics, meet up in relation to Brownian motion, *terra incognita* for both theories.

---

[69] *Marian Smoluchowski, His Life and Scientific Work*, S. Chandrasekhar, M. Kac, R. Smoluchowski, Polish Scientific Publishers, PNW (2000).

[70] S. G. Brush, *locus cit.*

In this context, by working in the same pragmatic spirit as Maxwell, Smoluchowski courageously showed how to use the theory of probability in physics as an efficient instrument, during an era when mathematicians looked down on it, and when physicists mostly ignored it. Without knowing it, Smoluchowski opened a new chapter of statistical physics, that nowadays bears the name Stochastic Processes[71].

1.5.2 Brownian motion and random walks

This probabilistic point of view is clearly present in Smoluchowski's first article on Brownian motion, *"Essay on the theory of Brownian motion and disordered media"*[72] published in 1906 (very likely under the pressure of Einstein's publication of his first two articles), as well as in another article, about the mean free path of molecules in a gas.[73] In these remarkable articles he was seemingly the first to establish the relation between *random walks* and Brownian diffusion, even though in 1900 Louis Bachelier had already introduced the model of a random walker in his thesis *The Theory of Speculation.* We shall return to this later.

Smoluchowski begins by citing Einstein's work from 1905 and writes that the latter's results *"completely agree with those I obtained a few years ago by an entirely different path of reasoning, and that since then I have considered an important argument in favor of the kinetic nature of these phenomena."* However, he adds further along that his own method *"seems more direct, simpler, and perhaps more convincing than that of Einstein."*

While Einstein (as Sutherland) avoids all treatment of collisions in favor of a general thermodynamic approach, Smoluchowski has a clear kinetic vision and treats the Brownian motion as a random walk or a game of heads or tails (see figure 2).

The newness and the originality of Smoluchowski's approach is the replacement of an incredibly difficult problem (a Brownian particle which collides within a gas or liquid) by a relatively simple stochastic process. Each dynamic event like a collision is considered as a random event similar to a game of heads or tails, or to the throw of a dice, where the elementary probabilities are (to a certain extent) determined by underlying mechanical laws. This way of reasoning plays a fundamental role in mechanics and statistical physics today and, as Marc Kac noticed, it is difficult for us today to imagine the degree of Smoluchowski's intellectual boldness for starting this subject during the early years of the last century.

---

[71]From the Greek word *stokhastikos*, "to aim well", "capable of making conjectures", already used by Jacques Bernoulli in 1713 in *Ars Conjectandi.*

[72]M. R. von Smolan Smoluchowski, *Rozprawy Kraków* **46A**, pp. 257–281 (1906); French translation: "Essai d'une théorie du mouvement brownien et de milieux troubles", *Bull. International de l'Académie des Sciences de Cracovie*, pp. 577–602 (1906); German translation: *Ann. d. Physik* **21**, pp. 755–780 (1906).

[73]M. R. von Smolan Smoluchowski, *Sur le chemin moyen parcouru par les molécules d'un gaz et sur son rapport avec la théorie de la diffusion, Bulletin International de l'Académie des Sciences de Cracovie*, pp. 202–213 (1906).

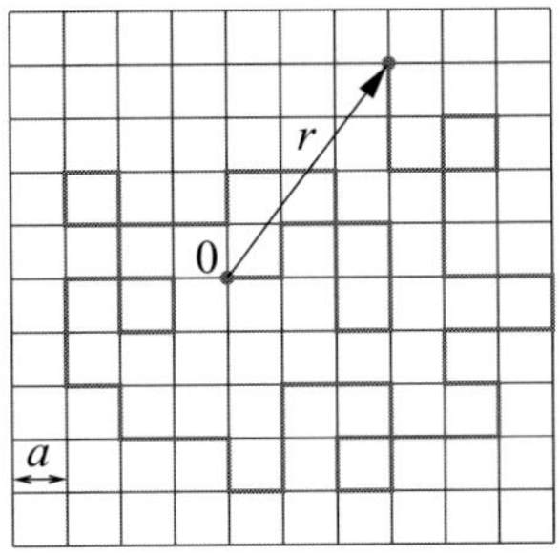

Figure 2: *Random walk on a square lattice with elementary lattice step a. We choose each step at random. In two dimensions, two equivalent methods exist. In the first one, we draw heads or tails (with a probability of 1/2) for a direction, vertical or horizontal, and next the orientation along the chosen direction. In the second method, we draw with the same probability (with probability of 1/4) one of the four possible directions. In the continuous limit where the lattice step goes to 0, a very long random walk will take the appearance of the Brownian motion of figure* 1.

1.5.3 Smoluchowski's contributions

Smoluchowski[74] knew about the most recent studies on Brownian motion and in particular the work of Felix Exner. He sent Smoluchowski diagrams made from memory, called "Krix-Krax" because of the several inter-crossing "jumps" apparently made by a Brownian particle observed under a microscope over a set of discrete instants of time.

Smoluchowski began by criticizing Nägeli's arguments who affirmed that a collision of a molecule of water with a sphere 0.001 mm in diameter would give a velocity of 3×10^{-6} cm/s, which would be impossible to observe under a microscope, and that the collision effects would cancel out on average. He compared this way of thinking to that of a player who believed to never be able to lose more than a single bet, despite repeated draws! By continuing the analogy further, he calculated for the heads or tails game how the positive (or negative) cumulated gains grow with the number n of draws ("time").

Let $p_{n,m}$ be the probability to have met m favorable outcomes in the total of n draws, with a net gain of $m-(n-m)=2m-n$. This probability can be written as

$$p_{n,m} = \frac{1}{2^n}\frac{n!}{m!(n-m)!} = \frac{1}{2^n}\binom{n}{m}, \tag{45}$$

where the number of combinations $\binom{n}{m}$ is the number of ways of choosing m out of n objects.

[74]In this section we follow Brush's presentation of Smoluchowski's work.

The positive or negative mean deviation from zero, δ_n, i.e., the average of the absolute value of a gain or of a loss after n turns, can be calculated as

$$\delta_n = \langle |2m-n| \rangle = 2 \sum_{m=n/2}^{n} (2m-n) p_{n,m} = 2 \sum_{m=n/2}^{n} (2m-n) \frac{1}{2^n} \binom{n}{m} = \frac{n}{2^n} \binom{n}{\frac{n}{2}},$$

where n is supposed an even number, to simplify the notation. For large n, we then use Stirling's formula $n! \simeq n^n e^{-n} \sqrt{2\pi}$, to evaluate δ_n:

$$\delta_n \simeq \sqrt{\frac{2n}{\pi}}, \quad n \gg 1.$$

The (arithmetical) average of successive gains (or losses) with respect to 0 increases then as $\sqrt{n}$, even when the total (algebraic) average is zero. The analogous number n of molecular collisions per second on a sphere, was estimated by Smoluchowski as 10^{16} in a gas and 10^{20} for a liquid. If the gain in velocity is of the order 10^{-6} cm/s at each collision, one obtains a mean cumulated velocity from 10^2 to 10^4 cm/s per second. However Smoluchowski immediately reduces this conclusion, remarking that each individual gain of velocity will fluctuate, and that a high velocity value decreases the probability of one more positive gain.

He shows next that a "true" velocity of displacement could be obtained from the equipartition of kinetic energy, and would give a velocity of 0.4 cm/s, again much too large in relation to experimental observations! In fact, Exner's diagrams in "Krix-Krax" gave a velocity of about 3×10^{-4} cm/s, an apparently irreconcilable disagreement. As Smoluchowski says, *"this contradiction, already seen by F. Exner, seems at first to be a decisive objection to kinetic theory. Nevertheless the explanation is very simple."*

He presents the following simple and clear explanation: such a velocity is too large to be observed with a microscope magnifying 500 times. What one observes is the average position of a particle having this velocity, but hit 10^{20} times per second, each time in a different direction, such that one cannot observe the instantaneous velocity. Each zig-zag displacement is incomparably smaller than the particle's size, and it is only when the geometric sum of these elements reaches a certain value that one can observe a displacement, which appears to us to be slower. It is clearly the substance of Einstein's argument, here supported by the concrete image of kinetic theory: the average displacement is the observable physical quantity, while velocity is not.

After such qualitative, but illuminating, considerations, Smoluchowski develops his model of random collisions. Let m and v (m' and v' respectively) be the mass and the velocity of a particle in suspension (of molecules in the liquid, respectively). From the equipartition of energy, one has on average:

$$\frac{v}{v'} = \sqrt{\frac{m'}{m}}. \tag{46}$$

He affirms that from *"the laws of collision of elastic spheres"*, the change of velocity of the sphere in suspension is, at a collision, on average given by a small transverse component $\alpha\, m'v'/m$, where $\alpha = 3/4$. The result is a random change of the velocity direction of a small angle $\varepsilon = \alpha\, m'v'/mv$. (According to (46), one also has $\varepsilon = \alpha v/v'$ on average.) He assumes also that the molecular impacts occur at equal intervals of time, which makes of the particle trajectory a chain made of constant-length segments.

In other words, Smoluchowski adapts the idea of mean free path of a molecule in a gas, even though here the persistence of motion is shortened by the presence of numerous molecules of the surrounding fluid.

The problem of Brownian motion is thus mathematically mapped onto the one of finding the end-to-end average distance Δ_n^2, of a chain of n segments, all of length ℓ, each randomly turned by a small angle ε with respect to the preceding one. He then obtains the general solution by a complicated recurrence relation, containing multiple angular integrals of trigonometric functions, of the form:

$$\Delta_n^2 = \ell^2 \left\{ \frac{2n}{\delta} + 1 - n - 2\frac{(1-\delta)^2 - (1-\delta)^{n+2}}{\delta^2} \right\}, \tag{47}$$

where $\delta = 1 - \cos\varepsilon \simeq \varepsilon^2/2$.

In the limit where $n\delta$ is small, one finds

$$\Delta_n = n\ell \left(1 - \frac{n\delta}{6}\right), \tag{48}$$

which represents a quasi-ballistic trajectory.

In the opposite case of a large number of collisions per unit of time $n\delta \gg 1$, the first term of (47) dominates and one finds the expected result:

$$\Delta_n^2 = \ell^2 \frac{2n}{\delta} = \ell^2 \frac{4n}{\varepsilon^2}. \tag{49}$$

If we call $\bar{n}$ the number of collisions per unit of time, such that there are $n = \bar{n}\, t$ collisions over the time t, we have for a free path $\ell = v/\bar{n}$, and by using $\varepsilon = \alpha v/v'$, we find an average quadratic displacement at time t,

$$\Delta_n^2 \equiv \Delta_t^2 = \frac{4}{\alpha^2} \frac{v'^2}{\bar{n}} t. \tag{50}$$

The momentum mv of the particle in suspension changes on average by a quantity $\alpha' m'v$ per collision, where, from Smoluchowski, $\alpha' = 2/3$, which means the friction force $F = -\bar{n}\alpha' m'v$, and thus the friction coefficient $\mu = \bar{n}\alpha' m'$. Substituting μ in $\bar{n}$ one obtains: $\Delta_t^2 = \frac{4\alpha'}{\alpha^2} \frac{m'v'^2}{\mu} t$. From the equipartition of kinetic energy of the molecules in the surrounding fluid: $\langle m'v'^2 \rangle = 3RT/\mathcal{N}$, and the result of Smoluchowski finally becomes:

$$\Delta_t^2 = \frac{2\alpha'}{\alpha^2} 6 \frac{RT}{\mu\mathcal{N}} t. \tag{51}$$

One finds again the Sutherland–Einstein result (15), (28), this time in three dimensions, with a supplementary numerical factor of kinetic origin $2\alpha'/\alpha^2 = (4/3)^3 = 64/27$. Because of the various physical and geometrical approximations involved, this factor should perhaps not come as a surprise! The experiments of Svedberg in 1907 seemed to support this result, but Langevin mentioned later in 1908, in his article to the Compte Rendus, that once these approximations were corrected, Smoluchowski's stochastic method gave the same formula (28) of Einstein. Smoluchowski himself adopted this formula in his subsequent articles.

Afterwards he gave the complete theory of density fluctuations within an ensemble of Brownian particles, as well as that of their sedimentation in a gravitational field and of the coagulation of colloids.[75] Smoluchowski's name is thus traditionally attached to the generalization of the diffusion equation (16) in a force field F:

$$\frac{\partial P}{\partial t} = D\Delta P - \frac{1}{\mu}\mathrm{div}(FP), \tag{52}$$

where μ is the same as in (14). This equation applies directly to the case of a uniform gravitational field. In one dimension it is simply written in the so-called Fokker–Planck[76] [77] form

$$\frac{\partial P(x,t)}{\partial t} = D\frac{\partial^2}{\partial x^2}P(x,t) + \frac{1}{\mu}\frac{\partial}{\partial x}\left(\frac{\partial V(x)}{\partial x}P(x,t)\right), \tag{53}$$

for a force field $F(x)$ derived from a potential $V(x)$.

We also mention that we owe to him (and to Einstein) the theory of critical opalescence as well.

1.5.4 Brownian motion and the second principle

Another aspect of Smoluchowski's work concerns the correct statistical formulation of the second principle of thermodynamics. With Theodor Svedberg's recent data on Brownian motion, Smoluchowski had experimental results which permitted him, armed with his own theory of fluctuations near-to-equilibrium, to estimate the recurrence and persistence time of a system slightly out of equilibrium, and to check the agreement with experiments. He used neither phase space, nor Liouville's theorem as in classical statistical mechanics *à la* Boltzmann. He introduced simply the calculus of probability.

By incorporating the theory of fluctuations he gave a correct formulation of the second principle of thermodynamics, where this principle appeared valid only in a *statistical* sense, and susceptible to multiple twists at the microscopic level.[78]

[75] M. von Smoluchowski, *Drei Vorträge über Diffusion, Brownsche Molekular Bewegung und Koagulation von Kolloidteilchen, Physikalische Zeitschrift*, **Jg. 17**, pp. 557–571, 585–599 (1916).

[76] A. D. Fokker, Thesis, Leiden (1913); *Ann. d. Physik* **43**, 810 (1914).

[77] M. Planck, *Sitzungsber. Preuss. Akad. Wissens.* p. 324 (1917); in *Physikalische Abhandlungen und Vorträge* II, p. 435, Vieweg, Braunschweig (1958).

[78] M. von Smoluchowski, *Phys. Z.* **13**, pp. 1069–1080 (1912); *Göttinger Vorträge über die kinetische Theorie der Materie u. Elektrizität*, Leipzig **S. 89–121** (1914).

A modern discussion of Smoluchowski's ideas was given by Richard Feynman in his famous lectures[79]. He compared Maxwell's demon with a ratchet and pawl and an electric rectifier, neither of which can systematically transform internal energy from a single reservoir to work. He wrote:

"If we assume that the specific heat of the demon is not infinite, it must heat up. It has but a finite number of internal gears and wheels, so it cannot get rid of the extra heat that it gets from observing the molecules. Soon it is shaking from Brownian motion so much that it cannot tell whether it is coming or going, much less whether the molecules are coming or going, so it does not work."

Modern day computer simulations strikingly reveal the fluctuation phenomena envisaged by Smoluchowski and Feynman.[80]

Smoluchowski's observation suggested that Maxwell's demon ought to be buried and forgotten. But that did not happen, apparently because Smoluchowski's approach left open the possibility that somehow, a perpetual motion machine operated by an "intelligent" being might be achievable. It was the fascinating idea of using intelligence that captured Leo Szilard's interest, in his classic 1929 paper, *"On the decrease of entropy in a thermodynamic system by the intervention of intelligent beings"*.[81]

The feature associated with intelligence that is needed by a demon is memory: it must remember what it measures, even so only briefly. Notably, Szilard discovered with his heat engine with a one-molecule working fluid, the idea of a "bit" of information with entropy $k_B \ln 2$, now central in computer science, and established the connection between entropy and information.

At this stage, rather than fully opening the Pandora box which contains the Protean Maxwell's demons, we prefer to suggest reading the survey, *Maxwell's Demon 2*, by H. S. Leff and A. F. Rex and in particular the thoughtful introduction of the second edition.[82] Let us only mention a few historical landmarks that are described in their presentation.

After a hiatus of 20 years, Léon Brillouin, assuming the use of (quantum) light signals in the demon's attempts to defeat the second law, concluded that information acquisition, like measurement, is dissipative. It led him to break new ground by developing an extensive mathematical theory connecting measurement and information. The impact of Brillouin's and Szilard's work was far reaching and the result was a proclaimed, but temporary, "exorcism" of the demon.

A new life began for the demon when Rolf Landauer made the important discovery that memory erasure in computers feeds entropy to the environment.[83]

[79] R. P. Feynman, R. B. Leighton and M. Sands, *The Feynman Lectures on Physics I*, Chap. 46, Addison-Wesley, Reading MA (1963).

[80] See, e.g., P. A. Skordos and W. H. Zurek, *Am. J. Phys.* **60**, 876 (1992).

[81] L. Szilard, *Z. Phys.* **53**, pp. 840–856 (1929); transl. reprinted in *The Collected Works of Leo Szilard, Scientific Papers*, B. T. Feld and G. Weiss Szilard, eds., The MIT Press, Cambridge, Mass. (1972).

[82] H. S. Leff and A. F. Rex, *Maxwell's Demon 2*, Adam Hilger, Bristol (2003).

[83] R. Landauer, *IBM J. Res. Dev.* **5**, pp. 183–191 (1961).

This is now called "Landauer's principle". It states that the erasure of one bit of information stored in a memory device requires sending an amount of entropy of at least $k_B \ln 2$ to the environment, i. e., a minimal heat generation of $k_B T \ln 2$.

Charles Bennett, after his important demonstration in 1973 that reversible computation, which avoids the erasure of information, is possible in principle, argued in 1982 that erasure of a demon's memory is the fundamental act that saves the second law because of Landauer's principle.[84] This was a turning point in the history of Maxwell's demon.

In his 1970 book *Foundations of Statistical Mechanics*, Oliver Penrose independently recognized the importance of "resetting" operations that bring all members of a statistical ensemble in the same observational state. Applied to Szilard's heat engine, this is nothing else than memory erasure, which sends an amount of entropy of at least $k_B \ln 2$ to the environment.

Among recent proofs of Landauer's principle we cite here, somehow arbitrarily, the one by K. Shizume, who uses a solvable model of memory based on a Brownian particle in a time-dependent potential well;[85] the one by M. Magnasco through a detailed analysis of Szilard's heat engine;[86] and the one by B. Piechocinska, who assumes the decoherence of the states of the thermal reservoir[87].

Let us finally mention that, despite several attempts to argue against its validity, the Landauer–Penrose–Bennett framework seems to be generally accepted as providing the solution to the Maxwell's demon-second principle puzzle, at least in classical mechanics, and in a thermodynamical limit of some sort.[88]

However, there are now indications that Landauer's, as well as the second principle, might not hold in the (strong) quantum regime. The source of the violation is *quantum entanglement* between the system and the constant-temperature reservoir, which then act as a single entity.[89]

In close relation to Brownian motion and the second principle, the topic of Brownian motors has recently received considerable attention.[90] C. Van den Broeck *et al.*[91] were able to find a solvable model for a *thermal Brownian motor*. They show that immersed in two different thermal baths, two rigidly coupled Brownian particles with a geometrical asymmetry, can function as a microscopic

[84] C. H. Bennett, *Int. J. Theor. Phys.* **21**, pp. 905–940 (1982).

[85] K. Shizume, *Phys. Rev. E* **52**, pp. 3495–3499 (1995).

[86] M. O. Magnasco, *Europhys. Lett.* **33**, pp. 583–588 (1996).

[87] B. Piechocinska, *Phys. Rev. A* **61**, 062314 (2000).

[88] For possible violations of Thompson's formulation of the second principle for a mesoscopic work source, see A. Allahverdyan, R. Balian and T. M. Nieuwenhuizen, *Entropy* **6**, pp. 30–37 (2004); see also *Europhys. Lett.* **67**, pp. 565–571 (2004).

[89] A. Allahverdyan and T. M. Nieuwenhuizen, *Phys. Rev. Lett.* **85**, pp. 1799–1802 (2000); *Phys. Rev. E* **64**, 056117 (2001); T. M. Nieuwenhuizen and A. Allahverdyan, *Phys. Rev. E* **66**, 036102 (2002); T. D. Kien, *Phys. Rev. Lett.* **93**, 140403 (2004).

[90] P. Reimann, *Phys. Rep.* **361**, 57 (2002); R. D. Astumian and P. Hänggi, *Phys. Today* **55**, 33 (2002); H. Linke (ed.), *Ratchets, Experiments and Applications*, *Appl. Phys.* A **75** (2002).

[91] C. Van den Broeck, R. Kawai and P. Meurs, *Phys. Rev. Lett.* **93**, 090601 (2004); C. Van den Broeck, P. Meurs and R. Kawai, *From Maxwell Demon to Brownian Motor*, *New Journal of Physics* **7**, 10 (2005).

engine able to rectify Brownian fluctuations. As expected, when the temperatures of the two baths are equal, the drift motion ceases, and one is left only with a standard Brownian displacement, which obeys Gauss' distribution law. The drift speed can be computed exactly for convex bodies, in the limit of dilute gases. Extemely precise molecular dynamics simulations with hard disks are possible, which confirm the theory. In effect, it is a microscopic and soluble Feynman's ratchet. In a work in progress, Van den Broeck and Kawai propose now a cooling mechanism, based on such a Brownian motor submitted to an external force. A heat flow is generated between the two components of the motor.

Such a marvelously simple microscopic model would have certainly greatly pleased Einstein, Smoluchowski and Sutherland!

It is necessary to note here that these discussions are current research topics of intense interest. In fact today there exist new theoretical results, known as the Gallavotti–Cohen fluctuation theorem,[92] Jarzynski's equality,[93] or Crooks' fluctuation theorem.[94] They quantify the spontaneous average work provided by a source of heat, during irreversible phenomena. The manipulations of single biological molecules like DNA and RNA, that are mesoscopic objects, allow to test these relations experimentally. The interpretations of these results and experiments are actually the topic of a lively debate,[95] just as at the dawn of Brownian motion![96]

1.5.5 Brownian motion and the mathematical aspects of irreversibility

Let us open here a brief mathematical parenthesis.[97] Einstein's and Smoluchowski's theories, based upon a Newtonian dynamics of the particles, in fact postulated the emergence of Brownian motion fom a classical non-dissipative reversible dynamics, a point of view which was far from being physically obvious or, a fortiori, mathematically rigorous. This precisely led to the heated controversy about the second principle. The key difficulty is similar to the justification of Boltzmann's molecular chaos assumption (*"Stosszahlansatz"*) standing behind the derivation of the Boltzmann equation. Mathematically, the dissipative character can only

[92]D. J. Evans and D. J. Searles, *Phys. Rev. E* **50**, 1645–1648 (1994); G. Gallavotti and E. G. D. Cohen, *Phys. Rev. Lett.* **74**, 2694–2697 (1995); *J. Stat. Phys.* **80**, 931–970 (1995); see also D. J. Evans, E. G. D. Cohen and G. P. Morris, *Phys. Rev. Lett.* **71**, 2401–2404 (1993); G. M. Wang, E. M. Sewick, E. Mittag, D. J. Searles and D. J. Evans, *Phys. Rev. Lett.* **89**, 050601 (2002).

[93]C. Jarzynski, *Phys. Rev. Lett.* **78**, pp. 2690–2693 (1997).

[94]G. E. Crooks, *Phys. Rev. E* **60**, pp. 2721–2726 (1999).

[95]See, e.g, E. G. D. Cohen and D. Mauzerall, *J. Stat. Mech.* P07006 (2004), and the reply by C. Jarzynski, arXiv:cond-mat/0407340.

[96]The interested reader can consult the texts by Ch. Maes and F. Ritort in the Poincaré Seminar on *Entropy* (2003), available on the website www.lpthe.jussieu.fr/poincare, and published in: J. Dalibard, B. Duplantier & V. Rivasseau *Eds.*, *Poincaré Seminar 2003*, Progress in Mathematical Physics, vol. 38, Birkhäuser, Basel (2004). See also C. Bustamante, J. Liphard and F. Ritort in *Physics Today*, July 2005, pp 43–48.

[97]The material of this section is borrowed from the introduction of the recent paper by L. Erdös, M. Salmhofer and H.-T. Yau, *Towards the quantum Brownian motion*, arXiv:math-ph/0503001 (2005).

emerge in a scaling limit, as the number of degrees of freedom goes to infinity.

As we shall see below, the first mathematical definition of Brownian motion was given only in 1923 by Wiener. But the derivation of Brownian motion from Hamiltonian dynamics was not seriously investigated until the end of the seventies. Kesten and Papanicolaou[98] proved that the velocity distribution of a particle moving in a random scatterer environment (so-called Lorenz gas with random scatterers) converges to Brownian motion in dimension $d \geq 3$. The same result was obtained in $d = 2$ dimensions by Dürr, Goldstein and Lebowitz.[99] A very recent work establishes the convergence to Brownian motion in position space as well.[100]

Bunimovich and Sinai proved the convergence to Brownian motion of the periodic Lorenz gas with a hard-core interaction.[101] The only source of randomness there is the distribution of the initial conditions. Finally, Dürr, Goldstein and Lebowitz[102] established that the velocity process of a heavy particle in an ideal gas converges to the Ornstein–Uhlenbeck process, that is a version of Brownian motion.

Brownian motion was discovered and theorized in the context of classical mechanics, and it postulates a microscopic reversible Newtonian world for atoms and molecules. Nowadays, it is thus natural to replace Newtonian dynamics with Schrödinger dynamics and investigate if Brownian motion still correctly describes the motion of a quantum particle in a random environment. For a very recent discussion of this fundamental and difficult question, we refer the reader to a recent work by Erdös, Salmhofer and H.-T. Yau[103] and to the references therein.

1.5.6 Smoluchowski's legacy

With Einstein, Smoluchowski shares the credit of having shown the importance of microscopic fluctuations in statistical physics, at the same time promoting the probabilistic approach. In this sense he appears as a great master inheritor in physics of the *Doctrine of Chance* from Abraham de Moivre.

Marian von Smoluchowski died prematurely in 1917, at the age of forty five, in Kraków.

1.6 Louis Bachelier

1.6.1 Bachelier and mathematical finance

Louis Bachelier is nowadays considered as having laid the foundation for mathematical finance, and is further credited with the first mathematical study of the

[98] H. Kesten, G. Papanicolaou, *Comm. Math. Phys.* **78**, pp. 19–63 (1980/81).

[99] D. Dürr, S. Goldstein, J. Lebowitz, *Commun. Math. Phys.* **113**, pp. 209–230 (1987).

[100] T. Komorowski, L. Ryzhik, *Diffusion in a weakly random Hamiltonian flow*, arXiv:math-phys/0505082 (2005); *The stochastic acceleration problem in two dimensions*, arXiv:math-phys/0505083 (2005).

[101] L. Bunimovich, Y. Sinai, *Commun. Math. Phys.* **78**, 479–497 (1980/81).

[102] D. Dürr, S. Goldstein, J. Lebowitz, *Commun. Math. Phys.* **78**, pp. 507–530 (1980/81).

[103] L. Erdös, M. Salmhofer and H.-T. Yau, *op. cit.*; see also arXiv:math-ph/0502025 (2005).

continuous Brownian process, including a random walk approach to the latter. A detailed and very interesting presentation of Bachelier's life and scientific achievements was given in 2000 in an essay, entitled *Louis Bachelier on the Centenary of Théorie de la Spéculation*, for the centenary of the publication of his thesis.[104] This section is essentially based on their presentation and a significant part of it incorporates material of the cited article.

The importance of Bachelier's work was not properly recognized during his time. As Benoît Mandelbrot writes nicely in *The Fractal Geometry of Nature*,[105] it was Kolmogorov in 1931 who re-discovered his name in an article in *Mathematische Annalen*.[106]

Bachelier was interested in the theory of speculation at the Paris stock market. He successfully defended his thesis, entitled *Théorie de la spéculation*, on 29 March 1900 at the Sorbonne, in front of a jury composed of Paul Appell, Joseph Boussinesq and Henri Poincaré, his thesis advisor. As a work of exceptional merit, stongly supported by Poincaré, his thesis was published in the *Annales Scientifiques de l'École Normale Supérieure*.[107]

1.6.2 The Thesis

Bachelier begins with the mathematical modeling of stock price movements, and formulates the principle that "the expectation of the speculator is zero", by which he means that the conditional expectation given the past information is zero. In other words, he assumes that the market evaluate assets according to a martingale measure. The further hypothesis is that the price evolves as a continuous Markov process (with no memory), homogeneous in time and space. Bachelier then shows that the density of one-dimensional distributions of this process satisfies an integral relation, now known as the Chapman–Kolmogorov equation. Bachelier, without addressing the question of uniqueness, shows the Gaussian density, with a linearly increasing variance, to solve the equation.

He also considers a discrete version of the problem, where the price process is the continuum limit of random walks, and where the binomial formula (45) appears. He then proceeds to show that the distribution functions of the process satisfy Fourier's heat equation, as in the similar eq. (22) in Einstein's article. Bachelier then introduces a novel expression: "the radiation of the probability".

One finds indeed many of the well-known results for Brownian motion: On p. 37 of his memoir, one reads that: *"On voit que la probabilité est régie par la loi de Gauss déjà célèbre dans le Calcul des probabilités"*; on p. 38, that *"L'espérance*

[104] J.-M. Courtault, Y. Kabanov, B. Bru, P. Crépel, I. Lebon and A. Le Marchand, *Louis Bachelier on the Centenary of Théorie de la Spéculation*, *Mathematical Finance*, Vol. **10**, No. 3, pp. 341–353 (2000).

[105] B. B. Mandelbrot, *The Fractal Geometry of Nature*, Freeman, New York (1982).

[106] A. N. Kolmogorov, *Über die analitischen Methoden in der Warscheinlichkeitsrechnung*, *Math. Annalen* **104**(3), pp. 415–458 (1931).

[107] L. Bachelier, *Ann. Sci. École Normale Supérieure* **17**, pp. 21–86 (1900).

mathématique

$$\int_0^\infty p x dx = k\sqrt{t}$$

est proportionnelle à la racine carrée du temps." Bachelier also calculates the probability that the Brownian motion does not exceed a fixed level and finds the distribution of the supremum of that motion.

He therefore developed in his first thesis a theory of continuous stochastic processes close to the modern mathematical theory of Brownian motion. As stressed by the authors of the essay *Louis Bachelier on the Centenary of Théorie de la Spéculation*, "more than one hundred years after the publication of the thesis, it is quite easy to appreciate the importance of Bachelier's ideas. The thesis can be viewed as the origin of mathematical finance, and of several branches of stochastic calculus such as the theory of Brownian motion, Markov processes, diffusion processes, and even weak convergence in functional spaces."

It is also quite interesting to read Poincaré's original report, translated in the essay cited above. Poincaré's report shows that Bachelier's thesis was highly appreciated by the outstanding mathematician. In contrast to the legend that the evaluation note *"honorable"* means somehow that the examinors were dissatisfied with the thesis, it can perhaps be argued that it might have been the highest grade possible for a thesis which was addressing a problem not in the realm of standard mathematics, and that in addition had a number of non-rigorous arguments.

The official report of the Thesis Committee states:

In the presentation of his First Thesis, M. Bachelier showed mathematical intelligence and insight. He has added some interesting results to those already contained in the printed version of the thesis, in particular an application of the image method.

As for the Second Thesis, he proved to possess a complete knowledge of Boussinesq's work on the motion of a sphere in an indefinite fluid.

The Faculty gave him the degree of Doctor with honors.

Paul Appell, President

It is indeed very intriguing that the Proposal given by the Faculty, subject of his Second Thesis, was entitled: *Resistance of an indefinite liquid mass with internal frictions, described by the formulae of Navier, to small translational motions of a solid sphere, submerged inside the fluid and adhering it.*

But there is of course no mention in his first thesis, published in 1900, about any link between the speculation problem and the motion of a sphere in a viscous fluid! However, we saw above Poincaré's early interest in Brownian motion in relation to Carnot's principle. We also saw that Einstein's (as well as Sutherland's) application of hydrodynamical laws to the motion of a sphere suspended in a fluid, was key to the solution of Brownian motion. We now observe the amazing coincidence that the thesis subject proposed by the Faculty, if joined with the subject of the first thesis, could just have led Poincaré and Bachelier to establish the quantitative theory of Brownian motion, before any Einstein, Sutherland or Smoluchowski! All necessary mathematical equations were indeed present for that, if only a little spark of physical intuition could have struck these eminent mathematicians!

1.6.3 Further Studies

Louis Bachelier continued to develop the mathematical theory of diffusion processes in a series of memoirs and books. In his 1906 memoir on the *Théorie des probabilités continues*,[108] he defined new classes of stochastic processes, which are now called processes with independent increments and Markov processes, and he derived the distribution of the Ornstein–Uhlenbeck process.

He was aware of the importance of his contributions. He wrote in his 1924 "Notice de Travaux" that *"this theory has no relation to the geometrical theory of probability, the range of application of which is quite limited. We are concerned here with a science of a different order of generality, compared to classical probability calculus. Among the new concepts, one can cite the assimilation to an energy of the probability which is an abstraction. That original concept was quite noticed by Henri Poincaré, and it made many progresses possible."* One also reads about his 1912 book *Calcul des probabilités*,[109] that *"it is the first that surpassed the great treatise by Laplace."*

We shall not describe in detail here the very unfortunate misunderstanding with Paul Lévy, which in 1926 prevented Bachelier to become a full professor at the University of Dijon. We refer the interested reader to the essay mentioned above for a thorough and well-documented analysis of this dramatic event.

Later, Lévy, under the influence of Kolmogorov's fundamental paper (1931) on diffusion processes, which referred to Bachelier's work, realized that a number of properties of Brownian motion had been discovered by Bachelier several decades earlier. He revised his opinion, and wrote him a letter with apologies.

Bachelier's ideas seem to receive nowadays a widespread recognition. Famous probability treatises, like the ones by W. Feller, *An Introduction to Probability Theory and its Applications* (1957), or by K. Itô and H. McKean, *Diffusion Processes and their Sample paths* (1965), refer to Bachelier's seminal work.

In the literature written by economists, one finds reference to him in Keynes (1921), and more recently in the work of other famous economists, like the 1997 Nobel laureates in Economic Sciences, Robert Merton and Myron Scholes. It is perhaps appropriate here to reproduce Merton's tribute to Bachelier:

"The origin of much of the mathematics in modern finance can be traced to Louis Bachelier's 1900 dissertation on the theory of speculation, framed as an

[108] *Théorie des probabilités continues*, J. Math. Pures et Appl., pp. 259–327 (1906).
[109] L. Bachelier, *Calcul des probabilités*, Gauthier Villars, Paris (1912).

option-pricing problem. This work marks the twin births of both the continuous-time mathematics of stochastic processes and the continuous-time economics of derivative-security pricing."

No doubt that today Bachelier would have been awarded a Nobel Prize in Economic Sciences for his work of 1900!

1.7 Paul Langevin

Knowing the great interest in the theory of Brownian motion, signalled by the works of Gouy, Einstein, and Smoluchowski, Langevin took the next steps in 1908. He first said that the factor of 64/27 of Smoluchowski's results, due to the approximations made, was erroneous and that the result coincided with Einstein's formula (28) after his correction. Next, he provided another demonstration of this fact, in which was contained the first mathematical example of a *stochastic equation.*

1.7.1 Langevin's equation

Langevin's argument is enlightening and we follow his demonstration faithfully.[110] The starting point is the Maxwell equipartition theorem of kinetic energy. It states that the energy of a particle in suspension inside a fluid in thermal equilibrium has, for instance in the x direction, an average kinetic energy $\frac{1}{2}\frac{RT}{\mathcal{N}}$, equal to that of any gas molecule, in a given direction, at the same temperature. This is directly related to van 't Hoff's law seen above, which affirms the identity between diluted solutions and perfect gases. If $v = \frac{\mathrm{d}x}{\mathrm{d}t}$ is the particle velocity in a chosen direction at a given moment, then the average over a large number of identical particles with mass m is

$$\frac{1}{2}m\langle v^2\rangle = \frac{1}{2}\frac{RT}{\mathcal{N}}. \tag{54}$$

A particle which is large compared to the molecules of a liquid, and is moving at speed v with respect to this liquid, experiences a viscous resistance force equal to $-6\pi\eta a v$, according to Stokes' formula. In reality this is only an average value, and because of the irregular shocks of the surrounding molecules, the action of the fluid on the particle fluctuates around the average value. The equation of motion along the direction x, given by Newtonian dynamics, is

$$m\frac{\mathrm{d}v}{\mathrm{d}t} = m\frac{\mathrm{d}^2x}{\mathrm{d}t^2} = -6\pi\eta a v + X. \tag{55}$$

The complementary force X, introduced by Langevin, is *random*, and also called *stochastic*. In reality we know little about it, apart from that it is indifferently positive or negative, and that its magnitude is such that it maintains the particle's agitation which, without it, would end by stopping because of the viscous resistance.

[110] P. Langevin, *C. R. Ac. Sci. Paris* **146**, 530 (1908).

By multiplying by x equation (55), one has[111]

$$\begin{aligned} mx\frac{\mathrm{d}v}{\mathrm{d}t} &= \frac{1}{2}m\frac{\mathrm{d}^2x^2}{\mathrm{d}t^2} - mv^2 \\ &= -\mu\, xv + xX = -\mu\frac{1}{2}\frac{\mathrm{d}x^2}{\mathrm{d}t} + xX, \end{aligned} \tag{56}$$

where the friction coefficient μ represents $\mu = 6\pi\eta a$ as before. If we consider a large number of identical particles and take the average of equations (56) written for each of them, then the average value of the term xX is "evidently" zero because of the irregularity of the random forces X, and one finds[112]

$$\frac{1}{2}m\frac{\mathrm{d}^2\langle x^2\rangle}{\mathrm{d}t^2} - m\langle v^2\rangle = -\mu\frac{1}{2}\frac{\mathrm{d}\langle x^2\rangle}{\mathrm{d}t}. \tag{57}$$

One puts $u = \frac{1}{2}\frac{\mathrm{d}\langle x^2\rangle}{\mathrm{d}t}$, and uses the equipartition theorem of kinetic energy (54) to get a simple differential equation of first order:

$$m\frac{\mathrm{d}u}{\mathrm{d}t} - \frac{RT}{\mathcal{N}} = -\mu u. \tag{58}$$

The general solution is

$$u = \frac{RT}{\mu\mathcal{N}} + C\exp\left(-\frac{\mu}{m}t\right), \tag{59}$$

where C is an arbitrary constant.[113] The exponentially decreasing term rapidly fades away, and the result goes to the constant value of the first term, in a limiting regime after a time τ of order $\frac{m}{\mu}$ or 10^{-8} seconds, for all Brownian particles.

Thus, we have

$$u = \frac{1}{2}\frac{\mathrm{d}\langle x^2\rangle}{\mathrm{d}t} = \frac{RT}{\mu\mathcal{N}}, \tag{60}$$

---

[111] Since $v = \frac{\mathrm{d}x}{\mathrm{d}t}$, we use the identities between derivatives $xv = x\frac{\mathrm{d}x}{\mathrm{d}t} = \frac{1}{2}\frac{\mathrm{d}x^2}{\mathrm{d}t}$, and $x\frac{\mathrm{d}v}{\mathrm{d}t} = x\frac{\mathrm{d}^2x}{\mathrm{d}t^2} = \frac{1}{2}\frac{\mathrm{d}^2x^2}{\mathrm{d}t^2} - v^2$.

[112] One should note that the force X disappears from the calculation thanks to that observation. The only under-lying role of X is therefore to ensure the physical possibility of a kinetic average $\langle v^2\rangle \neq 0$. On the other hand, the equality $\langle xX\rangle = 0$ does not appear as evident, because there could have existed a subtle correlation between the position x and the stochastic force X, as it exists between velocity and stochastic force. The existence of two types of stochastic calculations, *à la* Itô and *à la* Stratonovitch, illustrates this difficulty. (See for example N. G. van Kampen, *Stochastic Processes in Physics and Chemistry*, Elsevier, Amsterdam (1992).) Einstein made the same hypothesis in his third demonstration of Brownian motion; see in this volume the translation of his lecture on November 2, 1910 for the Zürich Physical Society.

[113] Here, there seems to be a contradiction between the existence of an exponential term and the hypothesis of the equipartition of energy, $m\langle v^2\rangle = \frac{RT}{\mathcal{N}}$, made *for every* t by Langevin, because it is only at large t that memory effects are exponentially suppressed. This hypothesis, as well as a solution of the form (59), can however be correct for all t, provided that one imposes the same condition for the initial velocity, which in fact fixes the value of the constant C to be equal to $C = -\frac{RT}{\mu N}$. We will come back to this point further along in a more detailed study of the solution of Langevin's equation.

from which, for the time interval t,

$$\langle x^2 \rangle = \frac{2RT}{\mu \mathcal{N}} t = \frac{RT}{\mathcal{N}} \frac{1}{3\pi\eta a} t, \tag{61}$$

if one supposes that the particle was observed at the origin $x = 0$ at time $t = 0$. Langevin's method indeed reproduces Einstein's result (28). In this paper (published in 1908 in the Comptes Rendus of the Academie de Sciences) Langevin introduced, without knowing it, the first element (the random force X) of what was to become stochastic calculus.[114]

1.7.2 Boltzmann's constant

Boltzmann's constant k_B is obtained by dividing the molar constant R of a perfect gas by Avogadro's number $\mathcal{N}$, such that one obtains a quantity which refers to a single molecule:

$$k_B = \frac{R}{\mathcal{N}} = 1.381 \times 10^{-23}\,\mathrm{J\,K^{-1}}. \tag{62}$$

The energy $k_B T$ gives the average thermal energy at the standard temperature: $k_B T = 4 \times 10^{-21}$ J. The constant k_B was not introduced by Boltzmann but by Planck in his famous presentation on December 14, 1900, on black body radiation, at the same time he presented Planck's constant h!

1.7.3 An analysis of the solution of Langevin's equation.

The method presented in section (1.7.1) is the one that Langevin gave in his original paper. A more modern formulation proceeds from the time-correlation functions of the stochastic force X in canonical form,

$$\langle X \rangle = 0, \; \langle X(t)X(t') \rangle = A\delta(t - t'), \tag{63}$$

where A is a coefficient to be determined and $\delta(t - t')$ is the Dirac distribution. The generalization to d dimensions is

$$\begin{aligned} \langle \vec{X} \rangle &= \vec{0}, \\ \langle X_i(t) X_j(t') \rangle &= A\delta_{ij}\delta(t - t'), \end{aligned} \tag{64}$$

where δ_{ij} is the Kronecker symbol and $i, j = 1, \cdots d$.

We can easily integrate the linear equation for the velocity

$$m \frac{\mathrm{d}\vec{v}}{\mathrm{d}t} = -\mu\vec{v} + \vec{X}. \tag{65}$$

[114] J. L. Doob, The Brownian Motion and Stochastic Equations, *Ann. of Math.*, **43**, pp. 351–369 (1942), reprinted in [Wax 1954, pp. 319–337], *op. cit.*

The solution is

$$\vec{v}(t) = \vec{v}(0)\, e^{-\frac{\mu}{m}t} + \frac{1}{m}\int_0^t \mathrm{d}t'\, \vec{X}(t')\, e^{-\frac{\mu}{m}(t-t')}. \tag{66}$$

Therefore by taking the square of the velocity and by using formula (64), we find the average value of kinetic energy at time t

$$\frac{1}{2}m\langle\vec{v}^2(t)\rangle = \frac{Ad}{4\mu}\left(1 - e^{-2\frac{\mu}{m}t}\right) + \frac{1}{2}m\vec{v}^2(0)e^{-2\frac{\mu}{m}t}. \tag{67}$$

We then see that this energy relaxes towards a constant value at large time, i.e., at equilibrium. From the theorem of equipartition of kinetic energy,

$$\frac{1}{2}m\langle\vec{v}^2(t)\rangle_{t\to\infty} = \frac{d}{2}k_B T, \tag{68}$$

we deduce the important identity

$$A = 2\mu k_B T. \tag{69}$$

We then have

$$\langle\vec{v}^2(t)\rangle = \frac{dk_BT}{m} + \left(\vec{v}^2(0) - \frac{dk_BT}{m}\right)e^{-2\frac{\mu}{m}t}. \tag{70}$$

A second stage consists of integrating equation (66) to obtain the displacement $\vec{r}(t) - \vec{r}(0)$. Then taking the square, and the stochastic average by means of formulae (64), we obtain after some calculation,

$$\begin{aligned}\langle[\vec{r}(t) - \vec{r}(0)]^2\rangle &= 2dD\left[t - \frac{m}{\mu}\left(1 - e^{-\frac{\mu}{m}t}\right)\right] \\ &+ \left(\vec{v}^2(0) - \frac{dk_BT}{m}\right)\left(\frac{m}{\mu}\right)^2\left(1 - e^{-\frac{\mu}{m}t}\right)^2,\end{aligned} \tag{71}$$

where $D = k_BT/\mu$, as before. The derivative u considered by Langevin is then given by

$$\begin{aligned}u &= \frac{1}{2}\frac{\mathrm{d}}{\mathrm{d}t}\langle[\vec{r}(t) - \vec{r}(0)]^2\rangle \\ &= d\frac{k_BT}{\mu} - \vec{v}^2(0)\frac{m}{\mu}e^{-\frac{\mu}{m}t} + \left(\vec{v}^2(0) - \frac{dk_BT}{m}\right)\frac{m}{\mu}e^{-2\frac{\mu}{m}t}.\end{aligned} \tag{72}$$

Notice first that these results at large t, or $t \gg \tau = m/\mu$, go asymptotically to those of thermal equilibrium and to the associated motion of diffusion, as expected. One remarks then the role played by the initial velocity in memory effects and in the approach to equilibrium. A very special value of $\vec{v}^2(0)$ is that of equipartition $\frac{dk_BT}{m}$. For this value only, the average quadratic velocity in (70) becomes *invariant* in time, $\langle \vec{v}^2(t)\rangle = \frac{dk_BT}{m}, \forall t$. The average quadratic displacement (71) then takes Ornstein's simple form (42), and the quantity u (72) takes the form predicted by Langevin in (59), with a determined value for C. One consistently obtains the same result by using for $\vec{v}^2(0)$ its most probable value, meaning its *thermal average at equipartition.* One can then understand (but only *a posteriori*) the consistency of Langevin's approach when he inserted the identity (54) in the middle of the derivation. That amounted to chosing the peculiar boundary condition $\vec{v}^2(0) = \frac{dk_BT}{m}$, which enforces *stationary* equipartition!

On the other hand, if one gives to the initial quadratic velocity $\vec{v}^2(0)$ a value which is different from that of equilibrium, the relaxation will occur in a bit more complex way, as we showed in the above results.

The regime at short times, $\frac{m}{\mu} \gg t$, also naturally depends on the initial conditions. In fact, by developing in series (71) one finds the expected ballistic regime

$$\langle [\vec{r}(t) - \vec{r}(0)]^2 \rangle = \vec{v}^2(0)\, t^2 + \mathcal{O}(t^3),$$

that naturally cross-checks with (44) if one takes once again the value at equipartition.

1.7.4 Microscopic model

The force proposed by Langevin, $-\mu v + X$, can only be an approximation to the underlying molecular reality, made up of innumerable collisions where multiple correlations, due to interactions between molecules, exist at very short time scales. The stochastic term X in (63,64) is a white noise without memory, i.e., it neglects temporal correlations.

As well, the hydrodynamic form of the friction term, $-\mu v$, is a description that pertains to the continuous limit, which requires extremely frequent collisions on a particle in suspension. The mass m of the particle must then be large enough so that the characteristic time $\tau = m/\mu$ is large compared to the inverse frequency of collisions.

To give an idea of the origin of Langevin's equation (55) and of its parameters μ and A (69), it is natural to consider the simplest model, where the collisions of a particle in suspension occur with a surrounding perfect gas, and thus without interaction.[115]

One can therefore consider a perfect gas of identical particles with mass m', a particle density n', at the temperature T, and colliding the particle of large mass m in suspension. To simplify, we consider the gas in one dimension, where the equations for the particle-gas elastic collisions are particularly simple. One then

[115] D. Durr, S. Goldstein, J. L. Lebowitz, *A Mechanical Model of Brownian Motion, Commun. Math. Phys.* **78**, 507 (1981).

finds that the equation for the momentum variation of the test particle is similar to Langevin's equation, with the explicit coefficients[116]

$$\mu = 4n'\sqrt{\frac{2m'k_BT}{\pi}}, \quad A = 8n'k_BT\sqrt{\frac{2m'k_BT}{\pi}}; \tag{73}$$

μ and A thus verify (69).

It is then particularly interesting to rewrite these terms as a function of *molar sizes* that characterize the perfect gas. One introduces as well the gas pressure[117] p', which responds to the equations of perfect gases $p' = n'k_BT$, which gives

$$\mu = 4p'\sqrt{\frac{2\mathcal{M}}{\pi RT}}, \quad A = \frac{2RT}{\mathcal{N}}4p'\sqrt{\frac{2\mathcal{M}}{\pi RT}}, \tag{74}$$

where $\mathcal{M} = \mathcal{N}m'$ is the molar mass of the gas.

1.7.5 Discontinuity in Nature and the existence of Brownian motion

The explicit results above, in their last form (74), rigorously state that the Sutherland–Einstein equation (13), $D = \frac{RT}{\mu\mathcal{N}}$, reflects the existence of molecules.

In fact, the friction coefficient μ can be expressed independently from Avogadro's number $\mathcal{N}$, and depends only on the ideal gas constant R and the *macroscopic* parameters of the surrounding gas, like the pressure p', temperature T, and molar mass $\mathcal{M}$. On the other hand, the variance A of the Langevin stochastic force, which controls diffusion, continues to depend on $\mathcal{N}$ and vanishes when Avogadro's number goes to infinity.

In the same way, the limit of the diffusion coefficient $D = \frac{RT}{\mu\mathcal{N}}$, when Avogadro's number goes to infinity, $\mathcal{N} \to \infty$, is of course *zero*, i.e., the *Brownian motion would cease immediately if Nature was continuous!* An entire branch of mathematics might perhaps never have seen the light of day.

1.8 Jean Perrin's experiments

1.8.1 The triumph of the "Molecular Hypothesis"

Jean Perrin is often cited as the one who established the Einstein–Smoluchowski–Sutherland theory by his beautiful experiments. He was also an outstanding promotor of atomistic ideas. His book, *Atoms*,[118] which contains a detailed description of his experiments on Brownian motion, is highly recommended. It begins:

[116] One calculates, in the process of discrete collision, the average momentum variation $\langle\frac{\mathrm{d}p}{\mathrm{d}t}\rangle = -\mu\langle v\rangle$ as well as the fluctuations $\langle\frac{\mathrm{d}p(t)}{\mathrm{d}t}\frac{\mathrm{d}p(t')}{\mathrm{d}t'}\rangle - \langle\frac{\mathrm{d}p(t)}{\mathrm{d}t}\rangle\langle\frac{\mathrm{d}p(t')}{\mathrm{d}t'}\rangle = A\delta(t-t') + \cdots$, and finds by comparison the values (73) of parameters μ and A for Langevin's equation. See the article from B. Derrida and É. Brunet in *Einstein aujourd'hui*, éds. M. Leduc and M. Le Bellac, Savoirs actuels, EDP Sciences/CNRS Éditions (2005).

[117] In one dimension, the pressure p' is equivalent to a force, because the boundaries of the "box" containing the gas are simple points.

[118] J. Perrin, *Les Atomes*, Félix Alcan, Paris (1913); réédition Champs Flammarion (1991); English translation: *Atoms*, transl. by D. Ll. Hammick, Ox Bow Press, Woodbridge (1990).

"Molecules: Some twenty-five centuries ago, before the close of the lyric period in Greek history, certain philosophers on the shores of the Mediterranean were already teaching that changeful matter is made up of indestructible particles in constant motion; atoms which chance or destiny has grouped in the course of ages into the forms or substances with which we are familiar. But we know next to nothing of these early theories, of the works of Moschus of Sidon, of Democritus of Abdera, or of his friend Leucippus. No fragments remain that might enable us to judge of what in their work was of scientific value. And in the beautiful poem, of a much later date, wherein Lucretius expounds the teachings of Epicurus, we find nothing that enables us to grasp what facts or what theories guided Greek thought."

He expounded in addition on the idea that non-differentiable continuous functions, such as the trajectory of Brownian Motion, were as completely natural as differentiable functions, objects of all prior studies. In the preface of *Atoms*, by considering the very irregular surface of a colloid and by making the analogy with the shape of Brittany's coast, he announced with a dazzling geometric intuition the ideas of Lewis Fry Richardson on Hausdorff anomalous dimensions, which would later be developed by Benoît Mandelbrot.[119]

Regarding Brownian Motion, we find as well:

"We are still in the realm of experimental reality when, under the microscope, we observe the Brownian movement agitating each small particle suspended in a fluid. In order to be able to fix a tangent to the trajectory of such a particle, we should expect to be able to establish, within at least approximate limits, the direction of the straight line joining the positions occupied by a particle at two very close successive instants. Now, no matter how many experiments are made, that direction is found to vary absolutely irregularly as the time between the two instants is decreased. An unprejudiced observer would therefore come to the conclusion that he was dealing with a function without derivative, instead of a curve to which a tangent could be drawn."

Further along we read:

"It is impossible to fix a tangent, even approximately, to any point on a trajectory, and we are thus reminded of those continuous functions[120] without derivative that mathematicians had imagined. It would be incorrect to regard such functions as mere mathematical curiosities, whereas Nature suggests them as much as differentiable functions."

These remarks would stimulate the research of the young mathematician Norbert Wiener.[121]

[119]B. Mandelbrot, *Fractal Objects*, (3ème éd.), followed by *A Survey of Fractal Language*, Flammarion, Nouvelle Bibliothèque scientifique (1989).

[120]"*Continuous* because it is not possible to regard the grains as passing from one position to another without cutting any given plane having one of those positions on each side of it." [original note]

[121]N. Wiener, *I am a Mathematician*, Doubleday, Garden City, NY (1956). He writes: "The Brownian motion was nothing new as an object of study by physicists. There were fundamental papers by Einstein and Smoluchowski that covered it, but whereas these papers concerned what was happening to any given particle at a specific time, or the long-time statistics of many particles, they did not concern themselves with the mathematical properties of the curve followed by a single particle.

Here the literature was very scant, but it did include a telling comment by the French physicist Perrin in his book *Les Atomes*, where he said in effect that the very irregular curves followed by particles in the Brownian motion led one to think of the supposed continuous non-differentiable curves of the mathematicians. He called the motion continuous because the particles never jump over a gap and non-differentiable because at no time do they seem to have a well-defined direction of movement."

The beautiful experiments of 1908–1909 by Perrin and his students, on emulsions of gum-resin ("gamboge") or of mastic, are described in detail in his review article *Brownian Motion and Molecular Reality,* which appeared in Annales de Chimie et Physique in 1909,[122] and the results are published in several *Notes aux Comptes Rendus.* The same material is also summarized in his book *Atoms.*

Perrin began by verifying the exponential distribution of the density of n particles in a suspension, as a function of the height h in a gravitational field g, a formula that generalizes the barometric formula for the atmosphere. Perrin writes it in the form

$$\frac{2}{3}W\ln\frac{n_0}{n} = \phi(\rho - \rho_0)gh, \tag{75}$$

where ϕ is the volume of each grain, ρ and ρ_0 are the mass per unit volume of the grains and of the inter-granular liquid, respectively, and last but not least, W is the average kinetic energy per particle (with $W = \frac{3}{2}\frac{RT}{\mathcal{N}} = \frac{3}{2}k_B T$).

He writes:

> "I indicated this equation at the time of my first experiments (*Comptes Rendus*, May 1908). I have since learned that Einstein and Smoluchowski, independently, at the time of their beautiful theoretical researches of which I shall speak later, had already seen that the exponential repartition is a necessary consequence of the equipartition of energy. Beyond this it does not seem to have occurred that in this sense, an *experimentum crucis* could be obtained, deciding for or against the molecular theory of the Brownian movement."

He continues:

> "If it is possible to measure the magnitudes other than W which enter into this equation, one can see whether it is verified and whether the value it indicates for W is the same as that which has been approximately assigned to the molecular energy. In the event of an affirmative answer, the origin of the Brownian movement will be established, and the laws of gases, already extended by van 't Hoff to solutions, can be regarded as still valid even for emulsions with visible grains."

He built as well an apparatus for fractioned centrifugation to produce emulsions of uniform size, a key element of his success. Using three independent processes to measure the radius of particles, one of which went via Stokes' law, he could verify the validity of the latter for particles in suspension. It was in fact one of the weak points of the theoretical proofs, because the continuity conditions required by hydrodynamics were far from being clearly fulfilled in the case of small spheres in very active Brownian motion.

Finally, by ingenious and patient observations, he could verify the law of rarefaction of density (75).[123] Thanks to the value of W (independent of all experimental conditions except the temperature), he verified the famous law of equipartition of energy, and obtained a first estimate of Avogadro's number, $\mathcal{N} = 7.05\times 10^{23}$, compared with the present accepted value $\mathcal{N} = 6.02 \times 10^{23}$.

[122] J. Perrin, *Ann. Chim. Phys.* **18**, 1 (1909); available online at http://gallica.bnf.fr/.
[123] J. Perrin, *C. R. Acad. Sci. Paris* **146**, 967 (1908); **147**, 475 (1908).

1.8.2 Einstein's formulae

Perrin turned next to Einstein's formulae for Brownian diffusion:

> "...another approach was possible and was proposed by Einstein, in conclusion to his beautiful theoretical works, of which I must now speak."

Further on, he adds:

> "It's fair to recall that, almost at the same time as Einstein and by another route, Smoluchowski arrived at a formula a bit different in his remarkable work on *A kinetic theory of Brownian motion* [Bulletin de l'Acad. des Sc. de Cracovie, July 1906, p. 577] where one finds, besides very interesting observations, an excellent history of work before 1905."

In *Atoms* he stresses that:[124]

> Einstein and Smoluchovski have defined the activity of the Brownian movement in the same way. Previously, we had been obliged to determine the "mean velocity of agitation" by following as nearly as possible the path of a grain. Values so obtained were always a few microns per second for grains of the order of a micron.[125]
>
> But such evaluations of the activity are *absolutely wrong*. The trajectories are confused and complicated so often and so rapidly that it is impossible to follow them; the trajectory actually measured is very much simpler and shorter than the real one. Similarly, the apparent mean speed of a grain during a given time varies *in the wildest way* in magnitude and direction, and does not tend to a limit as the time taken for an observation decreases [...].
>
> Neglecting, therefore, the true velocity, which cannot be measured, and disregarding the extremely intricate path followed by a grain during a given time, Einstein and Smoluchowski chose, as the magnitude characteristic of the agitation, the rectilinear segment joining the starting and end points; in the mean, this line will clearly be longer the more active the agitation. The segment will be the *displacement* of the grain in the time considered.

He begins his review by recalling the early work of Exner, anterior to the publication of Sutherland–Einstein–Smoluchowski formula for the average quadratic displacement (28), and in which one can see "at least one presumption of partial verification for the formula in question."

Soon after the publication of this formula, verification was quickly tried by Theodor Svedberg, who thought he achieved it.[126] Perrin made a sharp criticism of these results, and declared him "a victim of an illusion," regarding his description of Brownian trajectories "as regularly modulated in amplitude and with well defined wavelength!"[127]

Victor Henri's results, published in *Comptes Rendus* in 1908, were obtained from a better founded cinematographic study of Brownian motion of natural latex grains. The average displacement varied as the square root of time, but the coefficient was three times too large.[128]

---

[124] *Atoms, op. cit.*, chapter IV.

[125] "Incidentally this gives the grains a kinetic energy 10^5 times too small." [original note]

[126] Th. Svedberg, *Studien zur Lehre von den kolloidalen Lösungen, Nova Acta Reg. Soc. Sc. Upsaliensis*, **2**, 1907.

[127] One must add that Svedberg won the Nobel Prize for Chemistry in 1926 for his invention of the ultracentrifuge.

[128] Perrin then noted almost mischievously: "As far as I could judge from the conversation, a current of opinion was produced among the French physicists community that closely followed

Having prepared grains with known diameter, Perrin asked his student Chaudesaigues to verify the law of Brownian displacement by direct observation, sequenced every thirty seconds, with gamboge grains of radius 0.212 μm.[129] This was completed by similar measurements by M. Dabrowski[130] on mastic grains, and gave the famous diagrams of random positions that one can find in Jean Perrin's book. (See figure 3.)

Figure 3: *Brownian motion. Bottom left: Strong magnification, showing the discretized aspect of sequential recordings of the position of a particle in suspension, observed by Jean Perrin and his collaborators. Bottom right: Magnification showing the self-similarity of the continuous Brownian curve.*

The conclusion was "the rigorous exactness of the formula proposed by Einstein", and "that some unknown complication or unknown cause of systematic error oddly affected the results of Victor Henri." They then deduced a new average value of Avogadro's number, $\mathcal{N} = 7.15 \times 10^{23}$. A wonderful verification was at last made of "Maxwell's irregularity law", that is, of the Gaussian distribution

these questions, and which really shocked me, proving to me how much the credit that we give to theories is limited, and at what point we see them as instruments of discovery rather than as true demonstrations. Without hesitating, they admitted that Einstein's theory was incomplete or inexact. On the other hand, there was no reason to renounce placing the origin of Brownian motion in molecular agitation, because I just showed by an experiment that a diluted emulsion behaves as a very dense perfect gas in which the molecules had a weight equal to the grains of the emulsion. They limited themselves to assuming that a few unjustified complementary hypotheses slipped into Einstein's reasoning."

[129] M. Chaudesaigues, *C. R. Acad. Sci. Paris*, **147**, 1044 (1908); *Diplôme d'Études*, Paris (1909).

[130] J. Perrin and Dabrowski, *C. R. Acad. Sci. Paris*, **149**, 477 (1909).

(26) of the Brownian particle's position in a plane orthogonal to gravity.

Jean Perrin did not stop there, but turned to *rotational Brownian motion.* Einstein himself did not really think that his predictions (41) were experimentally verifiable, because the speed of rotation seemed to be too large to be observable. In fact, for grains of 1 μm in diameter, the rotation is about 8 degrees per hundredth of second. Perrin could then prepare spheres with larger diameter, from 10–15 μm up to 50 μm, and he succeeded in preparing them in suspension in a 27% solution of urea. In this case the angular speed falls to a few degrees per minute. The spheres carried inclusions of different refractive indices, which made their rotation observable under a microscope! The result was a spectacular verification of Einstein's second formula (41), this time for grains 100 000 times heavier than the small grains of gamboge first studied.[131] On 11 November 1909, Einstein wrote to Perrin: "I would not have considered a measurement of rotations as feasible. In my eyes it was only a pretty triffle".[132]

Perrin received the Nobel Prize in 1926 for his work on Brownian motion. His book, *Atoms*, one of the most finely written physics books of the 20th century, contains a postmortem, in the great classic style, about the fight for establishing the reality of molecules:

> "La théorie atomique a triomphé. Enore nombreux naguère, ses adversaires enfin conquis renoncent l'un après l'autre aux défiances qui, longtemps, furent légitimes et sans doute utiles. C'est au sujet d'autres idées que se poursuivra désormais le conflit des instincts de prudence et d'audace dont l'équilibre est nécessaire au lent progrès de la science humaine."
>
> "The atomic theory has triumphed. Its opponents, who until recently were numerous, have been convinced and have abandoned one after the other the sceptical position that was for a long time legitimate and likely useful. Equilibrium between the instincts towards caution and towards boldness is necessary to the slow progress of human science; the conflict between them will henceforth be waged in other realms of thought."

To conclude this section, let us return for a last time to Einstein. One reads in his autobiographical notes:[133]

> "The agreement of these considerations with experiment together with Planck's determination of the true molecular size from the law of radiation (for high temperatures) convinced the sceptics, who were quite numerous at that time (Ostwald, Mach) of the reality of atoms. The antipathy of these scholars towards atomic theory can undoubtedly be traced back to their positivistic philosophical attitude. This is an interesting example of the fact that even scholars of audacious spirit and fine instinct can be obstructed in the interpretation of facts by philosophical prejudices. The prejudice – which has by no means died out in the meantime – consists in the faith that facts by themselves can and should yield scientific knowledge without free conceptual construction. Such a misconception is possible only because one does not easily become aware of the free choice of such concepts, which, through verification and long usage, appear to be immediateley connected with the empirical material."

Let us finally mention Ostwald's magnanimous concession: In 1908 he refers to the Brownian motion results and says they *"entitle even the cautious scientist*

[131] J. Perrin, *C. R. Acad. Sci. Paris*, **149**, 549 (1909).

[132] Quoted in J. Stachel, *Einstein's Miraculous Year* (Princeton University Press, Princeton, New Jersey, 1998).

[133] *Albert Einstein: Philosopher-Scientist*, The Library of Living Philosophers, Vol. VII, Paul Arthur Schilpp Ed., Open Court, La Salle, Illinois, 3rd Edition (2000).

to speak of the experimental proof for the atomistic constitution of space-filled matter". In 1910, he is the first person to nominate Einstein for the Nobel Prize (for special relativity).

2 Measurements by Brownian fluctuations

Jumping ahead a century, we observe how the theory of Brownian fluctuations, whose construction we just described, today finds spectacular applications in physics applied to biology. We will give an example from the physics of DNA.

2.1 Micromanipulation of DNA molecules

2.1.1 The interest of DNA for physicists

Physicists are interested today in DNA for several reasons. First of all, it is a remarkable polymer for its length, reaching several centimeters, and for its monodispersity (the DNA of the virus bacteriophage-λ, for example, has always 48502 base pairs with identical sequence). DNA is an important subject in polymer physics because it can be easily shaped by bio-molecular tools and it can be directly observed and manipulated. A fluorescent intercalation placed between base pairs (such as ethidium bromide) permits the observation, under a microscope and by fluorescence, of *single* DNA molecules in solution.

2.1.2 Experimental realization of a micro-manipulation

One can also micro-manipulate molecules individually. The techniques of micro-manipulation of isolated bio-molecules have developed considerably during the past few years, thanks to an ever-growing number of tools: optical or magnetic "tweezers", atomic force microscopes, glass micro-fibers, and also hydrodynamic flow observations.

A recent example consists of pulling a single DNA molecule to measure its extension as a function of the force, which allows one to measure various important mechanical parameters of the DNA chain.

In "magnetic tweezers" (figure 4), a magnetic bead is placed in the field of a magnet; the bead is attracted towards regions with a high gradient field, and one can move the magnets or rotate them. This allows to pull the DNA or to twist it, creating as well torsions, or super-coilings, that are a part of topological configurations for biological functions.

We give a brief overview of forces playing a role in biology, and of the specific problems related to their smallness.

2.1.3 Biological interaction forces and thermal agitation forces

The interaction forces in biological systems are typically generated by hydrogen or ionic bonds, as well as by van der Waals interactions that shape nucleic acids

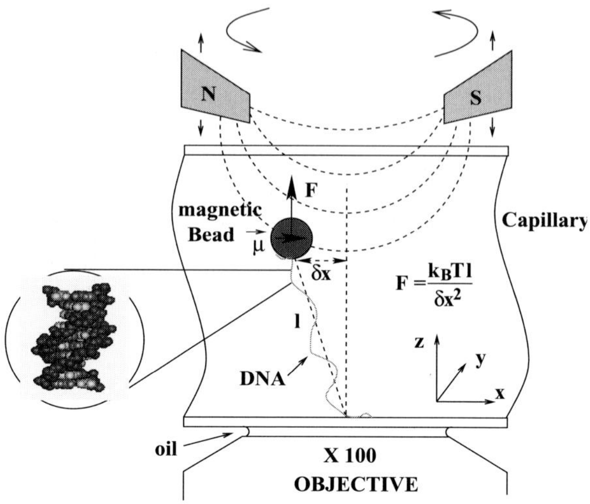

Figure 4: *Micro-manipulation of a DNA molecule by "magnetic tweezers".*

and proteins. Their order of magnitude is typically obtained by dividing k_BT, the order of magnitude of the "quantum of energy" provided by the hydrolyzation of the ATP in ADP[134] (10 k_BT), by the characteristic size of biological objects, of the order of a nanometer (nm). We then find the *picoNewton*:

$$\frac{k_BT}{1\text{nm}} = 4 \underset{10^{-12}\ \text{N}}{\underset{\parallel}{\text{p}N}} \quad .$$

Such a force is the one typically needed to stretch a DNA molecule. As it is extremely small, it is not easy to detect with standard measuring devices.

The smallest measurable forces are in principle limited by the *thermal agitation* of the measuring device (see figure 5). This thermal agitation generates Langevin's stochastic force seen above, whose value depends on the coefficient of viscous friction of the object, and also on the temporal window of observation. We have:

$$\langle X^2_{\text{Langevin}} \rangle = 2k_BT\, 6\pi\eta\, a\, \delta f,$$

where η is the medium's viscosity, a the radius of a spherical bead taken as an example, and δf the observed frequency range. For example, for $a = 1.5\ \mu$m, in water (viscosity $\eta = 10^{-3}$ Poise), the average force over a period of a second is

[134]ATP: adenosine triphosphate, universal biological "fuel", made of one sugar, ribose, and of one base, adenine, and of three phosphate groups; ADP: adenosine diphosphate, is the degraded version after losing a group of phosphates under enzymatic action and release of energy.

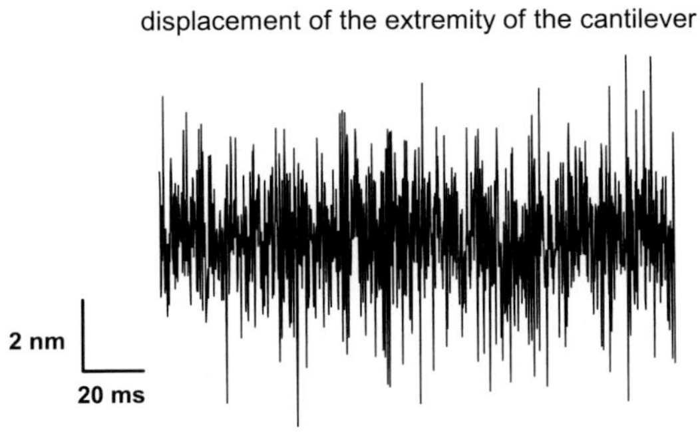

Figure 5: *The track of the random displacement, in a liquid, of the tip of the cantilever in an atomic force microscope. It executes a one-dimensional Brownian motion. (Kindly provided by Pascal Silberzan and Olivia du Roure, Curie Institute.)*

$X_{\text{Langevin}} \sim 15 \underset{(10^{-15}\ \text{N})}{\underset{\parallel}{f\text{N}}}$, i.e., 15 femtoNewtons.

Astonishingly, Brownian fluctuations can be used directly to measure forces of biological origin!

2.2 Measurement of force by Brownian fluctuations

This technique of measuring a force is largely inspired by the method proposed by Einstein[135] for measuring the elastic constant of a spring by means of *Brownian fluctuations*. When we apply a force upon a small magnetic bead in a gradient field, the stretched molecule and the bead form a miniscule pendulum of length ℓ (figure 4). The bead is animated by Brownian motion, connected to the thermal agitation of surrounding water molecules. The small magnetic pendulum is thus perturbed from its equilibrium position by Langevin's random force. It is then brought back towards equilibrium by the pulling force exerted by the DNA (figure 6).

As we will show in detail further along, the pendulum possesses a transverse elastic constant $k_\perp$ that is directly related to the pulling force F by $k_\perp = F/\ell$. If we call x the position of the bead with respect to its equilibrium position in the

[135] A. Einstein, *Investigations on the Theory of the Brownian Movement*, R. Fürth *Ed.*, A. D. Cowper *Transl.*, Dover Publications, p. 24 (1956).

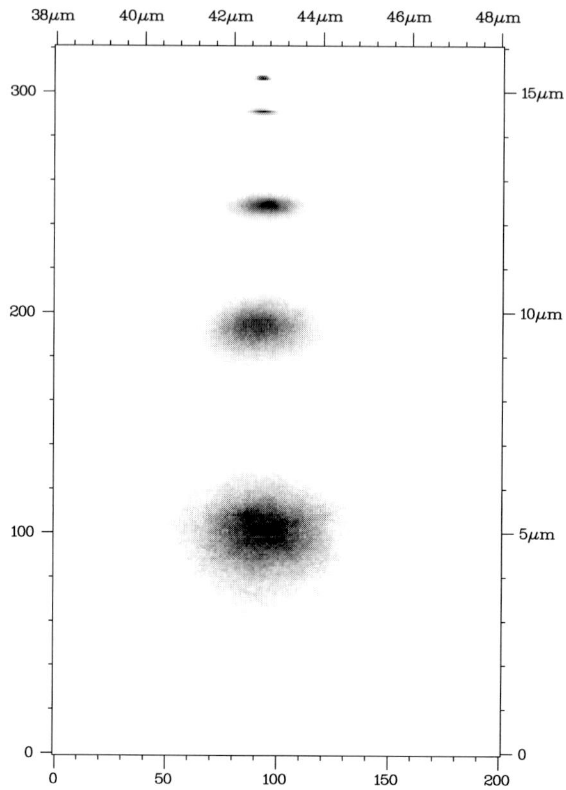

Figure 6: *Brownian cloud of a bead's fluctuating position in the vertical plane* (Ox, Oz), *for different applied forces. The larger the force, the more a molecule is stretched, and the more the Brownian fluctuations are constrained. (Figure kindly provided by Vincent Croquette, Statistical Physics Laboratory, ENS, Paris.)*

direction perpendicular to the force $\vec{F}$, the theory states

$$F = k_B T \ell / \langle x^2 \rangle,$$

where $\langle x^2 \rangle$ is the average quadratic fluctuation of x. To measure the pulling force on a DNA molecule, one simply measures the length ℓ and the average quadratic fluctuation $\langle x^2 \rangle$! This is reminiscent of Einstein's formula (28), as well as of the surprise of being able to deduce Avogadro's number from it.

To measure such fluctuations, one must follow the movements of the bead during a given amount of time, just as in Jean Perrin's experiments of 1908 on Brownian motion. Today, a computer program analyzes in real time the images on a video of the bead observed via a microscope, and determines its positions in a three-dimensional space with a precision of 10 nm (figure 6). Such precision is

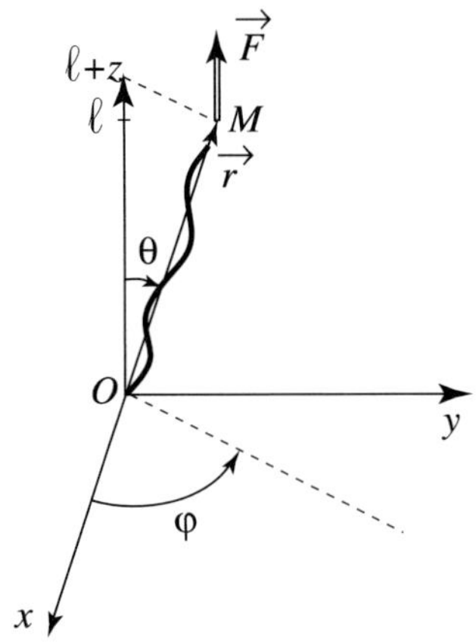

Figure 7: *Axis of a DNA chain fluctuating around the vertical position; the extremity M moves from the equilibrium position* $(0, 0, \ell)$ *in presence of F towards the random position* $(x, y, \ell + z)$.

obtained through a technique of image correlation.

This sort of Brownian measurement has several advantages:

- One gauges the force by absolute measurement of position fluctuations.
- There is no contact with the bead, therefore it is non-invasive.
- The range of values of x is between μm to nm, the force goes from a dozen femtoNewtons to hundreds of picoNewtons.

The drawback is its slowness: to accumulate sufficient fluctuations and to have reliable statistics, a minute of recording is needed for a force of 1pN, and more than an hour for 10fN.

We shall now describe the theory of measurement by Brownian fluctuations.

2.3 Theory

2.3.1 Equilibrium and fluctuations

One considers a DNA chain of length ℓ_0 with one extremity fixed at the origin 0, while the other extremity M is determined by $\overrightarrow{OM} = \vec{r}$ (see figure 7). A force $\vec{F}$ acts on the extremity M along the direction of the Oz axis. At equilibrium, the chain is parallel to the Oz axis and is elastically stretched up to a length ℓ dependent on F. The Brownian fluctuations, originating from the shocks between the bead that is attached to the DNA chain and the molecules of the solution, induce small displacements (x, y, z) that one can consider as perturbations of the macroscopic equilibrium position (0,0,ℓ). The extremity M is thus shifted from its equilibrium position $(0,0,\ell)$ (in the presence of F) to a random position $(x,y,\ell+z)$. Let $r = |\overrightarrow{OM}!!|$ be the radial distance between the extremities of the chain. Because of elasticity, the chain develops a *restoring radial force* $F_r(r)$. At equilibrium, one has $F_r(\ell) = F$, where F is the external force given experimentally.

In the presence of fluctuations, the radial distance is written

$$r = [(\ell+z)^2 + x^2 + y^2))]^{1/2}, \tag{76}$$

and the restoring force

$$\vec{F}_r = -\frac{\vec{r}}{r}F_r(r) = \begin{cases} F_{rx} = -\dfrac{x}{r}F_r(r) \\ F_{ry} = -\dfrac{y}{r}F_r(r) \\ F_{rz} = -\dfrac{\ell+z}{r}F_r(r). \end{cases} \tag{77}$$

2.3.2 Series expansions

One writes the series expansion of the distance r for x,y,z small compared to ℓ:

$$r = [(\ell+z)^2 + x^2 + y^2]^{1/2} = \ell + z + \cdots \tag{78}$$

An expansion to the first linear order in x,y,z will be sufficient, and from now on we will denote by $+\cdots$ all second order terms (of $O(x^2,y^2,z^2)$) in the expansions.

The radial force $F_r(r)$ of the DNA on the bead, depends only on the radial distance r; therefore, from (78), it has the series expansion:

$$F_r(r) \quad = \quad F_r\,[\ell+z+\cdots] = F_r(\ell) + z\frac{d\,F_r}{d\,r}(\ell) + \cdots . \tag{79}$$

One can now easily determine the components (77) of the radial force by using (78) and (79):

$$\begin{aligned} F_{rx} &= -\frac{x}{r}F_r(r) = -\frac{x}{\ell}F_r(\ell) + \cdots , \\ F_{ry} &= -\frac{y}{r}F_r(r) = -\frac{y}{\ell}F_r(\ell) + \cdots , \\ F_{rz} &= -\frac{\ell+z}{r}F_r(r) = -F_r(\ell) - z\frac{d\,F_r}{d\,r}(\ell) + \cdots . \end{aligned}$$

We finally note that at the equilibrium point the external force, $\vec{F} = F\vec{u}_z$, exactly cancels the term $-F_r(\ell)\vec{u}_z$ of the vertical component $F_{rz}\vec{u}_z$. Leaving aside terms of second order, our analysis leads us to a fluctuating resultant force on the DNA:

$$\vec{f} = F\vec{u}_z + \vec{F}_r = \begin{cases} -\dfrac{x}{\ell} F_r(\ell) \\ -\dfrac{y}{\ell} F_r(\ell) \\ -z\dfrac{dF_r}{d\ell}(\ell) \end{cases} = -\nabla_{\vec{r}} U. \tag{80}$$

2.3.3 Elastic energy

The beauty of this approach is that one can determine the elastic energy of the Brownian fluctuations of the DNA chain without even knowing the analytic form of the elastic force. In these expressions, it must be understood that the equilibrium length ℓ is determined by the external force, while the fluctuating force (80) is *linear* in x, y, z, as expected from an expansion to first order. A quadratic potential energy U is associated to the force by $\vec{f} = -\nabla_{\vec{r}} U$, given by the simple expression:

$$U = \frac{1}{2}\left(x^2 + y^2\right)\frac{1}{\ell}F_r(\ell) + \frac{1}{2}z^2\frac{d\,F_r}{d\,\ell}(\ell). \tag{81}$$

2.3.4 Elastic constants

One can write the energy U (81) as that of a three-dimensional anisotropic harmonic oscillator with two elastic constants, $k_\perp$ and $k_\parallel$, corresponding to the perpendicular and parallel directions, respectively, with respect to the force:

$$U = \frac{1}{2}k_\perp\left(x^2 + y^2\right) + \frac{1}{2}k_\parallel\, z^2, \tag{82}$$

with

$$\begin{cases} k_\perp &= \dfrac{F_r(\ell)}{\ell}, \\ k_\parallel &= \dfrac{d\,F_r}{d\,\ell}(\ell). \end{cases} \tag{83}$$

As one can intuitively imagine, the transverse elastic constant, which opposes lateral movements of the DNA molecule, is weaker than the longitudinal elastic constant, which opposes mechanical stretching of the DNA.

2.3.5 Energy equipartition

In *classical* statistical mechanics, we have seen the historically important result about the *equipartition of energy*. The theory simply states that each *quadratic* degree of freedom has average energy $\frac{1}{2}k_BT$ exactly, where k_B is Boltzmann's constant and T is the absolute temperature. In the case of the harmonic energy (82), the theorem immediately gives us:

$$\frac{1}{2}k_\perp \langle x^2\rangle = \frac{1}{2}k_\perp \langle y^2\rangle = \frac{1}{2}k_\parallel \langle z^2\rangle = \frac{1}{2}k_B T\,. \tag{84}$$

Therefore we find, with the help of (83)

$$\begin{cases} k_\perp & = \dfrac{F_r(\ell)}{\ell} = \dfrac{k_B T}{\langle x^2\rangle}\,, \\ k_\parallel & = \dfrac{d\,F_r}{d\,\ell}(\ell) = \dfrac{k_B T}{\langle z^2\rangle}\,. \end{cases} \tag{85}$$

Because of the difference between the elastic constants, $k_\perp < k_\parallel$, transverse fluctuations dominate over longitudinal ones: $\langle x^2\rangle = \langle y^2\rangle > \langle z^2\rangle$, as one can see in figure 6. One sees, for instance, that the fluctuations $\sqrt{\langle x^2\rangle}$ and $\sqrt{\langle z^2\rangle}$ are of the order of $2\,\mu$m and of less than $1\,\mu$m, respectively, for the second Brownian cloud from the bottom. Such Brownian fluctuations can be directly measured optically, as can the length ℓ, and equation (85) allows a truly ingenious direct measurement of the elastic force $F_r(\ell)$ and its derivative $F_r'(\ell)$! One can then compare the experimental results to the predictions of theoretical models for the statistical description of the DNA configurations (see figure 8).

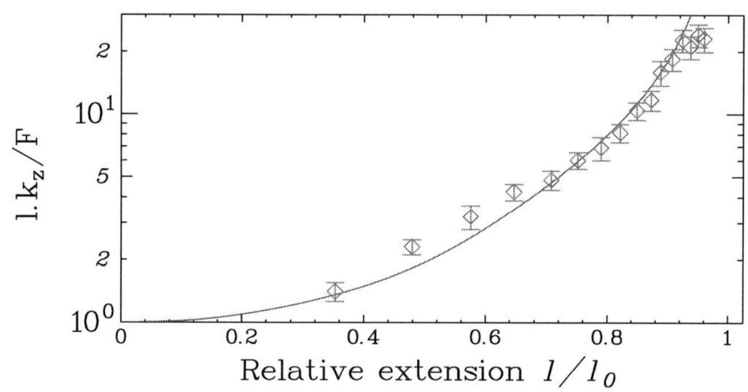

Figure 8: *Dimensionless ratio* $\frac{\ell k_z}{F} = \frac{k_\parallel}{k_\perp} = \frac{\langle x^2\rangle}{\langle z^2\rangle} = \frac{\ell}{F_r(\ell)}\frac{d\,F_r}{d\,\ell}(\ell)$*, plotted as a function of the length* ℓ *of the DNA chain (in units of maximum length* ℓ_0*). The points correspond to the ratio of experimental measurements of transverse (*$\langle x^2\rangle$*) and vertical (*$\langle z^2\rangle$*) quadratic Brownian fluctuations. The curve is theoretically predicted from the knowledge of* $F_r(\ell)$*, in a model of a semi-flexible DNA chain, also known as the Worm-like Chain Model. We stress the remarkable agreement between experiment and theory. (Figure kindly provided by Vincent Croquette.)*

3 Potential theory and Brownian motion

Et ignem regunt numeri[136]

3.1 Introduction

3.1.1 Laplace's equation

Potential theory concerns the equilibrium properties of continuous bodies, like the distribution of electrostatic charges on conductors, the distribution of the Newtonian potential in the classic theory of gravitation, the distribution of temperature in Fourier's theory of heat conduction, or in addition the distribution of positions of a stretched elastic membrane.[137]

A deep relation exists between potential theory and the theory of diffusion, and therefore also with Brownian motion.[138] We will first give an intuitive illustration within the framework of Fourier's theory of heat conduction.

The temperature of a body, $u(x, y, z; t)$ at the point x, y, z and at the instant t, follows the equation of heat

$$\frac{\partial u}{\partial t} = D\Delta u, \tag{86}$$

where, as in the case of Brownian motion, D is the diffusion coefficient, and Δ is the Laplacian in three-dimensions $\Delta = \frac{\partial^2}{\partial x^2} + \frac{\partial^2}{\partial y^2} + \frac{\partial^2}{\partial z^2}$. In general, the Laplacian in d dimensions is:

$$\Delta = \sum_{i=1}^{d} \frac{\partial^2}{\partial x_i^2}, \tag{87}$$

where x_i are d-dimensional Cartesian coordinates. When the temperature reaches equilibrium, the time dependence cancels, and the temperature field is described by Laplace's equation:

$$\Delta u = 0. \tag{88}$$

Any function with zero Laplacian is called *harmonic.*

Such a function, the *potential*, therefore can be seen as the equilibrium solution of a diffusion process (at infinite time), which is the first elementary relation we meet between potential theory and Brownian diffusion. To specify in our example the value of the temperature everywhere, we must fix the initial conditions in case one starts from an out-of-equilibrium situation.

In the case we will consider here, we want to directly study equilibrium and the associated harmonic functions, or more generally the potential. For that

[136] Joseph Fourier's major work, *La théorie analytique de la chaleur*, was published in 1822, with *Et ignem regunt numeri* as its motto (*Numbers rule fire*).

[137] One can cite O. D. Kellogg's classic work *Foundations of Potential Theory*, Springer-Verlag (1929); Dover Books on Advanced Mathematics (1969).

[138] See the article *Brownian Motion and Potential Theory*, by R. Hersch and R. J. Griego, Scientific American, **220**, March 1969; translated into French in *Le mouvement brownien et la théorie du potentiel*, appearing in 1977 within the first out-of-series of Pour la Science.

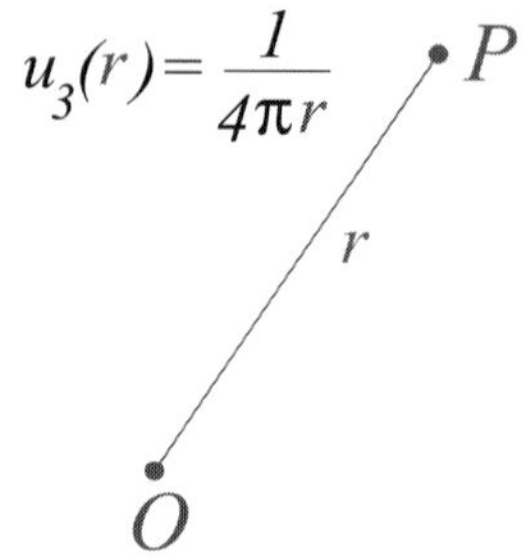

Figure 9: *Newtonian potential in three dimensions.*

purpose one must know either the position of the sources of the potential, or the boundary conditions on it, in a way that will be made more precise in the following.

Giving the position of the sources is natural in the well-known theory of the Newtonian or Coulomb potential, where the sources of the potential are masses or electrostatic charges. Imposing boundary conditions on the potential is also possible, as is natural in the case of heat conduction and temperature distribution, where one gives the temperature distribution on the surface of a body to determine the internal temperature distribution.

Such representations are mathematically equivalent. Let us first recall elementary properties of the Newtonian or Coulomb potential, that will be useful for obtaining the finer properties of harmonic functions. To fix the ideas, we will adopt the familiar language of a Newton or Coulomb potential created by masses or electrostatic charges, but the mathematical results of course will not depend on this choice.

3.2 Newtonian potential

3.2.1 The potential created by a point source

In order to consider the potential in a universal way, as for gravitation or electrostatics, the physical constants like the universal gravitational constant G, or the electric permeability of the vacuum, ε_0, are not indicated. In general, we will adopt the electrostatic language.

The potential at a point P in three dimensions created by a unit charge or mass placed at the origin O is

$$u_3(r) = \frac{1}{4\pi r}, \quad r = |\overrightarrow{OP}|, \tag{89}$$

where r is the distance between O and P (figure 9).

The associated electric (or gravitational) field is

$$\vec{E}_3(\vec{r}) = -\nabla_{\vec{r}}\, u_3(r) = \frac{1}{4\pi}\frac{\vec{r}}{r^3}, \tag{90}$$

where $\vec{r}$ is the relative position vector $\vec{r} = \overrightarrow{OP}$.

In d dimensions, the potential and field generalize to

$$u_d(r) = \frac{1}{(d-2)S_d} \frac{1}{r^{d-2}}, \tag{91}$$

and

$$\vec{E}_d(\vec{r}) = -\nabla_{\vec{r}}\, u_d(r) = \frac{1}{S_d} \frac{\vec{r}}{r^d}, \tag{92}$$

where $S_d = 2\pi^{d/2}/\Gamma(d/2)$ is the surface of the unit sphere in $\mathbb{R}^d$.

The two-dimensional case is more complicated, and leads to a logarithmic potential,

$$u_2(r) = \frac{1}{2\pi} \log \frac{1}{r}, \tag{93}$$

$$\vec{E}_2(\vec{r}) = -\nabla_{\vec{r}}\, u_2(r) = \frac{1}{2\pi} \frac{\vec{r}}{r^2}. \tag{94}$$

3.2.2 Laplace's equation and the Dirac distribution

The Laplacian of the above potential $u_d(r)$ vanishes identically everywhere in space, except at the origin: $\Delta u_d(r) = 0,\ r \neq 0$. At $\vec{r} = \vec{0}$ it is divergent, and its value is given by a distribution, namely

$$\Delta u_d(r) = \frac{1}{(d-2)S_d} \Delta \frac{1}{r^{d-2}} = -\delta^d(\vec{r}), \tag{95}$$

where $\delta^d(\vec{r})$ is the Dirac distribution in d dimensions, zero everywhere except at the origin $\vec{r} = \vec{0}$, where it is singular (infinite). This divergence is such that the integral

$$\int_{\mathbb{R}^d} f(\vec{r})\delta^d(\vec{r})\, \mathrm{d}^d r = f(\vec{0}) \tag{96}$$

yields the value at the origin of any test function $f(\vec{r})$.

Equation (95) is Poisson's equation, where the second term represents the mass or charge density, i.e., the source of the potential. In the case of a potential (89), (91) or (93), such a source is a point, at which a singular density appears.

In the elementary approach that follows, we shall not use this formalism. Rather, we will follow the elementary path that uses Gauss' theorem.[139]

3.2.3 Gauss' theorem

Gauss' theorem says that the flux of an electric (or gravitational) field across any closed surface Σ is equal to the total charge $Q(\Sigma)$ (or mass) enclosed by the surface:

$$\int_\Sigma \vec{E} \cdot d\vec{S} = Q(\Sigma). \tag{97}$$

[139] O.D. Kellogg, *op.cit.*

This theorem can be proved in two stages. By linearity, since the case of a distribution of charges can be treated by adding the fields, one can reduce it to the case of a point charge. Actually, if each one of these fields satisfies Gauss' theorem, their sum will as well.

Next, for a point charge enclosed by the surface, we notice that the flux of $\vec{E}$ is invariant when we deform the surface Σ without crossing the charge.[140] We can thus restrict attention to a sphere around the charge, for which Gauss' theorem is trivial. Actually, because of the form (90) of the $1/r^2$ field with spherical symmetry, the integral (97) on a sphere of a radius r is equal to the charge.

Gauss' theorem immediately generalizes to any number of dimensions.

3.2.4 Potential generated by a sphere

Let us consider the sphere $\mathcal{S}(a)$ of radius a centered at the origin O. Imagine that it carries a charge Q uniformly distributed over its surface.

The associated field $\vec{E}(r)$ is radial and with spherical symmetry. It satisfies Gauss' theorem (97). If one chooses the surface Σ like a sphere $\mathcal{S}(r)$ centered in O, of radius $r > a$, i.e., it is exterior to $\mathcal{S}(a)$, we have $Q(\Sigma) = Q$, and the flux of $\vec{E}(r)$ across Σ is simply, by spherical symmetry, $E(r)4\pi r^2 = Q$. We then deduce that $E(r) = \frac{Q}{4\pi r^2}$ is the same field that would be created by a charge as if it was concentrated at the center of sphere. If the surface Σ is chosen like a sphere $\mathcal{S}(r)$ of radius $r < a$, i.e., inside $\mathcal{S}(a)$, then $Q(\Sigma) = 0$ and the flux (97) is then zero. By symmetry, we then deduce that the field $\vec{E}$ is zero everywhere inside the sphere.

Let $u_{\mathcal{S}}(P)$ now be the potential created at a point P by the same sphere $\mathcal{S}(a)$ of radius a with total charge Q, uniformly distributed on the surface. This potential has a spherical symmetry, as does its associated field. Outside the sphere, the field is the same as that of a point charge Q placed at the center, while inside the sphere the field is zero. The potential outside the sphere is therefore the one, (89), created by a point charge placed at the center of the sphere, while inside the sphere it is constant, and by continuity equal to its value on the boundary. One thus has

$$u_{\mathcal{S}}(P) = \frac{1}{4\pi}\frac{1}{r}\vartheta(r-a) + \frac{1}{4\pi}\frac{1}{a}\vartheta(a-r), \quad r = |\overrightarrow{OP}|, \tag{98}$$

where ϑ is the Heaviside distribution $\vartheta(x<0) = 0$, $\vartheta(0) = 1/2$, $\vartheta(x>0) = 1$.

3.3 Harmonic functions and the Theorem of the Mean

3.3.1 Gauss' theorem of the Arithmetic Mean

The property that two bodies or two charges attract one another with equal and opposite forces, reflects itself in the potential. Actually the potential is symmetric

[140] We have, from the Green–Ostrogradski theorem, that $\int_\Sigma \vec{E}\cdot d\vec{S} - \int_{\Sigma'} \vec{E}\cdot d\vec{S} = \int_{\mathcal{D}} \mathrm{div}\vec{E}\, d^3v = -\int_{\mathcal{D}} \Delta u\, d^3v = 0$, where $\mathcal{D}$ is the domain between the two surfaces Σ and Σ', and u is the potential. Indeed, we have the identities $\vec{E} = -\vec{\nabla}u$ and $\mathrm{div}(\vec{\nabla}u) = \Delta u = 0$, because u is harmonic in the domain $\mathcal{D}$ without charges.

with respect to the coordinates of the two points, in such a way that the potential at P of a charge Q at S is the same as the potential at S of a charge Q at P. From such a simple fact follow theorems with important applications. We derive two of them, called Gauss' theorems of the Arithmetic Mean.[141]

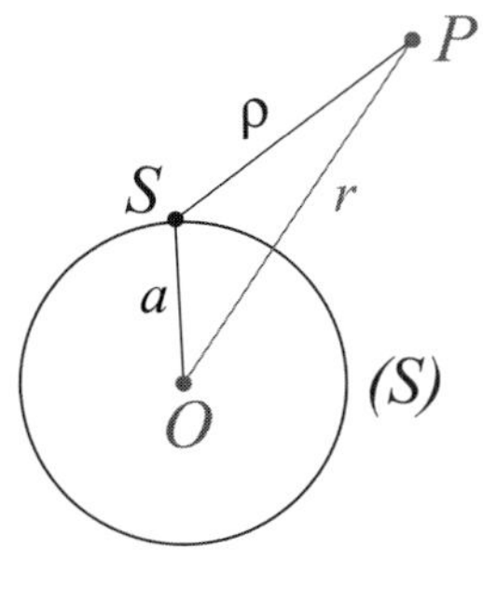

Figure 10: *Newtonian potential* (99) *created by a uniformly charged sphere of radius* a.

The potential

$$u_{\mathcal{S}}(P) \quad = \quad \frac{Q}{4\pi a^2}\int_{\mathcal{S}}\frac{d^2S}{4\pi\rho}, \quad \rho = |\overrightarrow{SP}| \tag{99}$$

is the one at point P, created by all points S on the surface of a sphere $\mathcal{S}$ of radius a, and with uniform charge density $\frac{Q}{4\pi a^2}$ (see figure 10). In (98) we just saw that outside the sphere the potential is equal to $\frac{Q}{4\pi r}$, where r is the distance $r = |\overrightarrow{OP}|$, while inside the sphere it is constant and equal to $\frac{Q}{4\pi a}$.

But because of the exchange symmetry which we just mentioned, the potential can also be interpreted as the arithmetic mean on the surface of a sphere of the potential created by the same charge Q, this time placed in P.

The equations (98) (99) therefore have the following interpretation:

a) The average on the surface of a sphere of the potential created by a charge situated *outside* the sphere, and at a distance r from its center, is equal to the value (varying as $1/r$) of the potential at the center of the sphere.

b) The average on the surface of a sphere of the potential created by a charge in any position *inside* the sphere, is equal to the value (varying as $1/a$) of the potential on the sphere, after concentrating the whole charge at the center of the sphere.

Now let us suppose that we have a group of charges placed either entirely on the outside of the sphere, or entirely on the inside. By adding up the above results for each elementary charge, we find the following two generalizations:

[141] O. D. Kellogg, *op. cit.*

a) **Gauss' Theorem of the Arithmetic Mean.** The average on a surface of a sphere of the potential created by charges situated entirely outside the sphere is equal to the value of the potential at the center.

b) **The Second Theorem of the Mean.** The average of the potential on a surface of a sphere, created by charges situated entirely inside the sphere, is independent of their distribution inside the sphere, and it is equal to the total charge divided by the radius of the sphere.[142]

3.3.2 Harmonic functions

Finally let us come back to harmonic functions, and consider a function u such that $\Delta u = 0$ in some domain $\mathcal{D}$. Such a harmonic function can be represented as a potential created inside the domain $\mathcal{D}$ by a distribution of charges outside $\mathcal{D}$. We can then apply the first of Gauss' theorems, and obtain the *mean-value theorem for harmonic functions*: *The average of a harmonic function u on a sphere S centered at a point P is equal to the value of u at P.* For instance, in three dimensions:

$$u(P) = \int_S u(S) \frac{d^2 S}{4\pi a^2}, \tag{100}$$

where a is the radius of the sphere; the theorem can be generalized to any number of dimensions.

The reciprocal is also true: any function that fulfills the Theorem of the Mean on every sphere inside a given domain, is harmonic inside that domain. This theorem is going to be the key relation between potential theory and Brownian motion.[143]

---

[142] One can find the first theorem in Gauss' complete works, *Allgemeine Lehrsätze*, vol. V, p. 222. The second theorem, less known, can be found there too.

[143] A proof of the Theorem of the Mean can be obtained by vectorial analysis. We write the average $\langle u \rangle_{\mathcal{S}}$ of u on the surface of the $(d-1)$-sphere $\mathcal{S}$ of radius a in $\mathbb{R}^d$, as the flux of the vector $u(\vec{r})\vec{r}/r^d$ on the surface of the sphere:

$$\langle u \rangle_{\mathcal{S}} = \frac{1}{S_d\, a^{d-1}} \int_{\mathcal{S}} u(S)\, d^{d-1}S = \frac{1}{S_d} \int_{\mathcal{S}} u(\vec{r}) \frac{\vec{r}}{r^d} . \vec{n}\, d^{d-1}S = - \int_{\mathcal{S}} u(\vec{r}) \vec{\nabla} u_d(r) . \vec{n}\, d^{d-1}S, \tag{101}$$

where S_d is the area of the unit sphere, $\vec{n}$ is the unit vector normal to the surface of the sphere (and directed towards the exterior), and where we used (92). We therefore use Green's theorem in the volume $\mathcal{D}$ inside the sphere:

$$\int_{\mathcal{D}} [u_d(r)\, \Delta u(\vec{r}) - u(\vec{r})\, \Delta u_d(r)]\, d^d r = \int_{\mathcal{S}} \left[u_d(r)\, \vec{\nabla} u(\vec{r}) - u(\vec{r}) \vec{\nabla} u_d(r)\right] . \vec{n}\, d^{d-1}S. \tag{102}$$

We have $\Delta u(\vec{r}) = 0$, because u is harmonic, and from (95) we have $\Delta u_d(r) = -\delta^d(\vec{r})$. From the definition (96) of Dirac distribution and by substituting (101) in (102), we have:

$$u(0) = \langle u \rangle_{\mathcal{S}} + \int_{\mathcal{S}} \left[u_d(r)\, \vec{\nabla} u(\vec{r})\right] . \vec{n}\, d^{d-1}S. \tag{103}$$

3.4 The Dirichlet problem

A classic problem of potential theory is the one of Dirichlet. One considers a domain $\mathcal{D}$ of the Euclidean space $\mathbb{R}^d$ and its boundary $\partial\mathcal{D}$. The potential u is given on the boundary by means of a given continuous function f:

$$\Delta u = 0 \text{ inside } \mathcal{D}, \tag{106}$$

$$u = f \text{ on } \partial\mathcal{D}. \tag{107}$$

For instance, the Dirichlet problem in the case of heat conduction is to determine the equilibrium temperature inside a conducting body $\mathcal{D}$, once the distribution f of the temperature along the boundary $\partial\mathcal{D}$ is given.

It is here that Brownian motion comes into play, to provide an entirely probabilistic representation of the solution.

3.5 Relation between potential theory and Brownian motion

3.5.1 Newtonian potential and probability density

The first relation, which contains the kernel of all the others, is obtained simply by considering the Gaussian probability density (26),[144] which represents the probability density of finding a Brownian particle at a point $\vec{r}$ at time t, knowing that the particle was at the origin at time $t = 0$. In d dimensions, formula (26) generalizes to

$$P(\vec{r};t) = \frac{1}{(4\pi Dt)^{d/2}} \exp\left(-\frac{r^2}{4Dt}\right), \tag{108}$$

where r is the distance from the origin.

By integrating $P(\vec{r};t)$ over the time variable t one obtains

$$D\int_0^{+\infty} P(\vec{r};t)\,\mathrm{d}t = \frac{1}{(d-2)S_d}\frac{1}{r^{d-2}} = u_d(r). \tag{109}$$

For a unit diffusion coefficient $D = 1$, the total Brownian probability density of arriving at $\vec{r}$ at any time is then exactly equal to the Newtonian potential created at $\vec{r}$ by a unit charge or mass.

Let us look now at the Dirichlet problem from a more general point of view.

As the Newtonian potential is constant on the sphere, $u_d(r) = u_d(a) = \frac{1}{(d-2)S_d a^{d-2}}$, the last flux integral is transformed into a volume integral and it yields

$$u_d(a)\int_{\mathcal{S}} \vec{\nabla}u(\vec{r}).\vec{n}\, d^{d-1}S = u_d(a)\int_{\mathcal{D}} \Delta u(\vec{r})\, d^d r = 0, \tag{104}$$

because u is a harmonic function by hypothesis. We have then obtained the Theorem of the Mean as expected:

$$\langle u\rangle_{\mathcal{S}} = u(0). \tag{105}$$

[144]For this subject one can consult the book of K.L. Chung, *Green, Brown, and Probability & Brownian Motion on the Line*, World Scientific, Singapore (2002).

3.5.2 Discrete random walks and the Dirichlet problem

This problem was considered in 1920s with the work of Phillips and Wiener[145], and of Courant, Friedrichs and Lewy.[146] They obtained a probabilistic representation of the solution of the Dirichlet problem (106, 107), in the form of an approximate sequence of random walks on a d-dimensional cubic lattice $\varepsilon\mathbb{Z}^d$, of lattice spacing ε.

More precisely, one considers random walkers $w = \{w_n, n \in \mathbb{N}\}$ on the lattice $\varepsilon\mathbb{Z}^d$, at discrete times $n = 0, 1, 2, \cdots$, all starting from the initial point $w_0 = P$ in domain $\mathcal{D}$ and diffusing away. When the walkers ultimately reach the boundary, one measures the value of the function f at that point on the boundary. One repeats the process and then takes the *average* of the values of the function f over all first contact points on the boundary reached by random walkers that started from P.

We can formally write the averaging operation as

$$u_\varepsilon(P) = \sum_{\{w:\ P \mapsto \partial\mathcal{D}\}} f(w_{\tau_\mathcal{D}}), \tag{110}$$

where the sum is over all random walks $w = \{w_n, n \in \mathbb{N}\}$ on the lattice $\varepsilon\mathbb{Z}^d$, at discrete times $n = 0, 1, 2, \cdots$, leaving the initial point $w_0 = P$ and diffusing towards the boundary. In (110), $\tau_\mathcal{D}$ is the first instant at which the boundary $\partial\mathcal{D}$ is reached by the random walker and $w_{\tau_\mathcal{D}}$ its position on the boundary at this instant. The sum must be normalized in a way to be a probability measure on the set of discrete random walkers.

To extend the result in the continuum, one next takes the limit of the lattice spacing ε to 0. The result $\lim_{\varepsilon\to 0} u_\varepsilon(P) = u(P)$ is then the value of u at point P, which is the solution of the Dirichlet problem in $\mathbb{R}^d$.

In the language of heat theory for instance, the temperature at point P is the average of the temperature at the boundary, evaluated after *random walking* towards it!

In mathematics, a standard notation of the average (110) is

$$u_\varepsilon(P) = \int f(w_{\tau_D})\Pi^\varepsilon_P(dw), \tag{111}$$

where Π^ε_P is the probability measure on discrete random walks in $\varepsilon\mathbb{Z}^d$ started at P.

3.5.3 Norbert Wiener

A first attempt to define integral calculus over a function space was made by Daniell (circa 1920).[147] A few years later, Norbert Wiener introduced a measure

[145] H. B. Phillips and N. Wiener, *J. Math. Phys.*, **2**, pp. 105–124 (1923).
[146] R. Courant, K. Friedrichs and K. Lewy, *Math. Ann.*, **100**, pp. 37–74 (1928).
[147] P.J. Daniell, *Ann. Math.* **21**, 203 (1920).

in function space which is rigorous from a mathematical point of view (it is a *bona fide* Borel measure), and which made it possible to define and calculate an integral over a space of functions.

Wiener had indeed known Einstein's theory since his visit to Cambridge in 1913. At 19, he came to study logic with Bertrand Russell, who suggested that he go listen to Hardy, the mathematician, and read Einstein!

So, motivated also by his reading of Perrin, Wiener constructed, in his fundamental article of 1923, "Differential Space",[148] a probability measure for Brownian paths in $\mathbb{R}$ (then in $\mathbb{R}^d$). The basic idea was to directly construct on the space of continuous functions $w(t)$ of a single real variable (representing the position as a function of time), a probability measure such that the changes of the positions $w(t_i) = x_i$, $i = 1, \cdots, n$, over disjoint time intervals, $[t_{i-1}, t_i]$, $i = 1, \cdots, n$, have a joint Gaussian probability distribution,

$$P(\{x_i\}; \{t_i\}) = \prod_{i=1}^{n} \frac{1}{[4\pi D(t_i - t_{i-1})]^{1/2}} \exp\left[-\frac{(x_i - x_{i-1})^2}{4D(t_i - t_{i-1})}\right]. \tag{112}$$

This is a direct generalization of the Brownian displacement distribution (25).

Wiener obtained his measure by using an explicit mapping of the space $\mathcal{C}$ of continuous functions into the interval (0,1) (minus a set of Lebesgue measure zero). This mapping allows to pull-back the ordinary Lebesgue measure on the space $\mathcal{C}$. In this language, the Brownian motion has the following probabilistic interpretation: a Brownian path corresponds to the random choice of an element of the measured set $\mathcal{C}$ (*i.e.*, a continuous function), endowed with the "Wiener measure".

Nowadays, this measure is indeed universally called *Wiener measure* in mathematical circles, while physicists prefer to speak of functional integrals, even though, like *Monsieur Jourdain*, they really calculate with the Wiener measure when they perform their formal calculations![149]

The integral over such a measure is called *Wiener average.* It is noted $\mathcal{W}(dw)$ here and more precisely $\mathcal{W}_P(dw)$ for a Brownian motion w started at P. It corresponds to the continuous limit for $\varepsilon \to 0$ of the measure $\Pi^\varepsilon(dw)$ on random walks on the discrete lattice $\varepsilon\mathbb{Z}^d$, introduced in the preceding section.

Once that construction was made, Wiener verified that the measure of the subset of differentiable functions vanishes, in agreement with Perrin's intuition, and that the support of the measure is given by Hölder functions (of order at least $1/2 - \epsilon, \epsilon > 0$). Along the years, he kept developing further the very broad ramifications of his theory.[150]

---

[148] N. Wiener, *J. Math. Phys.*, **2**, pp. 131–174 (1923).

[149] This is true in perturbation theory. See, e.g., in the case of polymer theory, B. Duplantier, *Renormalization and Conformal Invariance for Polymers*, in *Proceedings of the Seventh International Summer School on Fundamental Problems in Statistical Mechanics*, Altenberg, Germany, June 18–30, 1989, H. van Beijeren Editor, North-Holland, Amsterdam (1990).

[150] N. Wiener, *Acta. Math.* **55**, 117 (1930); ; R. E. A. C. Paley and N. Wiener, *Fourier Transforms*

In a study written in 1964 on Wiener and functional integration, Mark Kac highlighted the profound originality of Wiener during his time, and in counterpoint, the difficulty for mathematicians to understand his approach:[151]

"Only Paul Lévy, in France, who had himself been thinking along similar lines, fully appreciated their significance."

The next steps were indeed made by Paul Lévy, in his great work on Brownian motion, *Processus stochastiques et mouvement brownien* (1948).[152] Since then, the blooming of the subject in mathematics was such that one can only make an extremely partial citation list. We refer the interested reader to the introductory article of J.-F. Le Gall for a first journey into the Brownian world of mathematics,[153] and to D. Revuz and M. Yor's book for a more thorough visit.[154]

The connection between the Wiener path measure for Brownian motion and path integrals is perhaps best intuitively understood by considering the multiple distribution (112) for a set of successive equal time intervals, $t_i = \frac{i}{n}t, i \in \{1, n\}$. One conditions the path, normalized to start at the origin $x = 0$ at time $t = 0$, to be at times $t_i = \frac{i}{n}t, i \in \{1, n\}$ within intervals $\mathrm{d}x_i$ of the set of points x_i in $\mathbb{R}$, and one then takes the *formal* limit $n \to \infty$:

$$\begin{aligned}
\mathcal{W}(dw) &= \lim_{n\to\infty} \prod_{i=1}^{n} \mathrm{d}x_i P(\{x_i\};\{t_i\})) \\
&= \lim_{n\to\infty} \prod_{i=1}^{n} \mathrm{d}x_i \prod_{i=1}^{n} \frac{1}{(4\pi D\, t/n)^{1/2}} \exp\left[-\frac{(x_i - x_{i-1})^2}{4D\, t/n}\right] \\
&= \mathcal{D}w \exp\left(-\frac{1}{4D}\int_0^t \left(\frac{\mathrm{d}w(t')}{\mathrm{d}t'}\right)^2 \mathrm{d}t'\right), \qquad (113)
\end{aligned}$$

whith now a continuum "Lebesgue" measure on paths,

$$\mathcal{D}w = \lim_{n\to\infty} \prod_{i=1}^{n} \frac{\mathrm{d}x_i}{(4\pi D\, t/n)^{1/2}}.$$

This notation is marvellously appealing to physicists, since one recognizes in the exponential in (113) the Boltzmann–Gibbs weight associated with the classical kinetic energy of the particle. As Marc Kac noted[155],

in the Complex Domain, Amer. Math. Soc. Colloq. Publ., **19**, New York (1934); N. Wiener, *Generalized Harmonic Analysis and Tauberian Theorems*, MIT Press, Cambridge, Mass. (1964).

[151] M. Kac, *Bull. Amer. Math. Soc.*, **72**, pp. 52–68 (1964).

[152] Paul Lévy, *Processus stochastiques et mouvement brownien*, Gauthier-Villars, Paris (1965).

[153] J.-F. Le Gall, *Introduction au mouvement brownien*, Journées annuelles de la Société Mathématique de France, 28 janvier 1989, three exposés on Brownian motion (J.-F. Le Gall: *supra*, G. Ben Arous: *Grandes déviations et noyau de la chaleur*, B. Duplantier: *Le mouvement brownien en physique, les polymères et leur relation avec les phénomènes critiques*).

[154] D. Revuz and M. Yor, *Continuous Martingales and Brownian Motion*, Berlin-Heidelberg: Springer (1991); second edition, 1994.

[155] M. Kac, *Probability and related Topics in the Physical Sciences*, Interscience, New York (1959);

"The disadvantages of such an approach from the purely mathematical point of view are obvious, although it is appealing on formal grounds".

In d dimensions, the formal equivalence between Wiener's measure and functional integrals is simply obtained by using the d-dimensional Gauss distribution, so that

$$\begin{aligned} \mathcal{W}(dw) &= \mathcal{D}w \exp\left(-\frac{1}{4D}\int_0^t \left(\frac{\mathrm{d}\vec{w}(t')}{\mathrm{d}t'}\right)^2 \mathrm{d}t'\right), \qquad (114) \\ \mathcal{D}w &= \lim_{n\to\infty}\prod_{i=1}^{n}\frac{\mathrm{d}^d x_i}{(4\pi D\, t/n)^{d/2}}. \end{aligned}$$

The rigorous connection between the Wiener path integral and Brownian motion is further illuminated by the Feyman–Kac formula that allows one to write explicit path integral representations for the solutions of parabolic differential equations, corresponding to Brownian motion in presence of a general potential,[156] the case pioneered by Smoluchowski.

When formally continued to imaginary time, the Feyman–Kac formula provides an expression for the Green function of the Schrödinger equation, thus leading to the celebrated path integral representation of Quantum Mechanics invented by Feynman in 1948.[157]

3.5.4 S. Kakutani

The existence of the Wiener measure and Wiener integral allowed for some very important progress by S. Kakutani in 1944–1945.[158] He showed that by substituting an integral with the Wiener measure $\mathcal{W}$ in the formula (111) with the discrete measure Π^{ε} indeed solved the Dirichlet problem in continuous space $\mathbb{R}^d$. Thus we have Kakutani's formula

$$u(P) = \int f(w_{\tau_D})\mathcal{W}_P(dw). \qquad (115)$$

That means that *the potential at any point P is given by the average of the potential chosen at random on the boundary by a Brownian motion started at P* (figure 11).

In the following section we give an elementary demonstration of this result.

[156] M. Kac, *Probability and related Topics in the Physical Sciences, op. cit.*; L. S. Schulman, *Techniques and Applications of Path Integration*, John Wiley and Sons, New York (1981); F. W. Wiegel, *Introduction to Path Integral Methods in Physics and Polymer Science*, World Scientific, Singapore (1986); J. Zinn-Justin, *Quantum Field Theory and Critical Phenomena*, 4th Edition, Oxford University Press (2002).

[157] R. P. Feynman, *Rev. Mod. Phys.* **20**, 367 (1948); R. P. Feynman and A. R. Hibbs, *Quantum Mechanics and Path Integrals*, McGraw-Hill, New York (1965).

[158] S. Kakutani, *Proc. Imp. Acad. Japan*, **20**, pp. 706–714 (1944).

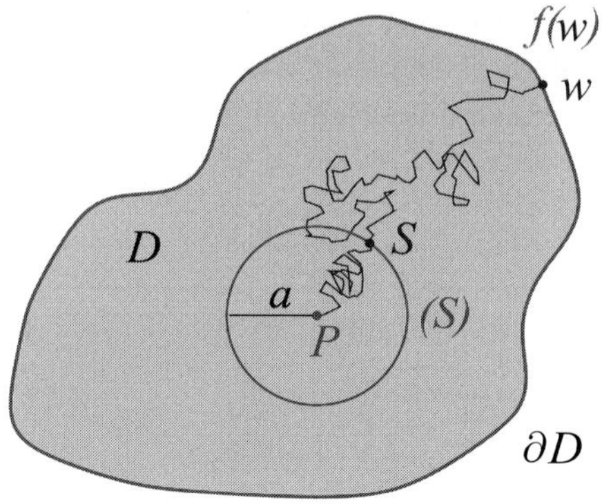

Figure 11: *The Dirichlet problem in a domain $\mathcal{D}$, and its Brownian representation. The point $w = w_{\tau_{\mathcal{D}}}$ is the point of first contact of a Brownian motion that started at P with the boundary $\partial\mathcal{D}$, at the instant $\tau_{\mathcal{D}}$ of first exit from the domain $\mathcal{D}$. The point S is the point of first passage across the surface of the sphere $\mathcal{S}$.*

3.5.5 Demonstration

In probability theory, the quantity $u(P)$ defined by equation (115) is called the *expectation value* associated to the point P, because it represents the expectation for a random sampling of the value f on the boundary, by a process of Brownian diffusion from P.

We want to verify that this expectation value fulfills the two conditions (106) and (107).

The second condition is easy to verify: if the point P is on the boundary $\partial\mathcal{D}$, any Brownian motion w coming from P is immediately stopped on the boundary at $w_{\tau_{\mathcal{D}}} = P$, therefore $u(P) = f(P)$ for P on $\partial\mathcal{D}$, as expected.

Moreover, if the Brownian motion leaves from an internal point P, close to a point P_0 of the boundary, it is (almost) certain (in a probabilistic sense) that the motion will meet the boundary in a neighborhood of P_0, and that the expectation value $u(P)$ will be close to the value $f(P_0)$ of f in P_0. Kakutani's solution has the right properties of regularity near the boundary, under the condition that the latter has a sufficiently regular geometry and that the "temperature" f on the boundary is a continuous function.

The continuity of the expectation value u, with respect to point P, is equally clear: a small displacement of P will only slightly modify the Brownian trajectories diffusing from P, as well as their subsequent exploration of the boundary.

We will now establish the first property, (106), i.e., that the expectation value $u(P)$ (115) is a harmonic function, by showing that it satisfies the equivalent property (100) on all spheres centered in P.

We draw a sphere of radius a $\mathcal{S}$ centered at P and contained inside the domain $\mathcal{D}$ (figure 11). The aim is to show that the Brownian expectation value $u(P)$ obtained by leaving from any point P is equal to the average of Brownian expectation values $u(S)$ obtained from any point S at the surface of the sphere $\mathcal{S}$.

In order to move beyond the boundary $\partial\mathcal{D}$ of the domain, a Brownian motion must cross the sphere $\mathcal{S}$ at least once. Calling S the first crossing point of the sphere (figure 11), and $u(P/S)$ the expectation value obtained for all Brownian motions coming from P and first crossing $\mathcal{S}$ at the point S.

As there is no preferential direction for Brownian motion, each point of $\mathcal{S}$ can be met first with equal probability. One distinguishes the average for Brownian motions starting at P in two steps: the choice of the point of first passage S, and diffusion across S, with the expectation value $u(P/S)$. By averaging the averages, one has the result that $u(P)$ must be equal to the average of $u(P/S)$ on the sphere, i.e., in mathematical terms:

$$u(P) = \int_{\mathcal{S}} u(P/S) \frac{d^2S}{4\pi a^2}. \tag{116}$$

The last thing to show is that the expectation value $u(P/S)$, obtained by leaving from P and passing through S, is the same as the expectation value $u(S)$, obtained by simply starting from S on the sphere. It is here that a very important property of Brownian motion comes into play: the motion at an instant t only depends on the position at that instant and not on previous motions. Somehow, there is an absolute loss of memory, where only the present instant and position are important: The Brownian motion is *Markovian*. In probability theory, one speaks generally as well of a *Markov process* when the future dynamic of a process is not influenced by its previous states. The future behavior of a Brownian particle leaving from S, or passing through S knowing that it began at P, does not differ. It follows that $u(P/S) = u(S)$, which ends the proof of the Theorem of the Mean (100).

3.6 Recurrence properties of Brownian motion

We give an illustration of a non-trivial probabilistic property of Brownian motion, which is deduced from potential theory, that is its *recurrence properties*.

3.6.1 Brownian motion in one dimension

Let us consider now the one-dimensional real line $\mathbb{R}$ and points x of a domain $\mathcal{D}$, here the line segment $\mathcal{D} = [0, R]$, where R is a positive number. Let us search for the harmonic function $u(x)$ that satisfies the simple Dirichlet problem: $u(0) = 0,\ u(R) = 1$. In one dimension, the Laplacian (87) is simply the second derivative, so the harmonic equation (106) becomes $\mathrm{d}^2u(x)/\mathrm{d}x^2 = 0$. The solution is simply linear in x: $u(x) = x/R$; it evidently satisfies the required conditions at the boundaries.

Let us consider Kakutani's solution for the Dirichlet problem by Brownian mathematical expectation. The boundary $\partial\mathcal{D}$ of the segment $\mathcal{D} = [0, R]$ is made up of two points: $\partial\mathcal{D} = \{0, R\}$. The function f with Dirichlet conditions (107), takes the values on the boundary: $f(0) = 0,\ f(R) = 1$. According to Kakutani's result, the value $u(x) = x/R$ of the harmonic function u is the average of the function f obtained from random sampling by means of a Brownian motion starting at x. The case of the first Brownian exit from the segment $\mathcal{D} = [0, R]$ at point $x = 0$ gives a value $f = 0$, and at point $x = R$ the value $f = 1$. The Brownian expectation of f is thus exactly the probability for the Brownian motion to first exit from the segment $[0, R]$ at the endpoint R rather than at 0, or else the probability, starting from x, to attain R before 0. The complementary probability to attain 0 before R is thus $p_R(x) = 1 - u(x) = 1 - x/R$.

Let us now keep the point x fixed while taking the limit $R \to \infty$, so that the segment $\mathcal{D}$ extends to the positive real axis $\mathbb{R}^+$. We see that $p_{R\to\infty}(x) \to 1$, and this is for all x. The probability $p_\infty(x)$, for a Brownian motion started at x, to reach the origin 0 before leaving to infinity is therefore identically equal to *one*.

The Brownian motion, wherever it leaves from, passes by the origin (quasi-) certainly[159]. Since spatial and temporal origins were arbitrary in our demonstration, the following property was established: *a Brownian motion in one dimension passes through all points on the real axis, infinitely many times.* One says it is *recurrent* in one dimension.

This property did not appear as evident a priori from the probability theory side. Thanks to the relation to potential theory, it has been obtained by simply resolving a second order differential equation! Einstein would surely not have thought of this in 1905, although who knows?

Now we will generalize the above study to two and then to d dimensions.

3.6.2 The two-dimensional case

This time, we consider the planar annular domain $\mathcal{D}$, which is that between two concentric circles, $\mathcal{C}_1$ and $\mathcal{C}_2$ centered at the origin O, of respective radii ρ_1 and ρ_2, with $\rho_1 < \rho_2$. The boundary of the domain $\mathcal{D}$ is then made of two circles, $\partial\mathcal{D} = \mathcal{C}_1 \cup \mathcal{C}_2$. We pose the Dirichlet problem in the annular domain $\mathcal{D}$:

$$\Delta u = 0 \text{ inside } \mathcal{D}, \tag{117}$$

$$u = 0 \text{ on } \mathcal{C}_1,\quad u = 1 \text{ on } \mathcal{C}_2. \tag{118}$$

By using the two-dimensional Newtonian potential $u_2(r)$ (93), it is easy to see that the solution to the Dirichlet problem is spherically symmetric and at a distance r

[159] In *continuous* probability theory, an event with probability 1 is only said to be "quasi-certain" or "almost surely true", contrary to the common language. The reason is that in the case of events forming a continuum, it can always exist a non-empty set of irreducible events where the prediction is not realized, which is still of zero measure in the sense of measure theory, and therefore of zero probability. One cannot forgo the consideration of zero-measure sets, hence go beyond the *"almost surely" (a.s.)* probabilistic description.

from the center evaluates to:

$$u_2(r;\rho_1,\rho_2) = \frac{u_2(r) - u_2(\rho_1)}{u_2(\rho_2) - u_2(\rho_1)} = \frac{\log r - \log \rho_1}{\log \rho_2 - \log \rho_1}, \quad \rho_1 \leq r \leq \rho_2. \quad (119)$$

Actually, this function obviously satisfies (118) and is harmonic in the annular domain $\mathcal{D}$, because the potential $u_2(r)$ (93) is harmonic too (except at the origin, which indeed does not belong to $\mathcal{D}$).

Let us come now to Kakutani's representation of the solution to the Dirichlet problem. In a manner similar to the one-dimensional case in the preceding paragraph, $u_2(r;\rho_1,\rho_2)$ (119) represents the probability that a Brownian motion, started at a distance r from the center, hits the outer circle $\mathcal{C}_2$ before hitting the inner circle $\mathcal{C}_1$.

As in the preceding paragraph, let us fix the distance r and the internal circle $\mathcal{C}_1$, and push the boundary of the outer circle $\mathcal{C}_2$ to infinity. By taking $\rho_2 \to \infty$ in formula (119), we see that by continuity the probability that the Brownian motion goes to infinity is $u_2(r;\rho_1,\infty) = 0$, for all r and ρ_1 finite. It means that the Brownian motion reaches the disk of radius ρ_1 with probability 1, whatever its point of departure outside of the disk. Since the initial departure time is arbitrary too, likewise the origin in the plane, one then concludes that *two-dimensional Brownian motion passes through neighboring points of any point infinitely often.* It is then *recurrent in two dimensions*, just as it is in one dimension.

It is equally interesting to fix r and ρ_2 in (119), and to take the limit of an infinitesimal circle around the origin, i.e., $\rho_1 \to 0$. We then find that by continuity $u_2(r;\rho_1 = 0,\rho_2) = 1$. The probability that a Brownian motion starting at a distance $r \neq 0$ from the origin, moves away from the origin up to a distance $\rho_2 > r$ without having visited the origin in $\rho_1 = 0$, is then equal to 1. In other words, a Brownian motion that does not leave from the origin O avoids the origin with probability 1, without ever being able to pass through it.

We deduce an apparently paradoxical result: *in two dimensions, any Brownian motion passes through a given point with zero probability, but it passes through immediate neighboring points infinitely often with probability* 1*!*

Such a double result is due to the presence in the expectation (119) of one function, the logarithm, that diverges both at short distance for $\rho_1 \to 0$, and at long distance for $\rho_2 \to \infty$. This is peculiar to two dimensions and announces exceptional properties known as *conformal invariance* in two dimensions, which will be described in the following section.

In $d > 2$ dimensions, a simple power law controls the Newtonian potential $u_d(r)$ (91), and only a divergence at short distance appears. We will see the consequences of such a divergence on the recurrence properties of Brownian motion.

Let us mention however that these properties only constitute the "tip of the iceberg": the singular character of the potential at short distances is the source of divergences in quantum field theories, which led to the creation of *renormalization theory*, whose consequences have been quite fruitful in the physics of elementary particles and in statistical mechanics.[160] Actually, the intersection of Brownian motions[161] provides the random geometric mechanism that underlies any interacting field theory.[162] This equivalence is fundamental in the theory of polymers[163] and also in the rigorous theory of second order phase transitions.[164] But *"Revenons à nos moutons."*[165]

3.6.3 The d-dimensional case

We are now well enough equipped to pass to the d-dimensional case, for $d > 2$. Let us consider two concentric hyperspheres, $\mathcal{S}_1$ and $\mathcal{S}_2$, centered at origin O, and of respective radii ρ_1 and ρ_2, with $\rho_1 < \rho_2$. The boundary of the domain $\mathcal{D}$ is then made of the two spheres $\partial\mathcal{D} = \mathcal{S}_1 \cup \mathcal{S}_2$. Let us state the Dirichlet problem (107)

$$\Delta u = 0 \text{ inside } \mathcal{D}, \tag{120}$$

$$u = 0 \text{ on } \mathcal{S}_1, \quad u = 1 \text{ on } \mathcal{S}_2. \tag{121}$$

Here again, in using this time the d-dimensional Newtonian potential $u_d(r)$ (91), it is easy to see that the spherically symmetric solution of the Dirichlet problem at a distance r from the center, is:

$$u_d(r;\rho_1,\rho_2) = \frac{u_d(r) - u_d(\rho_1)}{u_d(\rho_2) - u_d(\rho_1)} = \frac{r^{2-d} - \rho_1^{2-d}}{\rho_2^{2-d} - \rho_1^{2-d}}, \quad \rho_1 \leq r \leq \rho_2. \tag{122}$$

This function satisfies (121); it is harmonic in the annular d-dimensional domain $\mathcal{D}$, because the potential $u_d(r)$ (91) is harmonic too (except at the origin, which does not belong to $\mathcal{D}$).

Finally let us apply the probabilistic result: $u_d(r;\rho_1,\rho_2)$ (122) is the probability that a Brownian motion, starting from a given point at a distance r from the center, meets the outer sphere before the internal sphere.

First, let us take in (122) the limit $\rho_2 \to \infty$, at r and ρ_1 fixed. As the dimension d is here superior to 2, one has $\rho_2^{2-d} \to 0$. The probability for the Brownian motion to escape to infinity, $u_d(r;\rho_1,\rho_2 \to \infty)$, is by continuity the limit $u_d(r;\rho_1,\infty) = 1 - (\rho_1/r)^{d-2}$, which is *finite*.

[160] For this subject, one can consult the text *Renormalization* from Séminaire Poincaré 2002, *in* B. Duplantier & V. Rivasseau *Eds.*, *Poincaré Seminar 2002*, Progress in Mathematical Physics, vol. 30, Birkhäuser, Bâle (2003); see also the monograph by J. Zinn-Justin, *Quantum Field Theory and Critical Phenomena*, 4th Edition, Oxford University Press (2002).

[161] G. F. Lawler, *Intersection of Random Walks* (Birkhäuser, Boston, 1991).

[162] K. Symanzyk, in *Local Quantum Theory*, edited by R. Jost (Academic Press, London, New York (1969)).

[163] P.-G. de Gennes, *Phys. Lett.* **A38**, 339–340 (1972); J. des Cloizeaux, *J. de Physique* **36**, 281–291 (1975).

[164] M. Aizenman, *Phys. Rev. Lett.* **47**, 1–4, 886 (1981); *Commun. Math. Phys.* **86**, 1–48 (1982); D. C. Brydges, J. Fröhlich, and T. Spencer, *Commun. Math. Phys.* **83**, 123–150 (1982); G. F. Lawler, *Commun. Math. Phys.* **86**, 539–554 (1982).

[165] From *La farce de Maistre Pierre Pathelin* (c. 1460), meaning "Let's get back to our main subject".

This result shows that *in all spaces with at least three dimensions Brownian motion is not recurrent*, because the space is larger than that in one or two dimensions. We say that it is *transient*. Such a result, very important in the probability theory, was obtained in an elegant and simple manner via potential theory.

The complementary probability at a distance $\rho_1 \leq r$, $p_d(r;\rho_1,\infty) = 1 - u_d(r;\rho_1,\infty)$, that is, of visiting a neighborhood of the origin, is then equal to $(\rho_1/r)^{d-2}$. In the usual physical case, $d = 3$, one finds $p_3(r;\rho_1,\infty) = \rho_1/r$, for $\rho_1 \leq r$.

One can generalize the definition of $p_d(r;\rho_1,\infty)$ to the whole space, by giving it the value 1 inside the sphere of radius ρ_1, that is for $r \leq \rho_1$. Such a generalized function is called *potential capacity* of the sphere of radius ρ_1. The potential capacity of an ensemble $\mathcal{B}$ is an important concept in classic potential theory; it is a harmonic function outside $\mathcal{B}$, equal to 1 inside $\mathcal{B}$, and zero at infinity. It is then the probability that a single particle, animated by Brownian motion and leaving from a given point, will reach $\mathcal{B}$.

Research in this domain allowed the discovery of important generalizations, both for the theory of Brownian motion and potential theory. We have seen that the equivalence between them rests on the Markovian property of Brownian motion. Similarly, a generalized potential theory can be associated to any "standard" Markov process.

We see therefore the profound relation that exists between the mathematical theory of potential, invented in the 17th century by Newton, then developed by Laplace, Poisson and Green, and Brownian motion, observed during the same era, but understood only in the 20th century, thanks to Sutherland, Einstein, Smoluchowski, Perrin and Langevin in physics, then Bachelier, Wiener, Lévy, Kakutani and many others in mathematics.

4 The fine geometry of the planar Brownian curve

4.1 The Brownian boundary

In this last part, we are interested in the geometry of the Brownian curve in the plane. By Brownian curve, or Brownian path, we mean the random curve traced by a Brownian motion on the plane. We can see a typical representative in figure 1. In particular, we will consider the *boundary* of such a curve. It is the outer envelope of the Brownian curve. We observe that it is an extremely irregular curve, *fractal* in Mandelbrot's sense (figure 12).[166]

From a series of accurate numerical simulations, Mandelbrot made the conjecture in 1982 that such a boundary is the continuous limit of a particular random walk, the *self-avoiding walk (SAW)* (figure 13). That is a process where the random

[166] See the classic works of Benoît Mandelbrot, *Les objets fractals: forme, hasard et dimension, survol du langage fractal*, Champs, Flammarion (1999), and *The Fractal Geometry of Nature*, Freeman, New York (1982).

Figure 12: *Boundary or outer envelope curve of a planar Brownian path.*

walker cannot visit any point of his own path twice. To define it, one considers *a priori* the ensemble of all possible random paths of a given length (with and without self-intersections) on, say, a square lattice, and select among them the small subset of all the paths that do not self-intersect. Those are then weighted with a uniform measure.[167]

The resulting conjecture is that the fractal dimension or Hausdorff dimension of the Brownian boundary is equal to $D_H = 4/3$, like that which was calculated by the Dutch theoretical physicist Bernard Nienhuis in 1982 for the two-dimensional self-avoiding random walks.[168] The fractal dimension D_H is here defined in an non-rigorous way, as follows. We cover the fractal object of size R by small disjoint disks of radius ε, and we count the number n of these disks. In general, this number grows with a power law in R and ε, $n \propto (R/\varepsilon)^{D_H}$. We then see that D_H generalizes the notion of Euclidean dimension of regular sets to the case of very irregular sets.

Nienhuis used a representation of statistical mechanics, known as the Coulomb gas, a precursor to the methods of conformal invariance or of conformal field theories that in 1984 would break into the theory of two-dimensional critical phenomena, thanks to the work of Belavin, Polyakov, and Zamolodchikov.[169]

[167]See the monographs: P.-G. de Gennes, *Scaling Concepts in Polymer Physics*, Cornell University Press (1979); J. des Cloizeaux and G. Jannink, *Polymers in Solution, their Modeling and Structure* (Clarendon, Oxford University Press, 1989).

[168]B. Nienhuis, *Phys. Rev. Lett.* **49**, 1062–1065 (1982); *J. Stat. Phys.* **34**, 731–761 (1984); *Phase Transitions and Critical Phenomena*, edited by C. Domb et J. L. Lebowitz, (Academic Press, London, 1987), Vol. 11; see also M. den Nijs, J. Phys. A **12**, 1857–1868 (1979); Phys. Rev. B **27**, 1674–1679 (1983).

[169]A. A. Belavin, A. M. Polyakov and A. B. Zamolodchikov, *Nucl. Phys.* **B241**, 333–380 (1984). One can find an introduction in the book by C. Itzykson and J.-M. Drouffe, *Théorie statistique des champs*, tome 2, EDP Sciences/CNRS Éditions (2000); English version: *Statistical Field Theory*, Vol. 2, Cambridge University Press, Cambridge (1989). For further reading, see J. L. Cardy, in *Phase Transitions and Critical Phenomena*, edited by C. Domb and J. L. Lebowitz,

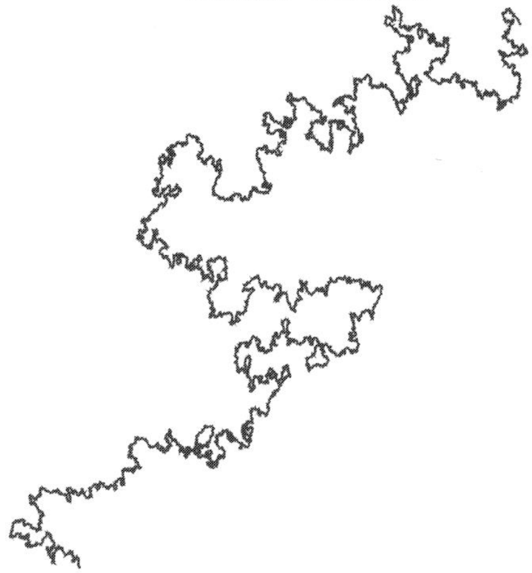

Figure 13: *A self-avoiding walk in the plane, made of 1 million steps! (Kindly provided by T. G. Kennedy, University of Arizona.)*

A heuristic demonstration of Mandelbrot's conjecture, inspired by some probabilistic results of conformal invariance by Lawler's and Werner's,[170] was given by the author in 1998 in the area of theoretical physics, by means of the formalism of *"2D quantum gravity"* in conformal field theory.[171]

Mandelbrot's conjecture was at last rigorously proved in the framework of probability theory in 2000 by Greg Lawler, Oded Schramm and Wendelin Werner,[172] by means of a conformally invariant stochastic process invented by Schramm, the SLE (*"Stochastic Loewner Evolution"*), which is itself based on Brownian motion.[173]

(Academic Press, London, 1987), Vol. 11; J. L. Cardy, *Conformal Invariance and Statistical Mechanics*, in "Fields, Strings, and Critical Phenomena", Les Houches Summer School 1988, edited by É. Brézin and J. Zinn-Justin, North-Holland, Amsterdam (1990); Ph. Di Francesco, P. Mathieu and D. Sénéchal, *Conformal Field Theory*, Springer-Verlag, New York (1997).

[170] G. F. Lawler and W. Werner, *Ann. Probab.* **27**, 1601–1642 (1999).

[171] B. Duplantier, *Phys. Rev. Lett.* **81**, 5489–5492 (1998); *ibid.* **82**, 880–883 (1999), arXiv:cond-mat/9812439.

[172] G. F. Lawler, O. Schramm, and W. Werner, *Acta Math.* **187**, (I) 237–273, (II) 275–308 (2001), arXiv:math.PR/9911084, arXiv:math.PR/0003156; *Ann. Inst. Henri Poincaré* PR **38**, 109–123 (2002), arXiv:math.PR/0005294; *Acta Math.* **189**, 179–201 (2002), arXiv:math.PR/0005295; *Math. Res. Lett.* **8**, 401–411 (2001), math.PR/0010165.

[173] O. Schramm, *Israel Jour. Math.* **118**, 221–288 (2000). The SLE_κ process, and its path, are generated by the Loewner equation, describing the evolution of the Riemann's conformal map which maps the unit disc, slit by the random path, onto itself. This map erases the path and maps its two sides onto the boundary of the unit disc, under the form of a Brownian motion characterized by a diffusion coefficient κ. This is the case of the so-called *radial* SLE. Another

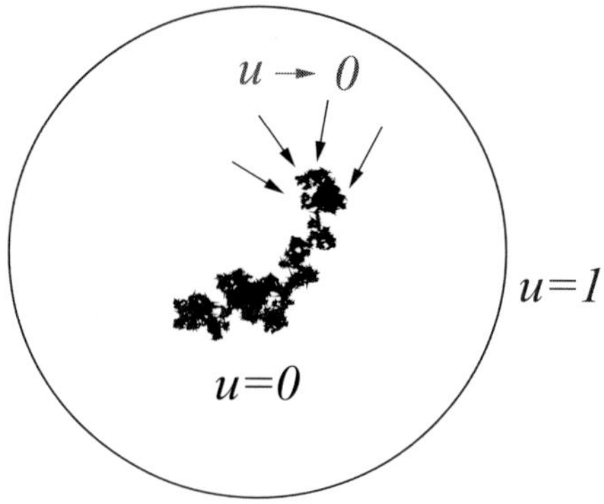

Figure 14: *Dirichlet problem associated to a planar Brownian path. The latter serves as an electrode where the potential vanishes.*

We are not going to describe this work in detail here[174], but we will look instead at the generalization of the results on the geometry of Brownian motion, and at the *multifractal* nature of its boundary. The latter actually reveals a structure

case is that of the *chordal* SLE, where the conformal map acts on the slit complex half-plane. See the recent book by G. F. Lawler, *Conformally Invariant Processes in the Plane*, Mathematical Surveys and Monographs, AMS, Vol. **114** (2005).

[174]For further details, see the article for the general public by Wendelin Werner, *Les chemins aléatoires*, published in Pour La Science in August 2001.

For the SLE process, consult: the notes from W. Werner's courses, *Random Planar Curves and Schramm–Loewner Evolutions*, Lectures Notes from the 2002 Saint-Flour Summer School, Springer L. N. Math. **1840**, 107–195, (2004), arXiv:math.PR/0303354; the book by G. F. Lawler, *Conformally Invariant Processes in the Plane, op. cit.*, as well as the article of W. Kager and B. Nienhuis, *A Guide to Stochastic Loewner Evolution and its Applications*, *J. Stat. Phys.* **115**, 1149–1229 (2004), arXiv:math-phys/0312056.

For the link for SLE with quantum gravity, see: B. Duplantier, *Conformal Fractal Geometry and Boundary Quantum Gravity*, in *Fractal Geometry and Applications, A Jubilee of Benoît Mandelbrot*, Proceedings of Symposia in Pure Mathematics, AMS, Vol. **72**, Part **2**, edited by M. L. Lapidus and F. van Frankenhuijsen, pp. 365–482 (2004); arXiv:math-phys/0303034.

For the link of SLE with conformal field theories, see in mathematics: R. Friedrich and W. Werner, *C. R. Acad. Sci. Paris Sér. I Math.* **335**, 947–952 (2002), arXiv:math.PR/0209382; *Commun. Math. Phys.*, **243**, 105–122 (2003), arXiv:math-ph/0301018; W. Werner, *Conformal restriction and related questions*, Proceedings of the conference *Conformal Invariance and Random Spatial Processes*, Edinburgh, July 2003, arXiv:math.PR/0307353; W. Werner and G. F. Lawler, *Probab. Th. Rel. Fields* **128**, 565–588 (2004), arXiv:math.PR/0304419; W. Werner, *C. R. Ac. Sci. Paris Sér. I Math.* **337**, 481–486 (2003), arXiv:math.PR/0308164; see also J. Dubédat, arXiv:math.PR/0411299; arXiv:math.PR/0507276; in physics: M. Bauer and D. Bernard, *Phys. Lett.* **B543**, 135–138 (2002), arXiv:math-ph/0206028; *Commun. Math. Phys.* **239**, 493–521 (2003), arXiv:hep-th/0210015; *Phys. Lett.* **B557**, 309–316 (2003), arXiv:hep-th/0301064; *Ann. Henri Poincaré* **5**, 289–326 (2004), arXiv:math-ph/0305061; Proceedings of the conference *Conformal Invariance and Random Spatial Processes*, Edinburgh, July 2003, arXiv:math-ph/0401019; M. Bauer, D. Bernard and J. Houdayer, *J. Stat. Mech.* P03001 (2005), arXiv:math-ph/0411038; M. Bauer and D. Bernard, arXiv:cond-mat/0412372; M. Bauer, D. Bernard and K. Kytölä, *J. Stat. Phys.* (to appear), arXiv:math-ph/0503024; K. Kytölä, arXiv:math-ph/0504057.

made of a continuum of fractal subsets that we will describe.

In continuity with the previous part, we will focus on the *potential theory associated with the neighborhood of a planar Brownian path.* We will show how the fine geometry of the Brownian boundary appears as an essential component of the solution to the Dirichlet associated electrostatic problem.

4.2 Potential theory in a neighborhood of a Brownian curve

4.2.1 Brownian Dirichlet problem

Let us then consider a planar Brownian path $\mathcal{B}$ enclosed by a large circle, and the associated Dirichlet problem where the potential u has the value $u = 0$ on the boundary $\partial\mathcal{B}$ of the Brownian curve, and $u = 1$ on the circle (figure 14). The Brownian path serves as an electrode creating the potential, and by electrostatic induction, its boundary will charge itself. This *a priori* appears as a rather complex problem, since the Brownian curve is completely random!

Far from the Brownian curve, the potential will depend on the global geometry of the system, and in particular on the presence of the outer circle that acts as an external electrode. Let us imagine for a moment that this circle is pushed towards infinity. Seen from intermediate regions located very far from the Brownian curve (and from the outer circle), the Brownian electrode would then appear to be confined to a point. Its potential will then coincide with that of a point charge equal to the total charge carried by the boundary of the Brownian curve, i.e the logarithmic Newtonian potential $u_2(r)$ (93).

On the other hand, close to the Brownian curve, the geometry of the boundary is crucial. The potential vanishes exactly on the boundary $\partial\mathcal{B}$, and the natural question here is its analytic behavior in the neighborhood of $\partial\mathcal{B}$, i.e the way in which it goes to 0. As the geometry of the boundary is particularly wild, the way the potential vanishes is as well.

However, the random Brownian curve hides at its heart a fundamental structural regularity connected to its *conformal invariance*, and one can in fact describe the potential close to the Brownian path in a way which is probabilistic, but *universal.*

4.2.2 Conformal invariance

A *conformal* map Φ of the plane is a bijection of the plane into itself that preserves *angles* between curve intersections. To any analytic function $\Phi(z)$ in the complex plane can be associated one such conformal map. Locally, i.e., infinitesimally close to the image $\Phi(z)$ of any point z in complex coordinates, a conformal map is the composition of a local dilation (by a factor of $|\Phi'(z)|$), and of a rotation around $\Phi(z)$ (by an angle $\arg \Phi'(z)$). This is why angles are locally conserved.

Let us come back for a moment to the Brownian representation of the general Dirichlet problem in a domain $\mathcal{D}$ (figure 11). An auxiliary Brownian motion issued from an arbitrary point P, stops upon touching the boundary $\partial\mathcal{D}$, and its Wiener integral represents the potential $u(P)$. Let us imagine the domain $\mathcal{D}$ to be transformed by a conformal map Φ into a domain $\mathcal{D}' = \Phi(\mathcal{D})$, while the Brownian trajectory $\mathcal{B}$ is transformed into a curve $\Phi(\mathcal{B})$, which is thus stopped upon touching the boundary $\partial\mathcal{D}' = \Phi(\partial\mathcal{D})$. Paul Lévy showed that $\Phi(\mathcal{B})$ is still the trajectory of a Brownian motion, after a time reparameterization: this is the property of *conformal invariance of planar Brownian motion.*[175]

Let us then consider the new potential $u'(P')$ at a transformed point $P' = \Phi(P)$, i.e. the solution to the Dirichlet problem in the transformed domain $\mathcal{D}'$. Since all geometric objects that represent the potential were transformed by Φ, and since the transformed auxiliary Brownian path is still Brownian, the result is that its Wiener integral, $u'(P')$, does not change. The potential $u'(P')$ is then equal to the potential $u(P)$, that is the solution to the Dirichlet problem in the original domain $\mathcal{D}$, and thus there is an *invariance of potential under a conformal map.*

In the case we are considering closely here, that is of Dirichlet problem of a potential $u(P)$ in the neighborhood of a planar Brownian curve (figure 14), the Brownian representation of the potential introduces a second auxiliary Brownian motion that diffuses from the point P, while avoiding the original Brownian curve (figure 15). As we just saw, the two Brownian paths are statistically conformal invariant and this probabilistic geometric problem is invariant under any conformal map in the plane.

4.2.3 The role of angles

Conformal maps preserve angles in the plane, and this is why the latter will play an essential role in the description of the potential close to the Brownian boundary.

Let us first consider the simple problem of a potential existing in an angular sector of the plane. More precisely, let us consider an open angle θ centered at a point w (figure 16). One easily shows, by using the singular conformal map of the complex plane that opens the angle θ into a flat angle, $\Phi(z) = z^{\pi/\theta}$, that the potential $u(z)$ varies at any point z close to w like

$$u(z) \approx r^{\pi/\theta}, \tag{123}$$

where r is the distance from w, $r = |z - w|$. For a flat angle, $\theta = \pi$, and we again find a linear behavior as a function of the distance, corresponding to a constant electric field close to a straight line.

4.3 Multifractality

4.3.1 Distribution of potential

Let us come back finally to the initial question of the distribution of the potential in the region close to a Brownian curve $\mathcal{B}$ (figures 14 and 15). Its boundary $\partial\mathcal{B}$

[175] Paul Lévy, *Processus stochastiques et mouvement brownien*, Gauthier-Villars, Paris (1965).

is a fractal curve without a microscopic scale, and the irregularities of this curve go down to the infinitesimally small. Among all these irregularities, it is natural, from the point of view of the potential theory and of conformal invariance, to look for those that are locally like "angles". Actually, such a distribution of angles and the distribution of the associated potential are invariant under a conformal map. They are then stable in the class of all Brownian curves which are obtained by conformal maps of a single realization of a Brownian curve.

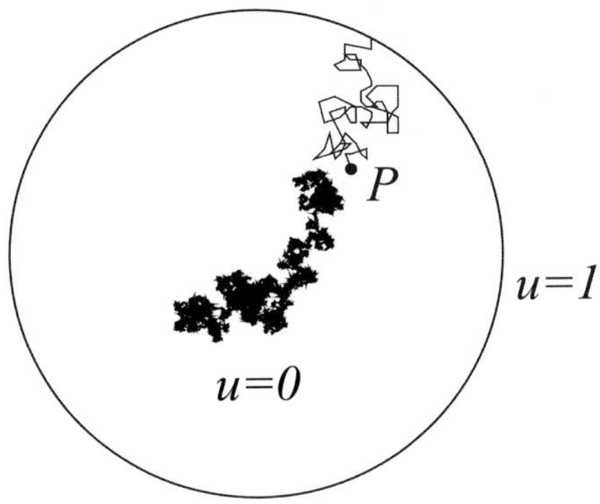

Figure 15: *The Dirichlet potential u created at point P by a Brownian curve (center), and vanishing on the boundary of the latter, is represented by a second auxiliary Brownian motion, that diffuses from P towards the exterior, while completely avoiding the first motion.*

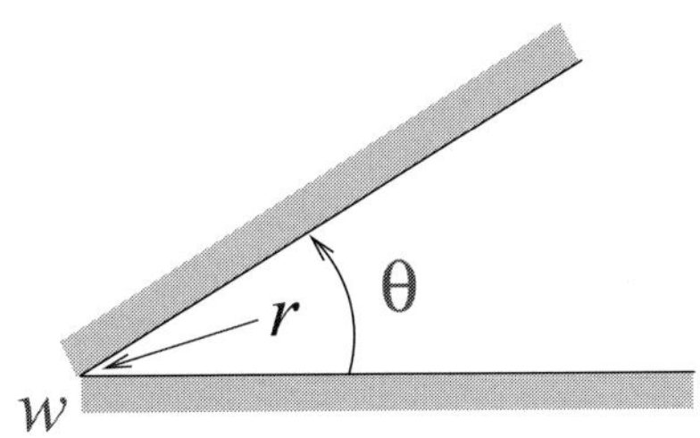

Figure 16: *Angular sector with apex w and angle θ.*

We can then classify the points w of the boundary $\partial\mathcal{B}$ according to the properties of variation of the potential $u(z)$ when a point P with complex coordinate z approaches w on the boundary. We say that a point w is of type α if

$$u\,(z \to w) \approx r^{\alpha}, \tag{124}$$

in the limit where the distance $r = |z - w| \to 0$. (Figure 17.)

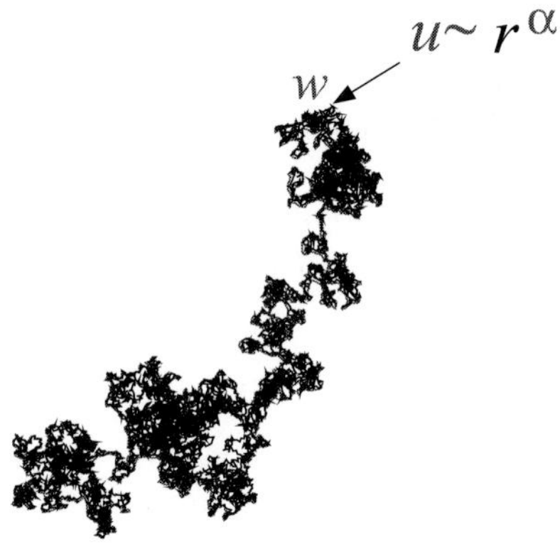

Figure 17: *Singular behavior in r^{α} of the potential close to a point w of type α.*

By comparing the property (124) to the form (123) of the potential of an angle, we see that an exponent α corresponds, from the point of view of the potential, to an *equivalent electrostatic angle* θ such that

$$\alpha = \frac{\pi}{\theta}. \tag{125}$$

The behavior is as if an angle $\theta = \pi/\alpha$ existed locally on the boundary.[176] The angular domain being such that $0 \leq \theta \leq 2\pi$, the domain of the exponents α is $1/2 \leq \alpha < \infty$, which is rigorously supported by a theorem of A. Beurling. The domain where α is close to $1/2$ corresponds to θ close to 2π, which is a completely open angular sector, and thus to the presence of an extremely thin *needle* on the boundary. The domain where α is very large corresponds to θ close to 0, thus to a very narrow angular sector, and one then speaks of a *fjord.*

Now, let $\partial\mathcal{B}_{\alpha}$ be the set of points of type α on the boundary. To measure the probability of finding such points of type α, we introduce the Hausdorff dimension of the set $\partial\mathcal{B}_{\alpha}$,

$$f(\alpha) = \dim(\partial\mathcal{B}_{\alpha}). \tag{126}$$

This defines the *multifractal spectrum* $f(\alpha)$ of the potential distribution. Such a spectrum is conformally invariant in two dimensions, because in any conformal map the local exponents $\alpha = \pi/\theta$ of the potential are themselves invariant.[177]

[176]The presence of a local singularity exponent α does not necessarily mean that $\theta = \pi/\alpha$ is a geometric angle, because the surroundings of a point w on a random fractal object will in general screen the potential, and reduce the equivalent electrostatic angle with respect to a possible geometric angle.

[177]The local definitions of the exponent α and of $f(\alpha)$ as given in (124) and (126) are only heuristic, since the way of taking limits was not explained. For any given point w on the boundary

From a historical point of view, the concept of *multifractality* was introduced by B. Mandelbrot in 1974,[178] about the phenomenon of turbulence in hydrodynamics, then by H. Hentschel, I. Procaccia, U. Frisch and G. Parisi.[179] It was then further developed at the University of Chicago by T .C. Halsey *et al.*[180] It corresponds to the existance of a continuous set of fractal dimensions $f(\alpha)$, that are functions of a continuum of exponents α.

4.3.2 The Brownian multifractal spectrum

One of the first properties is that the global Hausdorff dimension of a multifractal object is always the maximum of its multifractal spectrum. Thus, for the boundary of a Brownian curve,

$$D_H = \sup_{\alpha} f(\alpha) = \frac{4}{3}, \tag{127}$$

because of Mandelbrot's conjecture, which we mentioned above.

The complete spectrum $f(\alpha)$ for the Brownian curve was calculated in 1998 by a method called "quantum gravity".[181] One uses a representation of the same problem on a random surface where the metric fluctuates, instead of the normal Euclidean plane. The geometric and probabilistic laws are largely simplified by the "quantum" fluctuations of the metric, and the singular behavior of the Brownian Dirichlet problem is directly accessible!

Next, one can obtain the multifractal spectrum in the plane $\mathbb{R}^2$, thanks to a fundamental relationship between critical exponents in the plane and on a random surface, a formula known by the initials "KPZ", discovered originally in 1988 by

---

of a random fractal object, in general no stable local exponents α exist, such that they are obtained by a "simple limit" to the point. One then proceeds in another way. Define the *harmonic measure* $\omega(w, r)$ as the probability that the Brownian motion leaving from any point on the outer circle (therefore from infinity), touches the frontier $\partial\mathcal{B}$ for the first time inside a ball centered at w and of radius r. (This harmonic measure is similar to the Brownian representation of the potential $u(P)$, which is just the harmonic measure of the outer boundary of $\mathcal{D}$ seen from a point P). Next, we define the set $\partial\mathcal{B}_{\alpha,\eta}$ of points on the boundary $\partial\mathcal{B}$, $w = w_{\alpha,\eta}$, for which there exists a decreasing series of radii $r_j, j \in \mathbb{N}$ tending towards 0, such that $r_j^{\alpha+\eta} \leq \omega(w, r_j) \leq r_j^{\alpha-\eta}$. The multifractal spectrum $f(\alpha)$ is then globally defined as the limit $\eta \to 0$ of the Hausdorff dimension of the set $\partial\mathcal{B}_{\alpha,\eta}$, i.e.,

$$f(\alpha) = \lim_{\eta\to 0} \dim \left\{ w : \exists\, \{r_j \to 0,\, j \in \mathbb{N}\} : r_j^{\alpha+\eta} \leq \omega(w, r_j) \leq r_j^{\alpha-\eta} \right\}.$$

[178] B. B. Mandelbrot, *J. Fluid. Mech.* **62**, 331–358 (1974).

[179] H. G. E. Hentschel and I. Procaccia, *Physica D* **8**, 435–444 (1983); U. Frisch and G. Parisi, *Proceedings of the International School of Physics "Enrico Fermi"*, course LXXXVIII, edited by M. Ghil (North-Holland, New York, 1985) p. 84.

[180] T. C. Halsey, M. H. Jensen, L. P. Kadanoff, I. Procaccia and B. I. Shraiman, *Phys. Rev. A* **33**, 1141–1151 (1986); *ibid.* **34**, 1601 (1986).

[181] B. Duplantier, *Phys. Rev. Lett.* **82**, 880–883 (1999), arXiv:cond-mat/9812439.

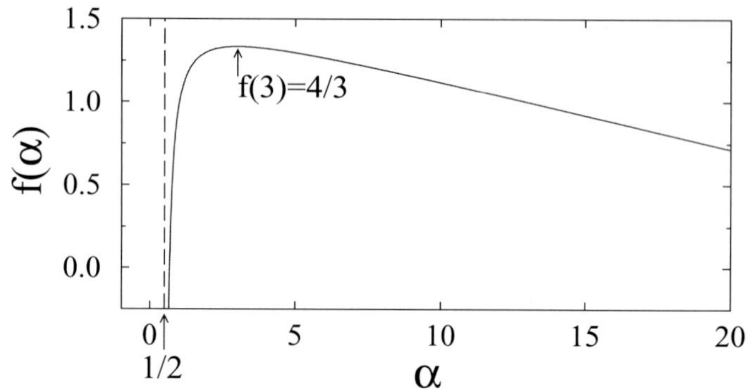

Figure 18: *Multifractal function $f(\alpha)$ of the Brownian frontier.*

three Russian physicists, V. G. Knizhnik, A. M. Polyakov, and A. B. Zamolodchikov.[182] We do not have space here to further develop this method.[183]

We find the formula

$$f(\alpha) = \alpha + b - \frac{b\alpha^2}{2\alpha - 1}, \quad b = \frac{25}{12}. \tag{128}$$

This curve is drawn in figure 18. The definition domain is the half-line $(1/2, +\infty)$. One verifies that the maximum of f is at $4/3$, in agreement with Mandelbrot's conjecture (127) for the fractal dimension of the boundary. It corresponds to a value of $\alpha = 3$, or to an typical electrostatic angle of $\pi/3$.

Moreover, one can calculate by the same method the multifractal spectrum of the potential close to a self-avoiding random walk,[184] and one finds a spectrum which is identical to that of a Brownian curve, fully confirming the identity of the Brownian frontier to a self-avoiding walk in the scaling limit.

One also predicts by this heuristic method that the spectra of a Brownian curve and of a critical percolating cluster are identical[185]. It then follows that

[182] V. G. Knizhnik, A. M. Polyakov and A. B. Zamolodchikov, *Mod. Phys. Lett.* **A 3**, 819–826 (1988).

[183] B. Duplantier, *Conformal Fractal Geometry and Boundary Quantum Gravity*, in *Fractal Geometry and Applications, A Jubilee of Benoît Mandelbrot*, Proceedings of Symposia in Pure Mathematics, AMS, Vol. **72**, Part **2**, edited by M. L. Lapidus and F. van Frankenhuijsen, pp. 365–482 (2004); arXiv:math-phys/0303034; see also V. Fateev, A. Zamolodchikov, Al. Zamolodchikov, *Boundary Liouville Field Theory I. Boundary State and Boundary Two-point Function*, arXiv:hep-th/0001012; I. K. Kostov, B. Ponsot and D. Serban, *Nucl. Phys.* **B683**, 309–362 (2004), arXiv:hep-th/0307189; I. K. Kostov, *Nucl. Phys.* **B689** 3–36 (2004), arXiv:hep-th/0312301; *Proceedings of the Conference "Lie theory and its applications in physics – 5"*, Varna, Bulgaria (2003), arXiv:hep-th/0402098, and references therein.

[184] B. Duplantier, *Phys. Rev. Lett.* **82**, 880–883 (1999), arXiv:cond-mat/9812439.

[185] B. Duplantier, *Phys. Rev. Lett.* **82**, 3940–3943 (1999), arXiv:cond-mat/9901008; M. Aizen-

both the Brownian frontier and the external perimeter of a critical percolation cluster coincide with the scaling limit of a self-avoiding walk, which further extends Mandelbrot's conjecture.

Let us mention that the works of Lawler, Schramm, and Werner contain also in principle the necessary information to calculate the spectrum of a Brownian potential. In a rigorous approach using SLE, these authors identify the boundary with that of the SLE_6 process, conjectured also to be an $\mathrm{SLE}_{8/3}$ and the scaling limit of a self-avoiding polymer.

This curve $f(\alpha)$, also called the harmonic measure spectrum, then solves the problem of the potential distribution close to a Brownian path in a probabilistic sense, since it gives the fractal dimension of the set of points where the potential varies in a specific way, namely as r^α.

Other values of b in (128) ($b = \frac{25-c}{12} \geq 2$, where $c \leq 1$ is the "central charge" of the associated conformal theory) generate the multifractal spectra of the potential or harmonic measure of conformally invariant random curves in the plane.[186] These are the SLEs describing the boundaries of critical clusters in two-dimensional statistical models, such as Ising or Potts models. For an SLE_κ, with $0 \leq \kappa < \infty$, one simply sets in (128)

$$c = \frac{1}{4}(6-\kappa)\left(6 - \frac{16}{\kappa}\right), \qquad b = 1 + \frac{1}{8}\left(\kappa + \frac{16}{\kappa}\right). \tag{129}$$

4.4 Generalized multifractality

4.4.1 Logarithmic spirals

Until now we have considered variations of the potential only. We can also study the form of the equipotential lines. As the potential follows the properties of conformal invariance of the Brownian curve, it is now necessary first to determine the geometric forms that are conserved by such invariance.

These are the *logarithmic spirals* that play a particular role in potential theory in two dimensions. One such spiral centered at the origin is defined by the logarithmic variation of the polar angle φ as a function of the distance r from the origin:

$$\varphi = \lambda \ln r \ ,$$

where λ is a real positive or negative parameter.

When we apply a conformal map Φ, around the center it is equivalent to a dilation $r \to |\Phi'(0)|\, r$, composed with a rotation. The dilation transforms the angle $\varphi = \lambda \ln r$ into $\lambda \ln r + \lambda |\Phi'(0)|$, which thus amounts to a local rotation of the spiral, whose geometrical shape is thereby locally conserved.

---

man, B. Duplantier et A. Aharony, *Phys. Rev. Lett.* **83**, 1359–1362 (1999), arXiv:cond-mat/9901018.

[186]B. Duplantier, *Phys. Rev. Lett.* **84**, 1363–1367 (2000), arXiv:cond-mat/9908314; *J. Stat. Phys.* **110**, 691–738 (2003), arXiv:cond-mat/0207743.

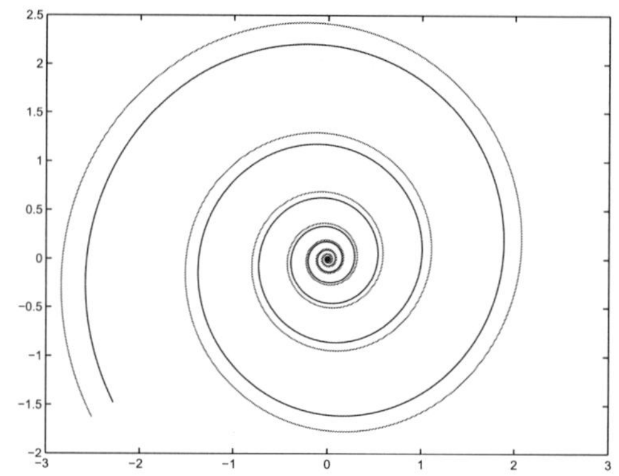

Figure 19: *Double logarithmic spiral.*

In the potential theory considered here, the Brownian frontier is equipotential by construction. There exists a multitude of points where such equipotential boundary will locally roll onto itself in a double logarithmic spiral, as shown in figure 19.

4.4.2 Mixed multifractal spectrum

We come then to Ilia Binder's idea from his thesis[187] in 1997 defining a generalized multifractality. One looks for a set $\partial\mathcal{B}_{\alpha,\lambda}$ of points w of the boundary $\partial\mathcal{B}$, where the potential varies like r^α, and the boundary spirals at a given rate λ. These conditions can be heuristically written for a point z close to w:

$$\begin{aligned} u\,(z \to w \in \partial\mathcal{B}_{\alpha,\lambda}) &\approx r^\alpha, \\ \varphi\,(z \to w \in \partial\mathcal{B}_{\alpha,\lambda}) &\approx \lambda \ln r\ , \end{aligned} \tag{130}$$

in the limit $r = |z-w| \to 0$. The Hausdorff dimension $f\,(\alpha,\lambda) = \dim\left(\partial\mathcal{B}_{\alpha,\lambda}\right)$ then defines the *mixed multifractal spectrum*, which is conformal invariant because under a conformal map both α and λ are locally invariant.

With Ilia Binder, we computed such a mixed spectrum for a Brownian motion, by the quantum gravity method.[188] It satisfies an exact scaling law

$$f(\alpha,\lambda) = (1+\lambda^2) f\left(\frac{\alpha}{1+\lambda^2}\right) - b\lambda^2\ , \tag{131}$$

which gives from (128)

$$f(\alpha,\lambda) = \alpha + b - \frac{b\alpha^2}{2\alpha - 1 - \lambda^2}, \quad b = \frac{25}{12}. \tag{132}$$

[187] I. A. Binder, *Harmonic Measure and Rotation of Simply Connected Planar Domains*, PhD Thesis, Caltech (1997).

[188] B. Duplantier and I. A. Binder, *Phys. Rev. Lett.* **89**, 264101 (2002); arXiv:cond-mat/0208045.

Its domain of definition is $\alpha \geq \frac{1}{2}(1+\lambda^2)$, according to a theorem of Beurling. Different spectra are represented in figure 20.

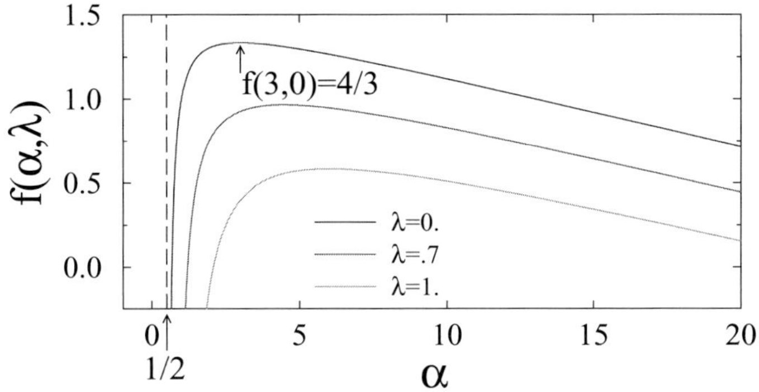

Figure 20: *Universal Multifractal Spectrum $f(\alpha, \lambda)$ of a Brownian path for different values of spiral rate λ. The maximum $f(3,0) = 4/3$ is the Hausdorff dimension of the Brownian frontier.*

Since this function does not depend on the sign of λ, spiral rotations in positive and negative directions are equiprobable, as expected. One recovers the harmonic spectrum $f(\alpha)$ as the maximum

$$f(\alpha) = f(\alpha, \lambda = 0) = \sup_{\alpha} f(\alpha, \lambda).$$

By symmetry, the most probable situation for a point on the boundary is the absence of spiral rotation, i.e., $\lambda = 0$.

One can then also consider only the fractal dimension $D_H(\lambda)$ of the points on the boundary, which are the tips of logarithmic spirals of type λ. For this, we take the maximum of the mixed spectrum with respect to the other variable, α:

$$D_H(\lambda) = \sup_{\alpha} f(\alpha, \lambda) = \frac{4}{3} - \frac{3}{4}\lambda^2.$$

This fractal dimension has then the form of a parabola as a function of λ, whose maximum is still the global Hausdorff dimension of the boundary, $D_H = 4/3$ (figure 21).

Let us add a few final remarks.

The quantum gravity calculations can be generalized to the whole class of conformally invariant curves on the plane, and to Schramm's SLE process. The spectra are given by the same formulae (128) and (132) for different values of the parameter b. For the SLE_κ process, one substitutes:

$$b = 1 + \frac{1}{8}\left(\kappa + \frac{16}{\kappa}\right) = \frac{1}{2\kappa}\left(2 + \frac{\kappa}{2}\right)^2, \quad \kappa \in \mathbb{R}^+.$$

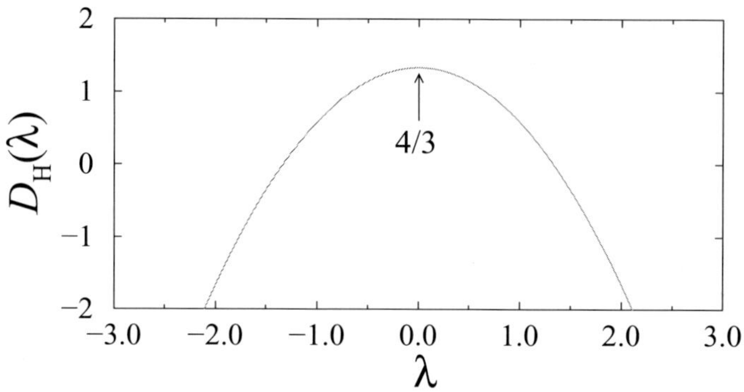

Figure 21: *Fractal dimension $D_H(\lambda)$ of spirals of type λ along the Brownian boundary.*

Lastly, these multifractal results, originally found heuristically in theoretical physics, can in principle be rigorously proved in the general probabilistic framework of SLE_κ.[189] The application of this general result to the case of the Brownian and percolation cluster frontiers is then obtained by identifying those boundaries to that of the SLE_6 process (thanks to works by Lawler, Schramm, and Werner and also by S. Smirnov[190], and V. Beffara [191]), while, from a rigorous point of view, the similar identification of the scaling limit of a self-avoiding walk to a $\mathrm{SLE}_{8/3}$ process, although certainly true, remains to be proven![192]

Here we pause in 2005 at the end of the path started in 1827 by Robert Brown with his observations at the microscope, and by Einstein in 1905 with his theory of Brownian fluctuations. The new paradigm of stochastic paths could be today the SLE, or Stochastic Loewner Evolution, generated itself by Brownian motion on the boundary of a planar domain, and its rather extraordinary conformal invariance properties in the Euclidean plane. This process brought us to the shores of two-dimensional quantum gravity, where the SLE stochasticity seems to call for fluctuations of the metric, hence "quantum gravity". In some sense, we are brought back to the work of Einstein, whose 1916 general relativity theory explained how

---

[189] I. A. Binder and B. Duplantier, in preparation; see also D. Beliaev, *Harmonic Measure on Random Fractals*, PhD thesis, KTH, Stockholm, Sweden, 2005.

[190] S. Smirnov, *C. R. Acad. Sci. Paris Sér. I Math.* **333**, 239–244 (2001).

[191] V. Beffara, arXiv:math.PR/0204208, *Ann. Prob.* **32**, (3) pp. 2606–2629 (2004); arXiv:math.PR/0211322.

[192] G. F. Lawler, O. Schramm and W. Werner, *On the Scaling Limit of Planar Self-Avoiding Walk*, in *Fractal Geometry and Applications, A Jubilee of Benoît Mandelbrot*, Proceedings of Symposia in Pure and Applied Mathematics, AMS, Vol. **72**, Part **2**, edited by M. L. Lapidus et F. van Frankenhuijsen, pp. 339–384 (2004), arXiv:math.PR/0204277.

gravitation is equivalent to a change of metric. Now it is Statistical Mechanics that stands in the breach, let us wish for fruitful developments!

Acknowledgements

I would like to thank John Stachel, Editor of the *Complete Papers of Albert Einstein*, for his kind permission to reproduce some material of the Editorial Notes.

I thank Bruce H. J. McKellar and Rod W. Home for kindly providing first-hand informations regarding the work of W. Sutherland, and Hector Giacomini for informations about the Besso–Einstein correspondence. I also wish to thank Jean-Pierre Kahane for stimulating exchanges about R. Brown and A. Brongniart, as well as André Rougé about L. Bachelier. I also thank Christian Van den Broeck and Chris Jarzynski for providing up-to-date information about their latest works, and for interesting discussions about Brownian motion, the Maxwell demon and the second principle.

I wish to thank Emmanuel Guitter for creating the figures, and Michel Bauer, Denis Bernard, Vincent Pasquier, and Jean-Louis Sikorav for useful comments.

Special thanks go to Thomas C. Halsey, David Kosower, and Kirone and Shamlal Mallick for their careful and critical proofreading of this article, to Thomas C. Halsey for introducing me to the concept of multifractality almost 20 years ago, and to Kirone Mallick and Ugo Moschella for their precious help during the completion of this article.

Last, but not least, I wish to express my gratitude to Emily Parks (and to Graziano Vernizzi!) who carried out the redoutable task of translating the original text from French into English.

Bertrand Duplantier
Service de Physique Théorique
Orme des Merisiers
CEA/Saclay
F-91191 Gif-sur-Yvette Cedex
France
email: Bertrand.Duplantier@cea.fr

gravitation is equivalent to a change of metric. Now it is Statistical Mechanics that stands in the breach, let us wish for fruitful developments!

Acknowledgements

I would like to thank John Stachel, Editor of the *Complete Papers of Albert Einstein*, for his kind permission to reproduce some material of the Editorial Notes.

I thank Bruce H. J. McKellar and Rod W. Home for kindly providing first-hand informations regarding the work of W. Sutherland, and Hector Giacomini for informations about the Besso–Einstein correspondence. I also wish to thank Jean-Pierre Kahane for stimulating exchanges about R. Brown and A. Brongniart, as well as André Rougé about L. Bachelier. I also thank Christian Van den Broeck and Chris Jarzynski for providing up-to-date information about their latest works, and for interesting discussions about Brownian motion, the Maxwell demon and the second principle.

I wish to thank Emmanuel Guitter for creating the figures, and Michel Bauer, Denis Bernard, Vincent Pasquier, and Jean-Louis Sikorav for useful comments.

Special thanks go to Thomas C. Halsey, David Kosower, and Kirone and Shamlal Mallick for their careful and critical proofreading of this article, to Thomas C. Halsey for introducing me to the concept of multifractality almost 20 years ago, and to Kirone Mallick and Ugo Moschella for their precious help during the completion of this article.

Last, but not least, I wish to express my gratitude to Emily Parks (and to Graziano Vernizzi!) who carried out the redoutable task of translating the original text from French into English.

Bertrand Duplantier
Service de Physique Théorique
Orme des Merisiers
CEA/Saclay
F-91191 Gif-sur-Yvette Cedex
France
email: `Bertrand.Duplantier@cea.fr`